Quantum
Mechanics

Quantum Mechanics

Sara M. McMurry

Trinity College, University of Dublin

Addison-Wesley Publishing Company
Wokingham, England • Reading, Massachusetts • Menlo Park, California
New York • Don Mills, Ontario • Amsterdam • Bonn • Sydney • Singapore
Tokyo • Madrid • San Juan • Milan • Paris • Mexico City • Seoul • Taipei

Cover designed by Designers & Partners, Oxford
and printed by The Riverside Printing Co. (Reading) Ltd.
Typeset by Keytec Typesetting, Bridport, Dorset.
Printed and bound in Great Britain by T.J. Press (Padstow) Ltd, Cornwall.

First printed 1993

ISBN 0-201-54439-3

British Library Cataloguing-in-Publication Data
A catalogue record for this book is available from the British Library.

Library of Congress Cataloging-in-Publication Data is available

Preface

Perhaps the best way for students of experimental physics to meet quantum theory for the first time is through a course in quantum physics, in which the emphasis is on a description of quantum phenomena and the general ideas (such as the uncertainty principle) that are essential for explaining them. The aim of such a course would be to give the students some feeling for the experimental phenomena described by quantum theory, and an understanding of how classical mechanics fails to explain them. They are then ready for a more mathematical presentation of quantum mechanics in the final two years of their undergraduate course. This text is intended to cover a course of the latter type, and the first chapter is a rapid overview of the important ideas that I assume students will have already met in an introductory course.

Maths and quantum physics

A course in quantum mechanics inevitably introduces some mathematics that is unfamiliar to most students. The reader of this text is expected to have a good knowledge of elementary calculus (including integration by parts), vectors and the solution of simple differential equations – that is, those of first order, and of second order with constant coefficients. Chapter 9 uses the expansion of a function in a Fourier series to introduce the superposition principle, and in Chapter 10 some elementary knowledge of matrix algebra is assumed. Appropriate references are provided for those not familiar with these last two topics. Concepts associated with the use of linear operators, their commutation relations and eigenvalue equations are introduced in the text, assuming no prior knowledge on the part of the reader.

It is important that students should try to come to grips with the basic mathematical ideas that are necessary for a proper understanding of quantum theory. However, this does not mean that they need to become experts at actually performing complex mathematical operations. In practice almost all quantum calculations on real systems can only be performed numerically, using a computer. What is necessary, therefore, is that the student should acquire an understanding of the overall strategy of solving the mathematical problems that arise in quantum mechanics – he or she should understand the nature of the predictions that it is possible to make about a given physical system, the type of mathematical calculation that is entailed, the way physical restrictions on the system are translated into mathematics, and what sort of results to expect.

Content and organization

It is traditional in textbooks on quantum mechanics to collect the basic mathematical ideas together in one chapter, with a title that is some variation of 'Mathematical foundations of quantum mechanics'. Although this is logical and concise, I feel that it is not the best approach for the student meeting the formal theory for the first time. In contrast I have spread the mathematical ideas throughout the text, using and expanding on one concept before introducing the next.

The fundamental assumptions on which quantum mechanics is based are encapsulated in four postulates, to be found in Chapters 2, 3, 9 and 12. Chapter 2 concerns the wave function and its probability interpretation, and Chapter 3 introduces the operator representation of dynamical variables, and the idea of an eigenvalue equation, in particular the time-independent Schrödinger equation. Chapters 4–8 continue the discussion of solutions of eigenvalue equations, in particular those for energy (the time-independent Schrödinger equation) and angular momentum. Commutators are introduced in Chapter 4, and are used throughout the book as an important tool, not only mathematically (as in Chapters 5 and 6, where they are applied in the solution of eigenvalue problems by algebraic methods) but also as an indication of the physical characteristics of a system (as in Chapter 7, where they are used to investigate the state of a particle in a central potential).

The description of an arbitrary state of a particle through the superposition principle is introduced in Chapter 9, and this mathematical formalism is necessary, to a greater or lesser extent, for developments in the last five chapters. It is impossible to cover all aspects of non-relativistic quantum mechanics in a single book. Important topics that are discussed briefly include the matrix formulation of the theory (Chapter 10), approximate methods for solving the Schrödinger equation (Chapter 11), time-dependent problems (Chapter 12), and the idea of a joint probability distribution for a system of many non-interacting particles (Chapter 13). Quantum scattering theory is omitted, but it would be artificial to avoid all mention of scattering experiments, since they provide such an important means of investigating quantum systems. Consequently scattering from a potential step and from a square well is discussed in Chapter 3, and the Born approximation is derived as an example of the application of time-dependent perturbation theory in Chapter 12. The final chapter is not intended to be part of an examined course, but aims to clarify the concepts of indeterminacy and non-locality that underlie quantum mechanics.

Examples

The physical examples used in the text have been chosen to reflect the enormously wide range of phenomena explained by quantum mechanics, and include some (in particular the scanning tunnelling microscope and quantum well structures) that are of importance in modern technology. Restrictions in the time allotted to a lecture course would usually make it impossible to cover all the examples mentioned, but many would be encountered in parallel physics courses – on spectroscopy or condensed matter physics, for example.

Problems

Problems are provided at the end of each chapter, some of which are fairly simple elaborations on the text, while others extend the theory that has been presented to different systems. Some problems (indicated by an asterisk) refer to programs on the accompanying computer disk, and provide detailed suggestions for using them. Briefly, the interactive programs allow students to solve energy eigenvalue equations for a range of systems and plot the associated wave functions and probability distributions, to examine the transmission of a plane wave through a one-dimensional square well or barrier, to watch the scattering of a Gaussian wave packet by a square barrier, and to construct wave packets for a particle in an infinitely deep one-dimensional square well or a one-dimensional simple harmonic oscillator and observe how they evolve with time.

The programs on the disk are described in detail in Appendix C. The full menu is as follows:

(1) The one-dimensional Schrödinger equation
  (a) Infinite square well
  (b) Finite square well
  (c) Harmonic oscillator
  (d) Anharmonic oscillator
  (e) Triangular well
(2) The Kronig–Penney model
(3) The Schrödinger equation: central potentials
  (a) Coulomb potential
  (b) Harmonic oscillator (three-dimensional)
(4) Orbital angular momentum
(5) Transmission
  (a) Plane wave, one-dimensional square well
  (b) Plane wave, one-dimensional square barrier
  (c) Gaussian wave packet, one-dimensional square barrier
(6) Wave packets (one-dimensional)
  (a) Infinite square well
  (b) Harmonic oscillator

The disk will run on any IBM-compatible pc, with at least a VGA monitor, though some of the programs will be rather slow on machines without a maths co-processor.

It should be noted that the software on the disk is free to qualified adopters only.

Acknowledgements

I should like to thank all my colleagues for useful discussions, advice, and technical information, and in particular Professor D.L. Weaire and Dr E.C. Finch for critical reading of the whole manuscript. I also wish to thank Andrew McMurry for writing the computer programs on the accompanying disc, and Fergal Toomey for the program demonstrating the scattering of a Gaussian wave packet. Finally, I am indebted to all those students on whom I have experimented with methods of teaching quantum mechanics over the last 25 years!

Figure 1.1a, taken from Preston T., *The Theory of Heat*, 1904, is reprinted by permission of Macmillan Ltd. Figure 1.1c is reprinted by permission of the COBE Science Working Group. Figure 1.2, is reprinted by permission of *Annalen der Physik*. **84**, 1927. Figures 1.3, taken from Millikan R. A., A direct photoelectric determination of Planck's *h*, *Physical Review* **7** 1916, 3.9, taken from Becker R. S. *et al* Electron interferometry at crystal surfaces, *Physical Review Letters* **55** 1985 and 7.5, taken from Moore K. J. Evolution of the electronic states of coupled (In, Ga) As-GaAs quantum wells into superlattice minibands, *Physical Review* B**42** 1990 and 8.8a, taken from von Klitzing K. The quantised Hall effect, *Reviews of Modern Physics*, **58** 1986 are reprinted by permission of the American Physical Society. Figure 1.5, taken from Tonomura A., *et al*, Demonstration of single-electron buildup of an interference pattern is reprinted by permission of the *American Journal of Physics* **57** 1989. Figure 4.4b taken from Gasiorowicz S. *Elementary Particle Physics*, 1966 is reprinted by permission of John Wiley & Sons Inc. Figure 4.4c, originally published in Mukhin A. I., Ozerov E. B. and Pontecorvo *Soviet Physics JETP* **4** 1957 and Figure 12.5, from Igo G., *High Energy Nuclear Structure*, ed. Nagle D., 1975 are reprinted by permission of the American Institute of Physics. Figure 4.6, from Goss Levi B., Super-deformed nuclei rotate so fast they make heads spin, is reprinted by permission of *Physics Today*. **41**, 1988. Figure 8.3a from E. Back and A. Lande, *Zeeman-Effekt und Multiplettstruktur der Spektrallinien* is reprinted by permission of Springer Verlag, Berlin. Figure 8.7, taken from de Haas W. J. and van Alphen P. M., *Proceedings of the Royal Academy of Sciences* **33**, 1930 is reprinted by permission of Elsevier Science Publishers. Table 4.2, from J. M. Hollas, *Modern Spectroscopy*, 1992 is reprinted by permission of John Wiley & Sons, Ltd.

Contents

List of Symbols and Physical Constants

Physical constants

α	$7.297 \times 10^{-3} = 1/137.04$	fine structure constant
c	$2.998 \times 10^8 \, \text{m s}^{-1}$	speed of light in vacuum
E_R	$13.61 \, \text{eV}$	Rydberg energy
e	$1.602 \times 10^{-19} \, \text{C}$	magnitude of electron charge
ε_0	$8.854 \times 10^{-12} \, \text{F m}^{-1}$	permittivity of free space
h	$6.626 \times 10^{-34} \, \text{J s}$	Planck's constant
$\hbar$	$1.055 \times 10^{-34} \, \text{J s}$	
k_B	$1.381 \times 10^{-23} \, \text{J K}^{-1}$	Boltzmann's constant
m_e	$9.11 \times 10^{-31} \, \text{kg}$	electron rest mass
m_p	$1.673 \times 10^{-27} \, \text{kg}$	proton rest mass
μ_B	$9.274 \times 10^{-24} \, \text{J T}^{-1}$	Bohr magneton
N_A	$6.022 \times 10^{23} \, \text{mol}^{-1}$	Avogadro's constant
R_H	$1.0968 \times 10^7 \, \text{m}^{-1}$	Rydberg constant for hydrogen
r_B	$5.292 \times 10^{-11} \, \text{m}$	Bohr radius

List of symbols

A	atomic mass number
A_n	Fourier expansion coefficient
$\hat{A}, \hat{A}^\dagger$	ladder operators
$\hat{A}, \hat{B}, \ldots$	linear Hermitian operators
$\mathbf{A, B, C}$	matrices
$A_{ij}, \hat{A}_{ij}$	matrix elements
$\mathbf{A}^T$	transpose of matrix

$\boldsymbol{A}^*$	complex conjugate of matrix
$\boldsymbol{A}^\dagger$	Hermitian conjugate of matrix
$\det(\boldsymbol{A})$	determinant of matrix
$\mathcal{A}, \mathcal{A}_x, \mathcal{A}_y, \mathcal{A}_z$	electromagnetic vector potential and its components
$\mathbb{A}$	area
$\hat{a}, \hat{a}^\dagger$	annihilation and creation operators
a_n, a	eigenvalues of $\hat{A}$
$\mathcal{B}, \mathcal{B}_x, \mathcal{B}_y, \mathcal{B}_z$	magnetic induction and its components
$\boldsymbol{b}, \boldsymbol{b}_n$	state vector and its components
$\boldsymbol{b}^\dagger$	Hermitian conjugate of state vector
$C(j, j_z; l, m_l, s, m_s)$	Clebsch–Gordan coefficient
c_s	velocity of sound in a crystal lattice
c_v	specific heat
$\mathcal{D}$	electric dipole moment
$\hat{\mathcal{D}}$	electric dipole moment operator
$\dfrac{\mathrm{d}\sigma}{\mathrm{d}\Omega}$	differential cross-section
$\mathrm{d}\Omega$	element of solid angle
E, E_{tot}	total energy
E_D	dissociation energy of diatomic molecule
E_F	Fermi energy
E_G	energy gap
E_G	ground state energy
E_R	Rydberg energy
$\hat{E}_{\text{tot}}$	total energy operator
$\mathbb{E}$	correlation function
$\mathcal{E}, \mathcal{E}_x, \mathcal{E}_y, \mathcal{E}_z$	electric field strength and its components
$\mathcal{E}$	electromagnetic energy density
$\mathcal{F}$	force
g	electron g-factor
$g(E)$	density of states
$\boldsymbol{g}$	reciprocal lattice vector
H_n	Hermite polynomial
$\hat{H}$	Hamiltonian operator
$\mathcal{I}$	electric current
$\mathcal{I}$	moment of inertia
J	total angular momentum quantum number
$\hat{\boldsymbol{J}}, \hat{J}_x, \hat{J}_y, \hat{J}_z$	angular momentum operator and its components
$\mathcal{I}_x, \mathcal{I}_y, \mathcal{I}_z$	components of the electric current density
j, j_z	angular momentum quantum numbers
$\boldsymbol{j}$	probability current density
$k, \boldsymbol{k}$	crystal momentum
$\boldsymbol{k}, k_x, k_y, k_z$	wave vector and its components
L	orbital angular momentum quantum number

$L_k^q(x)$	Laguerre polynomials
$\hat{\boldsymbol{L}}, \hat{L}_x, \hat{L}_y, \hat{L}_z$	orbital angular momentum operator and its components
$\mathbb{L}$	length
l	orbital angular momentum quantum number
M_J, M_L, M_S	total, orbital and spin angular momentum quantum numbers
$\boldsymbol{M}$	classical angular momentum vector
m	mass
m^*	effective mass
m_l, m_s	orbital and spin angular momentum quantum numbers
$\hat{N}$	number operator
N	number of unit cells
N	number density
N	number of events
n	principal quantum number
n_c	density of conduction electrons
$\boldsymbol{n}$	unit vector
$\hat{P}_r$	radial momentum operator
$P_l(\cos\theta)$	Legendre polynomial
$P_l^{m_l}(\cos\theta)$	associated Legendre function
$\hat{\boldsymbol{P}}, \hat{P}_x, \hat{P}_y, \hat{P}_z$	momentum operator and its components
$\mathbb{P}$	probability
p_r	radial momentum
$\boldsymbol{p}$	linear momentum
q	electric charge
$R(r), R_{nl}$	radial wave function
R_H	Rydberg constant for hydrogen
R_H	Hall coefficient
$\hat{\boldsymbol{R}}, \hat{X}, \hat{Y}, \hat{Z}$	position operator and its components
$\boldsymbol{R}_l$	lattice vector
$\mathbb{R}$	radius
$\mathbb{R}$	reflection coefficient
$\boldsymbol{r}, x, y, z$	position vector and its components
r	radial distance
S	spin quantum number
$\hat{\boldsymbol{S}}, \hat{S}_x, \hat{S}_y, \hat{S}_z$	spin operator and its components
s	spin quantum number
$\hat{T}$	transformation matrix
T	temperature in kelvin
$\hat{T}_l$	translation operator
T_{max}	maximum kinetic energy
$\mathbb{T}$	transmission coefficient
t	time
$u_k(x)$	periodic part of Bloch function

$V(x)$, $V(r)$	potential energy function
$\hat{V}_{mag}$	magnetic dipole interaction operator
V_{so}	spin–orbit coupling potential
$\mathcal{V}$	electromagnetic scalar potential
$\mathbb{V}$	volume
$\boldsymbol{v}$, v	velocity, speed
$\boldsymbol{x}$, $\boldsymbol{y}$, $\boldsymbol{z}$	unit vectors along Cartesian axes
$Y_{lm_l}(\theta, \phi)$	spherical harmonic
Z	atomic number

α, β	eigenvalues
γ	constant phase
Δx	uncertainty in x
$\delta(x - x')$	Dirac delta function
ϵ	dimensionless energy eigenvalue
θ	angle
λ	wavelength
λ, $\lambda_{\pm}$	eigenvalue of matrix
λ	variational parameter
$\boldsymbol{\mu}$	magnetic dipole moment
$\hat{\boldsymbol{\mu}}_l$, $\hat{\boldsymbol{\mu}}_s$	magnetic moment operators
v	frequency
ξ	dimensionless position variable
$\hat{\Pi}$	parity operator
ρ	dimensionless position variable
$\rho'(\lambda, T)$	spectral energy density as a function of wavelength
$\rho(v, T)$	spectral energy density as a function of frequency
ρ_{xx}, ρ_{xy}	resistivity
$\hat{\sigma}_x$, $\hat{\sigma}_y$, $\hat{\sigma}_z$	Pauli matrices
σ	total cross-section
σ	width of Gaussian
Φ	magnetic flux
Φ	wave function
$\phi(p, t)$	momentum space wave function
ϕ	work function
ϕ	azimuthal angle
X, χ	angular momentum eigenspinors
Ψ, ψ	wave function
ω	angular frequency
ω_c	cyclotron frequency
ω_L	Larmor frequency

CHAPTER 1

A Review of the Origins of Quantum Theory

1.1 ... and there was light!

Light is one of our most important sources of information about the physical nature of the world around us, and quantum theory arose from attempts to interpret observations of the light emitted by incandescent materials in terms of their atomic and molecular structure. Visible light brings us direct evidence of structures in our universe varying over an enormous range of sizes, from stars and galaxies to microscopic organisms. As infants we learn to interpret the signals detected by our eyes in terms of the size and shape of surrounding objects. This is the regime of geometrical optics, where the detected light has a wavelength much smaller than the scale of the features that it makes visible – the wavelength of visible light ranges from approximately 450 nm (1 nm = 10^{-9} m) at the blue end of the spectrum to 650 nm at the red end.

Light also carries information about features with dimensions comparable to its wavelength, though in this case the geometrical image becomes blurred by diffraction, and the shape and size of the features are not immediately recognizable. The information is carried in the **phase** of the light wave, and interference between light waves emanating from different parts of an object produces the diffraction pattern. This is the type of information recorded on a hologram, which cannot be understood until it is decoded and exhibited as a geometrical image by appropriate illumination. The wave theory is the key to this coded information, and only through this theory can we relate a diffraction pattern to the shape and size of the features that produced it. Objects a few hundred nanometres in width can produce diffraction patterns with visible light, and smaller features can be detected with light of wavelengths far below the visible. For example, the arrangement of atoms in a crystal can be deduced from

1

the diffraction patterns produced by X rays with wavelengths of the order of 10^{-1} nm, which is comparable to the spacing between the atoms.

The shape and size of an object determine the intensity distribution of light scattered from it. But light carries a wealth of further information that has only begun to be fully exploited in this century: this information is visible as the colour of a material (which results from the selective absorption of characteristic wavelengths, so that the reflected light is depleted in these wavelengths and appears coloured). The particular wavelengths that are absorbed by a material depend on its molecular and atomic structure, and the absorption is associated with an increase in the internal energy of the material. Similarly a loss of internal energy can be accompanied by the emission of light. Energy changes of the order of electron volts $(1\,\mathrm{eV} = 1.609 \times 10^{-9}\,\mathrm{J})$ are associated with the absorption or emission of wavelengths in the visible region, typical of many atomic spectra. Energy changes around a million times greater, of the order of $10^6\,\mathrm{eV} = 1\,\mathrm{MeV}$, occur in nuclear interactions, and can be associated with the absorption or emission of γ radiation with wavelengths of the order of 10^{-3} nm.

When Maxwell explained the nature of light as an electromagnetic wave, it became clear that light is absorbed through its interaction with electric charge distributions, and that the spectrum of wavelengths emitted by an incandescent material is coded information about the electromagnetic structure of matter. An incandescent gas of any particular element emits a unique series of spectral lines at discrete wavelengths, and this spectrum is characteristic of the chemical element concerned. Such a line spectrum must therefore contain information about the electromagnetic structure of an atom, and the way this differs from element to element. Incandescent solids, on the other hand, emit a continuous range of wavelengths, and the distribution of intensity with wavelength (which is what we mean by the spectrum of the radiation) is independent of chemical composition but varies with temperature. The ideal spectrum of this type, characteristic of the temperature but completely independent of the nature of the body producing it, is called the **black body radiation spectrum**, and it provides information about the thermal motion of atoms bound together in a solid.

By definition, a **black body** absorbs 100% of all thermal radiation falling on it [1]. A close approximation to this is a very small hole into a cavity in a solid maintained at some steady absolute temperature T. Any radiation falling on the hole will enter the cavity and be reflected back and forth inside, but very little will escape out through the hole again. Within the cavity the solid walls emit radiation whose spectrum depends on T, and this radiation is partly absorbed and partly reflected each time it is incident on a wall. If the cavity is in thermal equilibrium, the rate of emission per unit area of radiant energy at any particular wavelength must exactly balance the corresponding rate of absorption, and the spectrum of the radiation within the cavity will be characteristic of the temperature but independent of the material of the cavity walls. This is the closest we can come experimentally to the ideal black body radiation spectrum, and it can be measured by observing the small amount of radiation that escapes through the hole. The black body spectrum at several different temperatures is shown in Figure 1.1(a), where the energy density of the radiation, which is proportional to its intensity, is plotted as a function of its wavelength. The peak intensity moves to lower wavelengths λ, or higher frequencies $v = c/\lambda$ (where c is the velocity of light in vacuum), as the temperature is increased.

By the end of the nineteenth century the black body spectrum had been measured to a high degree of accuracy, and it was known that the total energy radiated by a black body at temperature T is proportional to T^4 (the **Stefan–Boltzmann law**). Similarly the atomic spectra of different elements were well documented: Lockyer and Ramsay had identified a new element – helium – from hitherto unknown spectral lines in the light from the Sun, and Balmer had discovered a mathematical formula to fit the wavelengths of a set of lines in the visible region of the hydrogen spectrum. But, in spite of much theoretical effort, the information contained in the wealth of experimental data on both black body and atomic spectra could not be deciphered, and the structure of matter on an atomic scale was still a mystery. It was Planck who discovered the key to the code in 1900, but it was not finally cracked until 1926, when Schrödinger and Heisenberg developed the wave and matrix formulations of quantum mechanics, respectively.

With the understanding provided by quantum mechanics we have now entered an era of engineering at a quantum level: the atomic and molecular structure of matter can be manipulated to construct new materials with properties useful for applications ranging from microelectronics and optical communications to space flight. At the subatomic level quantum mechanics is essential for an understanding of nuclear and subnuclear phenomena [2]. The electromagnetic force is not the only one of importance here, though it was the electromagnetic scattering of α particles by atoms that led Rutherford to his nuclear model of the atom, and more recently an experiment in which electrons were scattered electromagnetically by protons first demonstrated that protons themselves are composed of three even more elementary particles, called quarks. All theories that describe how particles behave under the influence of the strong and weak nuclear forces (which are responsible respectively for holding the nucleus together against the Coulomb repulsion of the protons, and for the β decay of nuclei) derive ultimately from the quantum mechanics first developed by Schrödinger and Heisenberg.

The study of the elementary particles of matter, such as electrons, neutrinos and quarks, relates the quantum world of the very small to the behaviour of matter on a cosmic scale, since in the early moments after

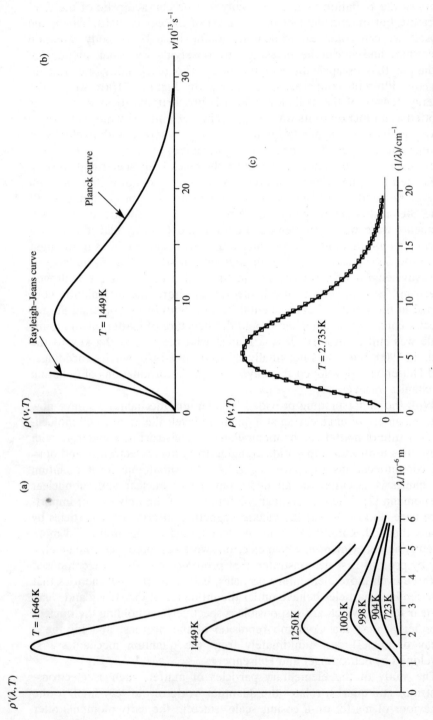

Figure 1.1 The spectral energy density of black body radiation: (a) as a function of wavelength λ, at the temperatures indicated [4]; (b) as a function of frequency ν, calculated from Planck's law (1.1) at $T = 1449$ K (the Rayleigh–Jeans prediction for the same temperature is also shown); (c) of the cosmic background radiation as a function of inverse wavelength $1/\lambda = \nu/c$. The experimental data (squares) are from the COBE satellite, and the solid curve shows the Planck prediction for $T = 2.735$ K [5].

4

the big bang the universe consisted of a hot 'soup' of subnuclear particles [3]. Until the universe was about three minutes old it was too hot for nuclei containing more than three or four nucleons (the generic name for protons and neutrons) to be stable. It took about 700 000 years more to cool sufficiently for electrons and nuclei to combine together in stable, electrically neutral atoms. Before this, the matter in the universe was a hot plasma containing charged particles such as electrons, positrons and nuclei, which continually emitted and reabsorbed electromagnetic radiation. The radiation was therefore in thermal equilibrium with the matter, with a black body type of spectrum. But when the matter condensed into neutral atoms the thermal connection between matter and radiation was broken, and the distribution of intensity with wavelength in most of the radiation was not modified by further interactions. This radiation therefore retained a black body spectrum, but, because the expansion of the universe produces a red shift, the peak was shifted to progressively longer wavelengths, corresponding to lower temperatures. This radiation is observable today as the cosmic background radiation, which reaches us almost uniformly from all directions in space, and has a black body spectrum characteristic of a temperature of just under 3 K (while the temperature of the universe when the thermal equilibrium with matter was broken was about 3000 K). The spectrum of the cosmic background radiation is illustrated in Figure 1.1(c). Not only the cosmic background radiation but also other fundamental characteristics of the universe as we observe it today, including for instance the overall ratio of helium to hydrogen, depend critically on the nature and interactions of the elementary particles in the primordial soup. So the development of the universe itself cannot be understood without an understanding of quantum mechanics.

1.2 The quantization of energy

The key to the decipherment of the coded information contained in optical spectra is the assumption that

> Oscillating charge distributions only emit and absorb electromagnetic energy in discrete units (quanta) which are integer multiples of $h\nu$, where ν is the frequency of the electromagnetic radiation and h is a new constant of nature – **Planck's constant**. (A)

It is impossible to account for either the black body radiation spectrum or atomic line spectra unless this is the case.

Planck had discovered that the spectrum of black body radiation could be fitted by his famous expression for the **spectral energy density** of the radiation in equilibrium with a black body at absolute temperature T:

$$\rho(v, T) = \frac{8\pi h}{c^3} \frac{v^3}{e^{hv/k_B T} - 1} \tag{1.1}$$

where $\rho(v, T) \, dv$ is the energy per unit volume in the frequency range from v to $v + dv$, k_B is Boltzmann's constant, c is the velocity of light and h is Planck's constant. The shape of the graph of $\rho(v, T)$ plotted as a function of v is shown in Figure 1.1(b). The substitution $v = c/\lambda$ gives $\rho(v, T) \, dv = \rho'(\lambda, T) \, d\lambda$, where $\rho'(\lambda, T)$ determines the energy density as a function of wavelength λ, and the resulting theoretical curves match the experimental data of Figure 1.1(a) very accurately. In Figure 1.1(c) experimental measurements of the cosmic background radiation spectrum made by the COBE satellite are shown, and compared with the Planck prediction for the temperature $T = 2.735$ K. Planck's expression provides a very good fit to the black body spectrum, but the only way he could derive it was by making an assumption essentially equivalent to (A).

There are two steps in the calculation of the spectrum of black body radiation. First the radiation in a black body cavity is treated classically. The wavelengths, and thus the frequencies, are limited to the set of discrete values allowed for standing waves, and electromagnetic theory gives the number of standing waves per unit volume in the frequency interval from v to $v + dv$ as [6]

$$N(v) \, dv = 2 \frac{4\pi}{c^3} v^2 \, dv \tag{1.2}$$

The factor 2 arises because there are two independent polarization states for each standing wave in the cavity. A **cavity mode** is a standing wave with a particular frequency and polarization. The second step is to find the average energy $\langle E \rangle$ in each cavity mode at temperature T. It was this second step that Planck modified.

Radiation in a cavity reaches thermal equilibrium by exchanging energy with oscillating charge distributions in the material of the walls. The oscillating electric field of the radiation in the cavity produces forced oscillations of the charge distributions, which, according to classical electromagnetic theory, will re-emit radiation at the frequency of their oscillation. This mechanism allows the radiation to exchange energy with the material (kept at a fixed temperature) and reach equilibrium. Planck showed that at equilibrium the average energy of a cavity mode of the radiation with frequency v is equal to the average energy of thermal oscillations at that same frequency v in the material. So the average energy of a cavity mode can be found by calculating the average energy of a charge distribution oscillating with natural frequency equal to the frequency of the mode.

According to classical statistical mechanics, the average energy of a particle in one-dimensional thermal oscillation should be $k_B T$, independent of its frequency [7]. Using this result for the average energy of a cavity mode in 1900 (just before Planck suggested (1.1)), Rayleigh found that the energy density of black body radiation, $\rho(v, T) = N(v)\langle E\rangle$, should increase with frequency as v^2. This implies that, whatever the temperature, more and more energy is emitted at larger and larger frequencies, and so the total energy emitted is infinite! This is the **ultraviolet catastrophe**. (Actually Rayleigh did not get his calculation quite right! He did not evaluate the numerical factor multiplying $v^2 \, dv$ in (1.2) until five years after his original paper on the subject, and when he did he got it wrong by a factor of 8. Jeans corrected this, so the classical law is called the **Rayleigh–Jeans law**.) Although this high-frequency behaviour cannot be correct, the Rayleigh–Jeans law agrees with experiment at low frequencies.

Planck replaced $k_B T$ as the average energy of a cavity mode of frequency v by

$$\langle E \rangle = \frac{hv}{e^{hv/k_B T} - 1} \tag{1.3}$$

The product of this with $N(v)$, (1.2), gives Planck's law (1.1). An expansion of the denominator of (1.3) shows that $\langle E \rangle \approx k_B T$ if $hv \ll k_B T$, which means that at sufficiently low frequencies Planck's law gives the same predictions as the Rayleigh–Jeans law, as can be seen in Figure 1.1(b). Assumption (A) is in flagrant contradiction to the classical belief that energy is always exchanged in a continuous manner, and it is this difference that leads to (1.3) instead of $\langle E \rangle = k_B T$. Both the classical expression and Planck's can be obtained by assuming that the probability of an oscillator having energy E at temperature T is given by the Boltzmann distribution, proportional to $e^{-E/k_B T}$. Classically, however, the energy E can vary continuously, so the calculation of $\langle E \rangle$ involves an integration over the continuous range of energies from 0 to ∞. On the other hand, Assumption (A) restricts energy changes to discrete units, and (1.3) is obtained by summing over the infinite but discrete set of energies $E = nhv$ ($n = 0, 1, 2, \ldots$) for fixed v [8].

Assumption (A) not only accounts for the black body spectrum but was also invoked by Bohr to make Rutherford's nuclear model of the atom [9] feasible, and to account for atomic line spectra. The chief argument against the picture of an atom as a positively charged nucleus orbited by negatively charged electrons was that such a structure could not be stable. Classically, an orbiting charge will radiate electromagnetic energy at the frequency of its orbit, and as it loses energy it will spiral down to orbits of smaller and smaller radius until it falls into the nucleus and the atom collapses. Since the frequency in an orbit increases as the radius decreases, the electron should emit radiation at frequencies in a continuous range above that of the original orbit, and so the observed line spectrum is completely inexplicable. Bohr saw that if an orbiting electron was only able to emit energy in discrete units, instead of continuously, this spiral-ling collapse would not occur. He pictured the electrons in an atom as confined to a set of discrete stable orbits, with discrete energy differences between them. Electromagnetic energy would only be emitted or ab-sorbed during a transition from one of these allowed orbits to another. A transition from an orbit with total energy E_i to one of lower energy E_f would result in the emission of a quantum of radiation of frequency

$$v = (E_i - E_f)/h \tag{1.4}$$

So an electron in an atom should have a discrete set of allowed energies $\{E_n; \; n = 1, 2, 3, \ldots\}$, and the separations between the levels may be determined experimentally by observing the frequencies of the lines in the atomic spectrum. The set of allowed energies in an atom is its **energy spectrum**, which must, like its line spectrum, be characteristic of the chemical element.

Balmer's formula for the wavelengths of a series of emission lines in the spectrum of hydrogen can be written in the form

$$\frac{1}{\lambda} = \frac{v}{c} = R_H \left(\frac{1}{2^2} - \frac{1}{n^2} \right) \quad (n = 3, 4, 5, \ldots) \tag{1.5}$$

where R_H is the Rydberg constant for hydrogen. Bohr recognized that this equation is of the same form as (1.4), provided that the energies of the allowed electron orbits in hydrogen are of the form

$$E_n = -hcR_H \frac{1}{n^2} \quad (n = 1, 2, 3, \ldots) \tag{1.6}$$

The Balmer series of spectral lines corresponds to transitions from energy levels with $n > 2$ to the level with $n = 2$. Other series of hydrogen lines exist, corresponding to transitions between levels with $n > m$ and the mth level, including a set of lines in the ultraviolet that represent transitions to the lowest energy level, or ground state, for which $n = 1$. Figure 1.2 shows a part of the emission spectrum of hydrogen, and indicates the lines of the Balmer series. The set of energy levels that accounts for this spectrum, and the transitions corresponding to the Balmer series, are also shown. Note that, according to (1.6), the energy levels get closer together as n increases, and approach the limit $E = 0$ as n becomes infinitely large. Experiment shows that a hydrogen atom in its ground state is ionized by radiation with frequency corresponding to an energy of 13.6 eV, and above this a continuous range of frequencies can be ab-sorbed. This indicates that the ground state is 13.6 eV below the energy $E = 0$ at which the electron escapes from the atom and becomes free. If a quantum of energy greater than the ionization energy is absorbed, the additional energy is converted to kinetic energy of the freed electron,

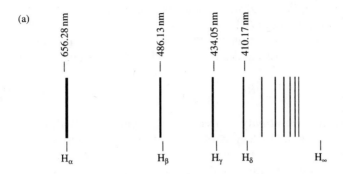

(a)

(b)

Figure 1.2 The hydrogen spectrum. (a) Lines from the Balmer series in the optical emission spectrum [10]. H_∞ marks the beginning of the continuous spectrum produced when an atom in the $n = 2$ level is ionized. (b) The energy spectrum calculated from (1.6), showing transitions responsible for the lines from the Balmer series illustrated in (a) in the enlargement on the right.

which can take any value in the continuous range from 0 to ∞. It is only when the electron is bound to the nucleus that its energy is restricted to the discrete set of values $\{E_n\}$.

Bohr derived an expression for the Rydberg constant in terms of the charge e and mass m_e of the electron, and Planck's constant:

$$R_H = \left(\frac{e^2}{4\pi\varepsilon_0}\right)^2 \frac{2\pi^2 m_e}{h^3 c} \tag{1.7}$$

(where ε_0 is the permittivity of free space). This agrees with the experimental value to a high degree of accuracy (see Problem 1.4). In this derivation he used classical mechanics to give the energy of an electron in the Coulomb potential due to the nucleus, and added a quantization condition in order to reproduce the form (1.6) for the energy levels. The necessary condition is that the magnitude of the orbital angular momentum can only take values that are integer multiples of $\hbar = h/2\pi$. This means that not only the total energy of the electron but also its angular momentum is quantized (though, as we shall see in Chapter 4, the correct, quantum mechanical, angular momentum quantization condition differs slightly from that suggested by Bohr).

With this concoction of classical and quantum ideas Bohr was able to describe atomic structure, explaining the differences between atoms of different elements in terms of differences in the charge on the nucleus and the number of electrons orbiting it. Various refinements were introduced, including the Pauli exclusion principle to explain the distribution of the electrons among the available energy levels in a multi-electron atom. But a complete and unambiguous derivation of atomic energy levels and spectra was not possible before the advent of quantum mechanics.

1.3 Particle/wave duality

The only new idea necessary to explain the black body radiation spectrum is Assumption (A). In Section 1.2 this was taken to mean that energy exchange with radiation of frequency v is restricted to integer multiples of hv because the thermal energy of charged particles in the solid is quantized. However, Einstein, in his famous paper on the photoelectric effect in 1905, made a different and even more controversial assumption:

> Electromagnetic radiation of frequency v can behave as if it consists of discrete units (quanta) of energy, of magnitude hv.   **(B)**

That is, electromagnetic energy itself is quantized. A solid (or an atom or any other quantum system) can absorb energy from a monochromatic beam of light of frequency v only in units of hv *because the light arrives in the form of discrete quanta, each with this energy.* The spacing between the energy levels of charged constituents of the material determines whether or not light quanta with this particular frequency *can* be absorbed.

Einstein's equation for the photoelectric effect can be derived as an energy conservation equation on the basis of Assumption (B). If an electron in the metal photocathode absorbs a quantum of electromagnetic energy hv, the electron will be emitted provided that this energy is sufficient to free it from the metal surface. The photoelectron will be emitted with a kinetic energy equal to the difference between hv and the energy necessary to overcome the potential holding the electron in the material. If the minimum amount of energy necessary for an electron to escape from the metal surface is ϕ, the **work function** of the metal, the photoelectrons will be emitted with a maximum kinetic energy T_{max}, where

$$T_{max} = hv - \phi \qquad (1.8)$$

The minimum frequency of light that will produce photoemission from a particular photocathode is $v_{min} = \phi/h$, when there is no surplus energy to provide the electrons with kinetic energy. Millikan's elegant series of experiments in 1915 verified Einstein's equation (1.8), which predicts that the maximum kinetic energy of the electrons should vary linearly with the frequency of the light, the slope of the line being Planck's constant. Some of Millikan's results are shown in Figure 1.3. The only effect of a change in light intensity is to change the rate of emission of photoelectrons. This is completely contrary to any classical calculations, which would link the energy absorbed by an electron to the intensity – proportional to the energy density – of the light beam. But, according to Assumption (B), the energy density of the beam is given by the product of the energy carried by each quantum, hv, with the number of quanta per unit volume in the beam. So the beam intensity determines the rate of arrival of quanta, which limits the rate at which photoelectrons are produced, but has no effect on their kinetic energy.

Einstein's idea inspired de Broglie to suggest that if light could sometimes behave like particles then particles should also show wave-like behaviour. He related the wavelength and frequency of the wave associated with a free particle to its momentum and energy respectively:

$$p = \frac{h}{\lambda}, \qquad E = hv \qquad (1.9)$$

The relationship between wavelength and momentum has been amply confirmed by electron and neutron diffraction experiments [12].

Einstein's Assumption (B) introduced the problem of wave/particle duality. Interference and diffraction phenomena can only be explained on

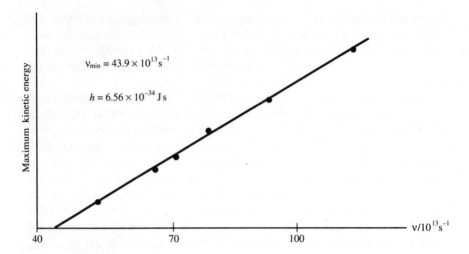

Figure 1.3 Millikan's data [11] for the maximum energy of photoelectrons emitted from a sodium surface as a function of the frequency of the light. The values he deduced for $v_{\min}$ (from the intercept) and h (from the slope) are given.

the basis of the wave theory, and in a classical wave the energy is spread continuously through a region of space. So how can Einstein's quantum picture of light be reconciled with the wave theory? And how do the wave-like properties of particles coexist with their particle-like characteristics (such as a finite size)? Although for many years most physicists felt that wave and particle characteristics are mutually incompatible, the ability of an atomic or subatomic particle to behave in some circumstances as a wave and in others as a particle is fundamental to quantum physics, and finds mathematical expression in quantum mechanics.

Since electromagnetic radiation exerts radiation pressure in the direction of propagation, it must carry momentum, so a quantum of light must have momentum as well as energy, just as any classical particle does. According to classical electromagnetic theory, radiation in which the electric field is $\mathcal{E}$ and the magnetic induction is $\mathcal{B}$ carries momentum of density $\mathcal{E} \times \mathcal{B}/c^2$ [13]. The magnitude of this is just $1/c$ times the energy density $|\mathcal{E} \times \mathcal{B}|/c$ carried by the radiation. Therefore we assume that each quantum of light with energy E carries momentum of magnitude p, where

$$p = \frac{E}{c} = \frac{hv}{c} = \frac{h}{\lambda} \tag{1.10}$$

For a free relativistic particle the total energy is $E = \sqrt{(p^2c^2 + m^2c^4)}$, and so (1.10) is what we should expect for a relativistic particle with zero mass. The Compton effect [14], in which X rays are scattered by electrons, with an accompanying change of wavelength, confirms that quanta of electromagnetic radiation can be scattered like massless particles. These particles are called **photons**, and travel with the velocity of light. Note that de Broglie's relationship between wavelength and momentum applies equally well to photons as to particles such as electrons and neutrons. However, the relationship between the wavelength and frequency is different for electromagnetic and for matter waves. For example, for a relativistic particle with $E = \sqrt{(p^2 c^2 + m^2 c^4)}$ we have

$$v = c \sqrt{\left(\frac{1}{\lambda^2} + \frac{m^2 c^2}{h^2} \right)}$$

In the quantum realm, then, there is no absolute distinction between particles and waves, and we shall use the term **quantum particle** to refer to photons as well as to electrons, protons, nuclei, atoms, and so on. A quantum particle exhibits wave-like or particle-like characteristics, depending on the experimental set-up used to observe it. However, the concept of the photon as a particle should be used with caution: not only is it a quantum rather than a classical particle, but it can never be treated even approximately as non-relativistic. This is because it travels with the velocity c, and, according to the special theory of relativity, all observers – no matter what their relative velocities – will agree on this. So there is no frame of reference in which a photon can be observed with a velocity less than c, and no frame of reference in which it is at rest. In contrast, an electron, even though it must be treated as a quantum particle, behaves non-relativistically at energies low compared with the electron rest mass energy 0.511 MeV. The non-relativistic quantum mechanics that is the subject of this book is not adequate for treating photons. Reconciling the quantum mechanics of Schrödinger and Heisenberg with relativity introduces a surprising number of completely new concepts (including the existence of antiparticles) and leads to quantum field theory. The quantum field theory of electrons, positrons and photons – **quantum electrodynamics (QED)** – is an amazingly successful theory, which predicts, for example, a value for the magnetic moment of the electron, in units of the Bohr magneton, that agrees with the experimental determination to 10 significant figures!

1.4 The two-slit diffraction experiment

In 1802 Thomas Young demonstrated very elegantly, in his famous two-slit interference experiment, that light behaves like waves. But if Einstein's Assumption (B) is correct, light can also be absorbed by matter as if it were as a stream of particle-like energy quanta. As de Broglie suggested, this duality is characteristic of all quantum particles. Indeed, beams of electrons or neutrons can produce an interference pattern just like that produced with light by Young's slits, provided some appropriate method is used to split the incident beam into two coherent beams, and then recombine them on a detector screen. We need to achieve a unified description of the wave and particle aspects of matter, but there are conceptual difficulties in doing this, which become very clear when we examine the production of the interference pattern by Young's slits from the point of view of the particle aspect of the incident beam.

The two-slit Fraunhofer diffraction experiment, shown schematically in Figure 1.4(a), is a modification of the original Young's slits experiment [15]. A coherent, plane, monochromatic beam of light falls on a screen S_1, in which there are two narrow closely spaced slits, and the resulting diffraction pattern is observed on a distant detector screen S_2. (In practice, lenses can be used to make the light from a point source parallel, and to put the screen S_2 at effectively an infinite distance from the screen S_1.) On basis of the wave theory it is easy to explain the diffraction pattern as the result of interference between the light passing through slit A and that passing through slit B. Provided the distance between the screens is sufficiently large, the optical path difference between waves from the two slits arriving at point P on S_2 is $d \sin \theta$, where d is the separation between the slits and θ the angular position of P. The waves interfere constructively, giving a maximum of the diffraction pattern, if this path difference is zero or an integer multiple of the wavelength λ of the light. Correspondingly, minima occur at angles such that the optical path difference between the two beams is an odd half-integer number of wavelengths:

$$d \sin \theta = (n + \tfrac{1}{2})\lambda \quad (n = 0, 1, 2, 3, \ldots) \tag{1.11}$$

According to de Broglie's relation (1.9), a parallel beam of quantum particles, all moving with the same momentum p and energy E, can be diffracted in just the same way as a plane monochromatic wave of wavelength $\lambda = h/p$. The diffraction pattern is predicted by the wave theory, and the position of the minima is determined by the same equation (1.11), whether the beam consists of light, electrons, neutrons, or any other quantum particle.

Now think of this experiment in terms of a stream of particles passing through the slits, and arriving at the detector screen. If the particles behave classically, the beam should spread out after passing through a slit, because of deflection at the edges, and a particle passing through slit A should arrive at any point in the strip R_A shown in Figure 1.4(b). Indeed, in the usual optical experiment we observe a fairly uniform distribution of light over this region if only slit A is open (though a close inspection will show fringes at the edge due to single-slit diffraction effects). Similarly, if only slit B is open, the strip R_B is illuminated.

(a)

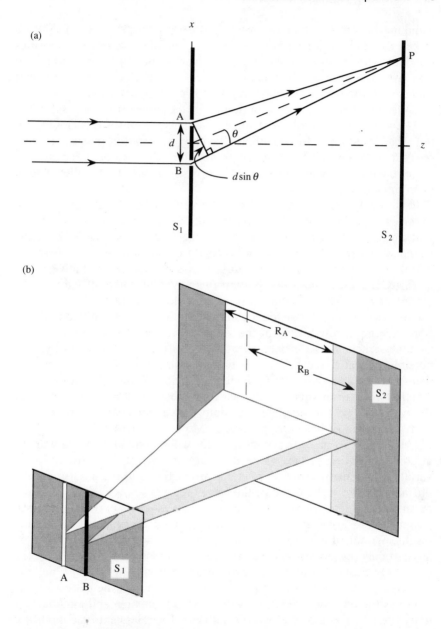

Figure 1.4 Two-slit diffraction: (a) Schematic representation of the experiment. The arrowed lines can either be interpreted as classical trajectories to point P through slits A and B, or as normals to the wave fronts reaching P from the two slits. The optical path difference is $d \sin \theta$. (b) The regions R_A and R_B that would be illuminated if only one slit were open in a two-slit diffraction experiment. When both slits are open the diffraction pattern is observed in the region where R_A and R_B overlap.

Thinking in terms of photons, we might expect the intensity of light on screen S_2 when both slits are open to be a simple sum of the intensities obtained with each slit separately. This would imply a uniform illumination over the area where regions R_A and R_B intersect, and could not account for the pattern of dark and light fringes that is actually observed. It is clear that there are certain points on the detector screen, satisfying (1.11), at which a photon from slit A can arrive when slit B is closed, but not when it is open. In fact, the behaviour of a photon passing through one slit is influenced by whether or not the other is open! Exactly the same remarks apply if the beam consists of electrons or any other type of quantum particle.

It is clearly difficult to interpret the diffraction process in terms of particles. However, the detection process at screen S_2 depends on the particle rather than the wave aspect of the beam. This can be demonstrated with light by reducing the intensity of the source: the diffraction pattern does not fade smoothly, but becomes granular, and at low enough intensity is seen to be a distribution of individual points of light. If the screen were replaced by an array of suitable photodetectors, we could observe the arrival of individual photons. Similarly, in an electron diffraction experiment the arrival of individual electrons can be recorded on a suitable screen. If the incident beam is of sufficiently low intensity, it is possible to watch the diffraction pattern being built up from points of light marking the arrival of individual particles. This is shown in Figure 1.5. Initially, when very few particles have been detected, it seems that their arrival at S_2 is completely random, but as more and more arrive a pattern gradually becomes apparent. Most particles arrive in regions corresponding to the diffraction maxima, while comparatively few arrive at points satisfying (1.11), where there are minima of the pattern. So the variation in density of arrival points is proportional to the intensity of the diffraction pattern predicted by the wave theory.

Any attempt to identify the slit through which a particular particle passed before arriving at some point on screen S_2 produces frustrating results. We might close one slit, B say, so that any particle arriving at S_2 has certainly passed through slit A. But in this case, as we know from the classical optical experiment, the two-slit diffraction pattern is lost, and the particles arrive almost uniformly over strip R_A. (In fact the pattern on the screen is the one-slit diffraction pattern, whose maxima and minima bear no relation to those of the two-slit pattern). Even less extreme modifications are doomed to failure, as discussed so clearly by Feynman [17]. Any measurement that can distinguish the slit through which a particular particle passed inevitably destroys the two-slit pattern, and if the two-slit pattern is visible then it is not possible to identify the slit through which any one of the particles contributing to it passed. A classical particle should follow a unique trajectory, through one slit or the other, and its behaviour can only be influenced by the slit through which it passes. But

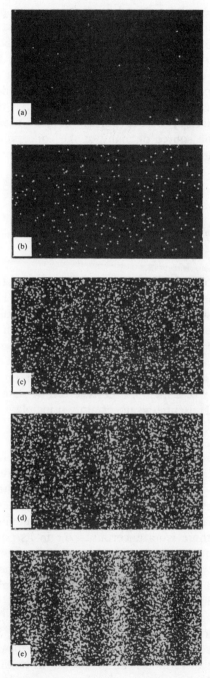

Figure 1.5 The gradual build-up of a two-slit diffraction pattern produced with electrons [16]. The numbers of electron arrival points recorded in sucessive photographs are (a) 10, (b) 100, (c) 3000, (d) 20 000 and (e) 70 000.

17

the diffraction pattern is produced by each quantum particle interacting with *both* slits, so that in a sense each particle must have passed through both slits if a two-slit diffraction pattern is observed. So we *cannot* think of each quantum particle following a unique, even though unknown, trajectory. Instead we have to admit that trajectories through both slits affect the arrival point of each particle on screen S_2. The path of a particle through the apparatus is not just unknown – there *is* no unique path, and the trajectory of the particle is said to be **indeterminate**. Indeterminacy is the source of many of the conceptual difficulties with quantum theory, and we shall return to it in Section 1.5 and again in Chapter 14.

1.5 Uncertainty and indeterminacy

The arrival point of a quantum particle on the detector screen in a diffraction experiment cannot be predicted precisely. The best we can do is to predict the statistical distribution of arrival points when a large number of particles is detected, and, as we saw in Section 1.4, this distribution is proportional to the intensity distribution of the diffraction pattern. This inherently statistical type of prediction is characteristic of quantum theory. There is a *fundamental* limitation on the precision with which the trajectory of a quantum particle can be determined, expressed mathematically by Heisenberg's uncertainty principle:

$$\Delta x \, \Delta p_x \gtrsim \tfrac{1}{2} \hbar \qquad (1.12)$$

This means, roughly speaking, that if a measurement finds the x component of the position of a quantum particle to lie somewhere between $x - \tfrac{1}{2}\Delta x$ and $x + \tfrac{1}{2}\Delta x$ then there is an inherent uncertainty $\Delta p_x \gtrsim \hbar/(2 \, \Delta x)$ in the value of the corresponding component of momentum. Conversely, if the value of the x component of momentum is measured to an accuracy Δp_x, there is an uncertainty Δx in its position along the x axis that is related to Δp_x by (1.12). If p_x is known precisely, that is, with uncertainty $\Delta p_x = 0$, then x is completely unknown, and vice versa.

The position and momentum of a particle are **dynamical variables**, that is, variables describing its motion. Other dynamical variables, such as energy and angular momentum, can be expressed in terms of position and momentum in both classical and quantum mechanics. So the uncertainty relation (1.12) places restrictions on the accuracy to which the values of these other dynamical variables can be known, as we shall see in detail in Chapter 4. In general in quantum mechanics it is not possible to know, simultaneously, precise values for all the dynamical variables relevant to a particular system.

The uncertainty principle implies that after a measurement of the position of a quantum particle its momentum will have changed by an *unpredictable* amount, such that the product of the uncertainties in position and momentum satisfies (1.12). If a position measurement has determined x to an accuracy Δx, it is not possible to predict the result of a momentum measurement to an accuracy better than Δp_x. Similarly, as we shall see in Chapter 4, if the total energy of a particle is precisely known, it is not possible to predict a unique value for the result of a momentum measurement, except in the case of a free particle (for which the potential energy is constant and independent of position). Conversely a momentum measurement on such a particle will produce an unpredictable change in its total energy.

This means that the uncertainty principle signals a breakdown of the classical assumption that measurements can be performed without disturbing the system on which they are made. When we look at the ordinary macroscopic objects around us we assume that our observation is completely passive – the reflection of light from an object into our eyes had no measurable effect on its motion. We take it for granted that a moving billiard ball would have moved in exactly the same way (assuming an identical impulse set it in motion) even if the lights were switched off and the ball moved in total darkness. But this is not true for a quantum particle. Scattering light (or a beam of any other type of quantum particle) from it to determine its position inevitably produces changes in its momentum, as illustrated by the Heisenberg microscope through experiment [18].

But there is another, more subtle, implication of the uncertainty principle, which tends to be obscured by discussions like that of the last paragraph concerning the effects of measurements. Changes produced in the value of one variable by the measurement of another are not simply unknown and unpredictable but also *indeterminate*. The uncertainty represents more than a lack of information on the part of the observer, it represents a fundamental lack of precision in the value of the variable concerned. For example, if there is a non-zero uncertainty Δx in the position of a particle then the particle does not have a unique, even if unknown, position: within the region of width Δx the wave aspect is dominant, so that in a sense the quantum particle is spread over the whole of this region. We have already met this indeterminacy in discussing two-slit diffraction. There, observation of the diffraction pattern is only possible if each particle contributing to the pattern has the wave-like characterisitic of interacting, not with one slit or the other, but with both. The position of such a particle as it passes through the slit screen (S_1 in Figure 1.4a) is indeterminate, and has an uncertainty at least as great as the spacing between the slits ($\Delta x \geqslant d$ in the direction perpendicular to the slits). If we make a measurement that determines through which slit a particle passed, the uncertainty in position at the slit screen must be reduced to $\Delta x < \frac{1}{2}d$. Suppose that $\Delta x \approx d/4\pi$; then, according to (1.12), $\Delta p_x \gtrsim h/d$. The wavelength, and therefore the magnitude of the momentum $p = h/\lambda$, is unchanged on diffraction. Only the direction of the momentum is changed, as we can see from Figure 1.4(a), and a particle diffracted through an angle θ must have acquired a component of momentum parallel to the screen given by $p_x = p \sin \theta$. The uncertainty in p_x means that the angle through which any particular particle is diffracted is indeterminate within the range $\Delta \theta \approx \sin \Delta \theta \gtrsim \lambda/d$ (and we have assumed that $\lambda/d \ll 1$). This means that the particles will arrive completely randomly at the detector screen S_2 over a region of angular width $\Delta \theta \gtrsim \lambda/d$, and there will be a more-or-less uniformly bright central region of this angular width. But the angular width θ of the central diffraction peak satisfies $\theta \approx \lambda/d$ (according to (1.11) with $n = \pm 1$), so that the uncertainty in direction of the diffracted particles will result in a loss of visibility of the first diffraction minima. As the uncertainty in the x component of position of each particle at the slit screen is decreased, the angular width of the bright band at the centre of the pattern on the detector screen will increase, and the two-slit diffraction pattern will become invisible. The uncertainty principle therefore predicts the loss of the two-slit pattern whenever the position of the particles at the slit screen is sufficiently well defined.

This indeterminacy in quantum physics accounts for the phenomena of **zero-point energy** and **barrier penetration**. Classically all molecular motion ceases at the absolute zero of temperature, so that a classical particle is at rest, with zero kinetic energy and a precise position, at zero Kelvin. But this cannot be true for a quantum particle. Suppose a quantum particle is known to be within a region of width Δx. (We can make the discussion easier by referring to only one dimension without altering the conclusions.) Then the momentum of the particle cannot be precisely zero, and is not defined to better than $\Delta p_x \approx \hbar/(2\,\Delta x)$, according to (1.12). Thus the kinetic energy of a non-relativistic particle cannot have a value less than

$$\frac{1}{2m}(\Delta p_x)^2 = \frac{1}{2m}\left(\frac{\hbar}{2\,\Delta x}\right)^2$$

For an electron in an atom, with radius of the order of 10^{-10} m, this gives a minimum kinetic energy of the order of an electron volt. The potential energy holding the electron in the atom must be of a similar order of magnitude, and indeed, as we saw in Section 1.2, the ionization energy of a hydrogen atom is 13.6 eV. The ionization energies corresponding to the removal of an outermost, or valence, electron from atoms of other elements are of the same order of magnitude. So the uncertainty principle gives the correct order of magnitude for the depth of the ground state energy level of a valence electron in an atom. The **zero-point energy** of a particle is its minimum total energy, which is always greater than the classically expected total energy for the particle at zero Kelvin.

Barrier penetration is the wave-like spreading of a quantum particle into a region of space that would be inaccessible to a classical particle. Classically, if a particle has total energy E and a potential energy (varying with position) $V(x)$, it can only be found in regions where $E \geqslant V(x)$, since $E - V(x)$ is the kinetic energy at the point x, which cannot be negative. So wherever the potential energy becomes greater than the total energy there is a barrier that the classical particle cannot penetrate. For classical waves the corresponding situation is incidence on a material in which they cannot propagate, such as a metal in the case of visible light. But for waves such a barrier is not completely impenetrable, though the intensity decreases rapidly with penetration depth. Quantum waves are like classical waves in this respect.

In Figure 1.6(a) the classically allowed regions for a particle with the total energy E shown are $0 \leqslant x \leqslant x_1$ and $x \geqslant x_2$. For points between x_1 and x_2 the potential energy curve lies above the total energy, forming a potential barrier through which a classical particle would be unable to pass. A classical particle would decelerate as it approached one of the points $x = x_1, x_2$, would come momentarily to rest at that point, and then turn and accelerate away from the point again. But, according to the uncertainty principle, a quantum particle cannot have both a precise position and a precise momentum, or a precise trajectory. Even if the particle has a precisely determined total energy E, there will be inherent uncertainties Δx and Δp_x in its position and momentum. So the point at which a quantum particle turns around is not well defined, but spread over a region of width equal to the appropriate uncertainty Δx centred on the classical turning point. The particle shows the wave-like characteristic of spreading a little way into the classically forbidden region.

Barrier penetration accounts for such phenomena as nuclear α decay [19]. If we take a slice across the centre of a large nucleus, the potential energy of an α particle within it would vary with distance from the centre in a similar way to the potential shown in Figure 1.6. If the uncertainties Δx_1 and Δx_2 in the turning points x_1 and x_2 are sufficiently large that they overlap then there is a possibility that the α particle can 'tunnel through' the potential barrier and escape from the nucleus.

Numerous electronic devices depend on barrier penetration. For example, the most accurate magnetometers are based on the Josephson effect [20], which involves the tunnelling of electrons through a thin layer of electrically insulating material separating two superconducting components of an electrical circuit. Another application of barrier penetration is in the scanning tunnelling microscope (STM) [21], which can be used to detect and manipulate individual atoms on a metallic surface, by means of the current produced when electrons tunnel through an insulating layer of air from a sharp metallic probe into the metallic target.

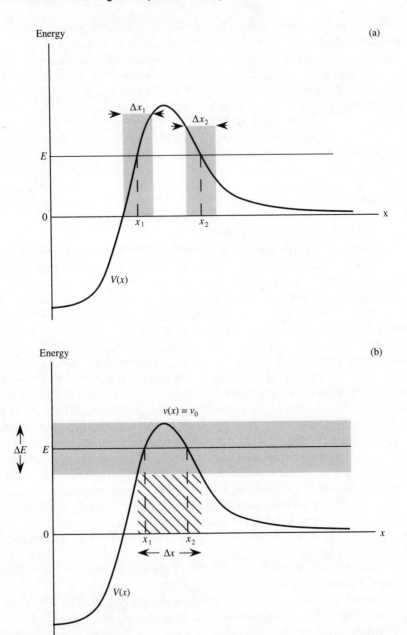

Figure 1.6 Uncertainty in the position of a particle moving in the potential $V(x)$. (a) The uncertainties Δx_1 and Δx_2 in the positions of the turning points for a particle with precise total energy E. (b) The minimum uncertainty in position, Δx, for which the particle may be found on either side of the potential barrier; this might be the result of an uncertainty ΔE in total energy.

An alternative way of looking at barrier penetration is in terms of an energy/time uncertainty relation:

$$\Delta E \, \Delta t \geqslant \tfrac{1}{2}\hbar \qquad (1.13)$$

If the total energy of a particle is not defined to better than the uncertainty ΔE then, as shown in Figure 1.6(b), the boundary of the classically allowed region is not precisely defined, and the particle can penetrate to points on the x axis within the shaded region. We can see that if $E + \tfrac{1}{2}\Delta E \geqslant V_0$, the particle can penetrate right through the barrier between x_1 and x_2 and escape. Equation (1.13) implies that during a time interval of the order of Δt the energy of a quantum particle is indeterminate, with an uncertainty ΔE depending on Δt. The characteristic lifetime of, for example, an α-decaying nucleus is related through (1.13) to the increase in energy that the α particle needs in order to escape.

Conversely, any energy measurement of duration Δt cannot give a result accurate to better than the corresponding value ΔE. This accounts for the natural width of spectral lines: excited atomic anergy levels decay spontaneously with a lifetime Δt of the order of 10^{-8} s, so that, according to (1.13), the energy of the transition is spread over a range $\Delta E \approx 10^{-8}$ eV, and the resulting spectral line has a minimum width (in terms of frequency) of the order of 10^7 s^{-1}. This is the **natural linewidth**, and is very small compared with the frequency of such a line, which is of the order of 10^{15} s^{-1}, so that the lines in optical atomic spectra appear very sharp. Notice that only a completely stable energy level – a ground state – can have a precise energy, with zero uncertainty associated with an infinitely long lifetime.

1.6 Non-classical phenomena

As the previous sections have indicated, quantum particles behave very differently from the ordinary macroscopic objects we can observe directly with our eyes. Planets orbiting the Sun, an apple falling from a tree and a billiard ball all behave classically. So what is the fundamental criterion that determines whether a particle should be treated classically or quantum mechanically?

The difference between classical and quantum mechanical behaviour is mainly one of scale: a billiard ball has a mass of the order of 100 g, and it is quite possible to measure its position to the nearest millimetre, and its velocity to an accuracy of 10^{-3} m s^{-1}. There are then uncertainties in its position and momentum of order of magnitude $\Delta x \approx 10^{-3}$ m and $\Delta p_x \approx 10^{-4}$ kg m s^{-1} respectively. The product of these is 10^{27} times the magnitude of Planck's constant. So these uncertainties could be reduced by a factor of 10^{12} on each measurement before we would even come within a factor of a thousand of the ultimate quantum limit, set by the uncertainty principle (1.12). But of course a reduction of the uncertainty in position even by only a factor of a million, to $\Delta x \approx 10^{-9}$ m, would indicate that we are making measurements on a scale where the molecular structure of the surface becomes apparent. While this might be interesting from the point of view of a materials scientist, it is completely irrelevant for a billiards player! So if we are concerned with the normal macroscopic motion of the billiard ball, the most accurate measurements that are of interest are on an enormously large scale compared with measurements for which quantum limitations on accuracy become important. The uncertainties in the macroscopic measurements are due to experimental limitations, and could conceivably be reduced to values so small as to be negligible on a macroscopic scale without becoming small enough for quantum effects to appear.

If we are interested in motion on a macroscopic scale, any changes in position or momentum due to the scattering of light from the object are completely insignificant. A photon of visible light carries a momentum of about 10^{-27} kg m s^{-1}, so that even the combined effect of thousands of millions of photons would be undetectable on a macroscopic scale. This means that it is possible to watch the motion of a macroscopic body through detecting light scattered from it, and make the usual classical assumption that this observation has no effect on the motion. Even the Brownian motion [22] of pollen grains suspended in a liquid is classical, though these random changes in direction are due to the effects of bombardment by the molecules of the liquid. In fact the motion, which is observable under a microscope, is the average effect of collisions with a large number of individual molecules, and is on too large a scale for quantum effects to be apparent. The light scattered from a pollen grain to make it visible produces changes in position and momentum on a very much smaller scale, and the observable Brownian motion is not affected by it.

What determines whether or not a particle behaves classically, then, is the scale of the position and momentum changes characteristic of the motion in which we are interested. If the product of the changes in momentum and position is very much larger than h then the uncertainty principle does not place any significant limitations on the motion. But if this product is of the same order as h then quantum effects become apparent. The product of momentum and spatial displacement is called **action**, and this quantity is important in classical as well as quantum mechanics. (In classical mechanics the **principle of least action** states that the actual path of a particle in a conservative field of force is such as to minimize the action.) During any interaction between quantum particles, the uncertainty principle tells us that changes in action smaller than h are impossible to observe. So the action can only change in discrete units of

Planck's constant. In other words, action is quantized and the funda-mental quantum of action is Planck's constant. To observe quantum behaviour, we must be able to detect changes in action of the order of magnitude of Planck's constant. The changes in action that occur in the motion of macroscopic bodies are always enormous compared with h, and so such bodies behave classically. Typically the scale of the changes in action associated with the motion of a body decreases with the size and mass of the body, and the motion of objects of molecular or atomic size may involve action changes of the same order of magnitude as h. In this case the act of observation produces changes in action of a similar magnitude, and we are at the limit of precision with which the motion can be observed.

When looking at theoretical formulae there is a useful rule of thumb for deciding whether or not the phenomenon described by a particular equa-tion is a classical or a quantum mechanical effect: any equation containing Planck's constant must be a quantum mechanical formula. There is no way to derive such an expression classically, because the presence of h indicates that it describes processes in which changes of action are quant-ized. For example, Planck's formula for the black body radiation spec-trum, (1.1), introduced h. If we expand the exponential, for temperature T and frequency v satisfying $hv/k_B T \ll 1$, the equation becomes

$$\rho(v)\,\mathrm{d}v = \frac{8\pi}{c^3}v^2 k_B T\left(1 - \frac{1}{4}\frac{hv}{k_B T} + \dots\right)\mathrm{d}v \qquad (1.14)$$

The first term is much the largest at frequencies that are low compared with $k_B T/h$, and h does not appear in it. This term is, in fact, the classical Rayleigh–Jeans formula mentioned in Section 1.2. The expan-sion (1.14) provides quantum corrections to the classical formula that are proportional to powers of h.

This criterion identifies as quantum phenomena the diffraction of elec-trons and neutrons, where the wave-like characteristics are related to the energy and the momentum of the particles by the de Broglie relations (1.9), and the photoelectric effect, described by Einstein's equation (1.8). Similarly, although the Balmer formula (1.5) seems to be classical at first sight, it contains the Rydberg constant, for which Bohr deduced the expression (1.7), which includes h. Whenever it is possible to take the limit $h \to 0$ in a quantum formula we approach the classical picture of continuous rather than discrete energy exchange, and the resulting ex-pression will have a classical derivation, as in the case of the Rayleigh–Jeans law.

Quantum theory not only predicts non-classical expressions, involving Planck's constant, for familiar variables such as energy, position and momentum, but it also introduces a completely new variable with no classical counterpart. This is the **spin** of an electron, which was intro-duced to explain details of the interaction of atoms with a magnetic field.

Classical mechanics predicts the spreading of the spectral lines emitted by atoms placed in a magnetic field (the classical Zeeman effect), while quantum mechanics, as we shall see in Chapter 8, predicts a splitting of the line into discrete components, due to a corresponding splitting of the atomic energy levels. Stern and Gerlach's experiment, which is also discussed in Chapter 8, was designed to test whether the magnetic field induced a (classical) continuous spreading or a (quantum) discrete splitting. Both the Stern–Gerlach experiment and the Zeeman effect show evidence of a discrete, non-classical, splitting of atomic levels. But the detailed structure of the splitting cannot be explained in terms of the interaction of the magnetic field with the atomic magnetic moment due to the orbital motion of the electrons, even using quantum mechanics. To explain the anomalous Zeeman effect, and Stern and Gerlach's original results concerning a beam of silver atoms, it must be assumed that the electron has an intrinsic magnetic dipole moment – it behaves like a point charge that is also a point bar magnet! The spin of the electron is defined to be proportional to this intrinsic magnetic moment.

Although the term 'spin' conjures up a picture of a tiny sphere of charge rotating about its own axis, this picture is not correct. The spin of the electron is *not* a rotation in physical space, even though it has the mathematical characteristics of an angular momentum, and it cannot be expressed in terms of position and momentum variables as ordinary orbital angular momentum can. In fact there is no completely satisfactory way of visualizing the spin in classical terms. Spin is an intrinsic characteristic of all quantum particles: the spins of protons and neutrons in the nucleus are necessary to explain the hyperfine structure of atomic spectral lines, the subnuclear particles observed in high-energy scattering experiments all have characteristic spins, which can influence the way they interact with one another, and the spin of the photon is associated with the polarization of light.

References

[1] Zemansky M.W. and Dittman R.H. (1981). *Heat and Thermodynamics*, Section 4-14. New York: McGraw-Hill.
[2] A good introduction to the world of subnuclear particles is Close F. (1983). *The Cosmic Onion*. London: Heinemann Educational.
[3] The early moments of the universe are described in Weinberg S. (1977). *The First Three Minutes*. London: Andre Deutsch.
[4] Preston T. (1904). *The Theory of Heat* 2nd edn. London: Macmillan. See Figure 181 on p. 596, which shows experimental results reported by Lummer

and Pringsheim in *Annalen der Physik* (1901).

[5] By courtesy of the COBE Science Working Group, National Aeronautics and Space Agency.

[6] Eisberg R. and Resnick R. (1985). *Quantum Physics of Atoms, Molecules, Solids, Nuclei and Particles* 2nd edn, Section 1.3. New York: Wiley.

[7] Flowers B.H. and Mendoza E. (1970) *Properties of Matter*, Section 5.3. London: Wiley.

[8] Reference [6], Section 1.4, Problem 1-4.

[9] Blin-Stoyle R.J. (1991). *Nuclear and Particle Physics*, Section 1.3. London: Chapman & Hall.

[10] Herzberg G. (1944). *Atomic Spectra and Atomic Structure* 2nd edn, p. 5. Figure 1. New York: Dover.

[11] Millikan R.A. (1916). *Physical Review* 7, 355.

[12] Reference [6], Section 3.1, and Greenberger D.M. (1983). The neutron interferometer as a device for illustrating the strange nature of quantum systems. *Reviews of Modern Physics* 55, 875.

[13] Feynman R.P. Leighton R.B. and Sands M. (1964). *The Feynman Lectures on Physics* Vol. II, Section 27-6. Reading MA: Addison-Wesley.

[14] Reference [6], Section 2-4.

[15] Young H.D. (1992). *University Physics* 8th edn, Section 37-2. Reading MA: Addison-Wesley.

[16] Tonomura A. *et al.* (1989). *American Journal of Physics* 57, 117.

[17] Reference [13], Vol. III, Section 1.6.

[18] Reference [6], Section 3.3 (Here the microscope thought experiment is ascribed to Bohr. In fact the original idea was Heisenberg's, but Bohr made the argument more rigorous.)

[19] Reference [9], Section 6.3.

[20] Some applications of the Josephson effect are presented in two articles in *Physics Today* (March 1986): Clarke J., Squids, brains and gravity waves, p. 36; Hayakawa H., Josephson computer technology, p. 46.

[21] Quate C.F. (1986). Vacuum tunneling: a new technique for microscopy. *Physics Today* 39, 26.

[22] Reference [13], Vol. I, Section 41-1.

[23] Weisskopf V.F. (1985). *American Journal of Physics* 53, 109. The other articles on quantum physics by Weisskopf in this series are also well worth reading: they appear as editorials in *American Journal of Physics* 53 (1985) (except for issue no. 8) and 54 (1986), nos 1 and 2.

Problems

1.1 Use (1.2) and the classical value for the average energy of an oscillator, $\langle E \rangle = k_B T$, to find the Rayleigh–Jeans prediction for $\rho(v, T)$. Sketch the curve of this function against frequency, and show that it has no maximum point.

The spectral energy density as a function of wavelength, $\rho'(\lambda, T)$, is defined by $\rho'(\lambda, T)\,d\lambda = \rho(v, T)\,dv$. Substitute $v = c/\lambda$ in (1.1) to find the Planck prediction for $\rho'(\lambda, T)$. At a temperature of 1000 K find the wavelength at which the maximum of $\rho'(\lambda, T)$ occurs, and the frequency at which the maximum of $\rho(v, T)$ occurs. (The solution to $(3 - x)e^x = 3$ is $x \approx 2.8214$, and that to $(5 - x)e^x = 5$ is $x \approx 4.9651$.)

1.2   If the probability of an oscillator having energy E is determined by the Boltzmann distribution Ze^{-E/k_BT}, show that, according to classical physics, its average energy is $\langle E \rangle = k_B T$, while if its energy is restricted to the set of discrete values nhv $(n = 0, 1, 2, 3, \ldots)$, its average energy is given by (1.3). (Ze^{-E/k_BT} is the probability of a particle having thermal energy E at temperature T, and Z is chosen to make the total probability of the particle having *some* energy between 0 and ∞ equal to 1. You may find the expansions

$$\frac{1}{1 - e^{-x}} = \sum_{n=0}^{\infty} e^{-nx}, \qquad \sum_{n=0}^{\infty} n e^{-nx} = -\frac{d}{dx} \sum_{n=0}^{\infty} e^{-nx}$$

useful.)

1.3   Assume that at temperature T the molecules in a solid vibrate independently with the same angular frequency v, and each molecule has three degrees of freedom. If the average vibrational energy of each degree of freedom is given by Planck's expression (1.3), find the average vibrational energy per mole of the solid, $\langle E \rangle_s$. Prove that the specific heat $c_v = \partial \langle E \rangle_s / \partial T$ is

$$3N_A k_B \left(\frac{hv}{k_B T} \right)^2 \frac{1}{(e^{hv/k_B T} - 1)^2}$$

and show that this reduces to the Dulong and Petit value $c_v = 3N_A k_B$ at sufficiently high temperatures. (This is the Einstein model for the specific heat of a solid.)

1.4   Use the wavelengths for the lines in the Balmer series given in Figure 1.2(a) to find a value for the Rydberg constant, and compare the result with the values calculated from (1.7), using (a) the free electron mass m_e and (b) the reduced mass $m = m_e m_p/(m_e + m_p)$. (The best spectroscopic measurements give $R_H = 1.096\,775\,76 \times 10^7 \text{ m}^{-1}$.) Calculate the wavelengths of the first three lines in the Lyman series (produced by transitions between an excited state of hydrogen and the ground state).

1.5   Find the frequency, wavelength and momentum associated with a photon of energy 13.6 eV. What would be the momentum of a free electron with this kinetic energy? Compare the frequency and de Broglie wavelength of the electron with that of the photon.

1.6   Find the maximum wavelength of light that can produce photoelectrons from sodium, for which the work function is 2.3 eV. What is the maximum kinetic energy of the photoelectrons emitted from sodium illuminated by light of frequency $1.5 \times 10^{15} \text{ s}^{-1}$?

1.7   According to Newton's second law, the Coulomb force on an electron orbiting a proton with speed v in a circular orbit of radius r is balanced by the centrifugal force:

$$\frac{mv^2}{r} = \frac{e^2}{4\pi\varepsilon_0 r^2}$$

Show that the total energy of the electron (according to classical physics) is

$$E = -\frac{1}{2}\frac{e^2}{4\pi\varepsilon_0 r}$$

Bohr made the assumption that the orbital angular momentum of the electron, mvr, can only take values that are integer multiples of $\hbar$. Show that this assumption implies that

$$r = \frac{\hbar^2 4\pi\varepsilon_0}{me^2} n^2 \quad (n = 1, 2, 3, \ldots)$$

and that this restricts the energy of the electron to the values determined by (1.6) and (1.7).

1.8   Find the frequency of the light emitted in a transition between the nth level and an adjacent one in hydrogen. Show that as n becomes very large, this approaches the frequency at which, according to classical physics, the electron rotates in the nth orbit.

1.9   The radial momentum of an electron in an orbit of radius r in a hydrogen atom is $\geqslant \hbar/r$, according to the uncertainty principle. If the momentum of the electron is put equal to $\hbar/r$ (assuming the angular momentum to be zero), its total energy becomes

$$E = \frac{\hbar^2}{2mr^2} - \frac{e^2}{4\pi\varepsilon_0 r}$$

Find the radius that minimizes this energy, and the minimum value of E. (*Answer*: one Bohr radius, and $-hcR_H = E_0$, (1.6)! See [23].)

1.10 If the electron in a hydrogen atom is assumed to follow a circular orbit, show that the requirement that its de Broglie wavelength be equal to the wavelength of a standing wave around the orbit is equivalent to Bohr's assumption that $mvr = n\hbar$ $(n = 1, 2, 3, \ldots)$ (Problem 1.7).

1.11 Find the de Broglie wavelength of each of the following:
   (a) a 70 kg man travelling at 6 km h^{-1};
   (b) a 1 g stone travelling at 10 m s^{-1};
   (c) a 10^{-6} g particle of dust moving at 1 m s^{-1}.

1.12   A neodymium (Nd) laser operates at a wavelength $\lambda = 1.06 \times 10^{-6}$ m. If the laser is operated in a pulsed mode, emitting pulses of duration 3×10^{-11} s, what is the minimum spread in (a) frequency and (b) wavelength of the laser beam?

1.13   Scattering of one nucleon (proton or neutron) by another can occur through the strong nuclear interaction. Consider a scattering process between a proton and a neutron in which the momentum of the proton changes from p to $p' = p - q$, and there is a corresponding change $E - E' = -\Delta E$ in its energy. The scattering process conserves energy and momentum, so that the momentum and energy of the neutron are changed by q and ΔE. According to Yukawa's model, the energy and momentum transfer from the proton to the neutron is due to the exchange of a particle (called a **meson**). But the emission of a particle by a proton (or its absorption by the neutron) cannot conserve both energy and momentum. (To prove this, use the relativistic expression $E = \sqrt{(m^2 c^4 + p^2 c^2)}$ for the energy of a particle of rest mass m and momentum p.) According to Yukawa, the conservation of energy may be violated by an amount ΔE provided that this energy is reabsorbed within a time Δt related to ΔE through (1.13). This limits the distance over which the meson can transfer energy, and accounts for the finite range of the interaction, which is of the order of 10^{-15} m. Estimate the rest mass energy of Yukawa's meson.

CHAPTER 2

The State of a Quantum System

2.1 The classical description of the state of a particle

The fundamental components of classical mechanics are first a mathematical description of the state of the system at any instant, and secondly an equation, or set of equations, of motion that will predict the state of the system at any future instant. These are usually differential equations, such as Newton's second law, or the Lagrangian or Hamiltonian equations of motion. Similarly, in quantum mechanics we need a mathematical description of the state of the quantum system at any time, which we shall introduce in this chapter, and an equation of motion to predict the evolution of the state with time, which will be discussed in Chapter 12.

How is the dynamical state of an object described classically? Think of a billiard ball. It has intrinsic properties such as colour and shape, but we can assume that these do not change and, in the case of the colour at least, are irrelevant to the motion. (Though even the colour affects the ball's trajectory indirectly – the first ball to be set in motion in a game of billiards is inevitably white!) The dynamical state of a billiard ball at any instant is specified by the position and momentum of its centre of mass, and its angular momentum about the centre of mass. These are the dynamical variables characterizing its motion. In general, the dynamical state of a classical object at time t is specified by giving the precise values of all relevant dynamical variables at that instant. In the case of a free point particle the relevant variables are just the position and momentum. In classical mechanics, if we specify the state of an object at some initial instant then we can attempt to find its state at any later (or earlier) instant using the classical equations of motion.

In practice, the accuracy of any measurement is limited, so that the values of the dynamical variables, and thus the state of an object, can never be determined absolutely precisely. But in classical mechanics we assume that it is possible, in principle, to measure variables to any required degree of accuracy. (This is a good approximation, as we saw in

Section 1.6, provided the motion in which we are interested involves changes in action on a scale very much larger than h.) So we feed approximate values for the initial state into the equations of motion, on the assumption that small inaccuracies in the initial conditions will only produce small errors in the predicted trajectory.

Actually this last assumption is not always justified. Indeed, it is only true for a very small subset of all possible trajectories, but until recently all except this small subset of stable, predictable trajectories was ignored. In the last 20 years, however, methods of dealing with non-predictable, chaotic motion have been developed [1]. In chaotic motion a slight change in the initial value of a dynamical variable, such as the position of a particle, is magnified exponentially with time, so that two states that are initially very similar can rapidly become totally different. This makes it impossible, in practice, to predict the trajectory of an object whose motion is chaotic, because small inaccuracies in the initial values of the dynamical variables fed into the equations of motion will lead to enormous inaccuracies in the values calculated for a later time.

But this classical chaotic unpredictability is completely different from the quantum unpredictability that we mentioned in Chapter 1. For example, some asteroids have chaotic orbits, and it is not possible to predict precisely where such an asteroid will be at any instant a year or so from now. But it is possible to find out, simply by watching and waiting. We cannot *predict* a particular future position and momentum for it, but we can *observe* the asteroid continuously without affecting its orbit significantly, and we can measure its position and momentum at any later time with any required degree of accuracy. Similarly, the random movements of pollen grains in Brownian motion cannot be predicted, but can be observed continuously under a microscope. The motion of the asteroid or the pollen grain is on a macroscopic scale (as defined in Section 1.6), and so we can assume that the object has a precise position and momentum at every instant, even if these cannot be measured with 100% accuracy in practice, or predicted in advance.

In the case of motion that can be described classically, whether it is predictable or chaotic, it is always possible to assume that at each instant a particle has precise values for the complete set of relevant dynamical variables, and these can all, in principle, be measured to any required degree of accuracy simultaneously. This is not true for the motion of a quantum particle. For example, Heisenberg's uncertainty principle tells us that it is not possible to know precise values for both the position and momentum of a particle simultaneously, and we cannot even assume that the quantum particle *has* unique, though unknown, values of both position and momentum. In general, both position and momentum will be indeterminate to some degree, depending on the previous history of the particle. So the classical concept of a state, described by a unique set of values for all the relevant dynamical variables at some instant, is useless for describing quantum behaviour. We need a new way to define the dynamical state of a particle in quantum mechanics.

2.2 The wave function for a single particle

The clue to constructing a description of the state of a quantum particle can be found in the way in which the two-slit diffraction pattern is built up from the arrival of individual particles: according to the discussion in Section 1.4, the diffracted beam of quantum particles shows its particle aspect when it is detected, but the distribution of particle arrival points at the screen S_2 in Figure 1.4 produces the diffraction pattern, which is determined by the wave aspect of the beam. Since the pattern is built up from individual particles even when the intensity of the beam is so low that only one particle at a time passes through the apparatus, the wave aspect of the beam must constrain the behaviour of each individual particle.

In order to describe the quantum behaviour of a single particle, we must restrict ourselves to the case of non-relativistic motion. There are serious difficulties in dealing with relativistic quantum particles individually, and, as we mentioned in Section 1.3, the relativistic version of quantum mechanics is beyond the scope of this book. Since photons are intrinsically relativistic, non-relativistic quantum mechanics cannot be applied to them. However, exactly the same type of diffraction effects as with light are also observed with non-relativistic particle beams. For instance, the energy of a beam used in an electron diffraction experiment is typically of the order of hundreds of electron volts, corresponding to a velocity no greater than about $0.03c$. So from now on our discussion should be understood to apply only to non-relativistic particles.

In a diffraction experiment it is not possible to predict the precise point at which any individual particle will arrive on the detector screen S_2. However, each particle has the greatest chance of arriving at the centre of a diffraction maximum, and (in the ideal case) has no chance at all of arriving at a diffraction minimum. The chance of arrival at any particular point varies from place to place in exactly the same way as the intensity of the pattern. So for each non-relativistic quantum particle we define a **probability distribution** over the screen S_2, which is proportional to the intensity distribution of the diffraction pattern, and which determines the relative chances of the particle's arrival at different points on the screen. Since the diffraction pattern can be predicted as the resultant intensity of a wave at points on the detector screen, we associate a **probability wave** with each (non-relativistic) particle in the beam. The intensity of this wave at any point determines the probability of detecting the particle at that point. In order to predict the two-slit diffraction pattern in the usual way, we assume that the probability wave associated with a particle in the incident beam is plane and monochromatic, and is split at screen S_1, with part passing through one slit and part through the other. The two parts of the wave then interfere at the screen S_2 to produce a probability distribution of the same form as the diffraction pattern.

This identification of the quantum wave intensity with a probability distribution applies everywhere, not only at the screen S_2 where the diffraction pattern is observed. At screen S_1 the wave is split, and the intensity of the wave at slit A, say, is proportional to the probability that the particle will pass through that slit. We do not have to imagine that the particle splits into two at the slits, but the wave that determines the relative probability of its passage through one slit or the other *does* split. The indeterminacy of the position of the particle on S_1, which we discussed in Section 1.5, is a wave-like characteristic, and is described mathematically in terms of the probability wave. So in quantum mechanics wave and particle aspects are combined through the introduction of a probability wave: the wave limits the possible positions at which the particle aspect may be detected, and assigns a statistical weight to each.

In the diffraction experiment each particle in the incident beam is in a particular dynamical state in which it has a precise momentum and energy (to correspond, through de Broglie's relations (1.9), to a plane monochromatic wave), but we shall extend the idea of a probability wave to arbitrary dynamical states. We assume that any (non-relativistic) state of the particle can be represented by a **wave function** $\Psi(x, y, z, t)$, which is the amplitude of a wave whose intensity at time t is the probability density for the spatial distribution of the particle at that instant. This means that the probability of finding the particle in volume V at time t is

$$\mathbb{P}_V = \int_V d^3r \, |\Psi(x, y, z, t)|^2 = \int_V d^3r \, \Psi^*(x, y, z, t) \Psi(x, y, z, t) \quad \textbf{(2.1)}$$

The intensity of a classical wave is just the square of its amplitude, but we shall find that the probability wave must be allowed to be complex rather than purely real. So we have used the square of the modulus of the amplitude as the wave intensity in (2.1).

If $\mathbb{P}_V$ is a probability, it can only take values in the range from 0 to 1. This imposes a restriction on the overall magnitude of the wave function. Since $\Psi(x, y, z, t)$ represents the state of a single particle, which exists *somewhere* at each instant, the probability of finding the particle somewhere in space should be 1 for each value of t:

$$\int_{\text{all space}} d^3r \, |\Psi(x, y, z, t)|^2 = 1 \quad \text{(2.2)}$$

(In the case of a plane wave, however, this is not feasible, and we have to use a slightly different condition, as discussed in Section 2.4.) If (2.2) holds for all t, this means that the probability is a conserved quantity.

Probability wave amplitude

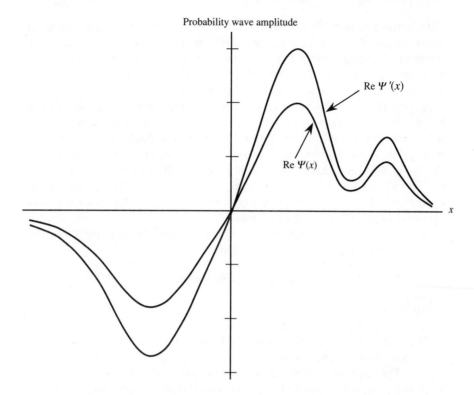

Re $\Psi'(x)$

Re $\Psi(x)$

x

Figure 2.1 The real parts of two wave functions, $\Psi(x)$ and $\Psi'(x) = \frac{3}{2}\Psi(x)$, which differ only in overall magnitude, and therefore represent the same quantum state. (A corresponding diagram could be drawn for their imaginary parts.)

The total probability for finding the particle in a finite volume V, P_V of (2.1), can only change with time if there is a flow of probability across the surface V, due to a change of the particular volume of space in which the particle is to be found. So probability should satisfy a continuity equation like that for electric charge. In Chapter 3 we shall find that the derivation of such an equation is intimately related to the possibility of the wave function being complex, rather than purely real.

But if $\Psi(x, y, z, t)$ is a wave function satisfying (2.2), what significance can we attach to the wave function $\Psi'(x, y, z, t) = C\Psi(x, y, z, t)$, where C is some constant, either real or complex, such that $|C| \neq 1$? The functions $\Psi(x, y, z, t)$ and $\Psi'(x, y, z, t)$ have the same shape, but differ in overall magnitude, as illustrated in Figure 2.1, and the intensity of the wave with amplitude $\Psi'(x, y, z, t)$ cannot be directly interpreted as a probability density because

$$\int_{\text{all space}} d^3r\,|\Psi'(x, y, z, t)|^2 = |C|^2 \neq 1 \qquad (2.3)$$

Clearly the intensity $|\Psi'(x, y, z, t)|^2 = |C|^2 |\Psi(x, y, z, t)|^2$ is *proportional* to the probability density at each instant and at each point in space. This constant of proportionality can be assigned no fundamental physical significance, so the wave functions $\Psi'(x, y, z, t)$ and $\Psi(x, y, z, t)$ describe exactly the same dynamical state of the particle.

A wave function that satisfies (2.2) is said to be **normalized**, and it must be possible to normalize any wave function representing a dynamical state, by multiplying it by an appropriate constant. In the case of $\Psi'(x, y, z, t)$ we can use C^{-1} as the normalization constant. However, this is not a unique choice. We could equally well choose $|C|^{-1}$, or indeed $C^{-1}e^{i\gamma}$, where γ is an arbitrary, real, constant phase. The phase of the normalization constant is therefore completely arbitrary, and we may always choose the constant to be real.

We are now in a position to define more formally how the state of a particle is to be represented in quantum mechanics.

Postulate 1

The state of a non-relativistic quantum particle at time t is described by a continuous, non-singular, complex wave function $\Psi(x, y, z, t)$, which can be normalized so that the square of its modulus is equal to the probability density for the results of a position measurement.

The wave function is required to be *continuous* because if there were a discontinuity at any point there would be two values of $\Psi(x, y, z, t)$ at that point, and thus no unique value for the probability density. A *non-singular* wave function remains finite everywhere, which is essential if it is to satisfy

$$\int_{\text{all space}} d^3 r \, |\Psi(x, y, z, t)|^2 = \text{finite number} \qquad (2.4)$$

Equation (2.4) must be satisfied for the wave function to be normalizable.

Later on in this book we shall be interested in **orthogonal sets** of wave functions. Two wave functions Ψ_1 and Ψ_2 are said to be **orthogonal** if they satisfy

$$\int_{\text{all space}} d^3 r \, \Psi_1^*(x, y, z, t)\Psi_2(x, y, z, t) = 0 \qquad (2.5)$$

You are probably already familiar with the orthogonality property of sets of trigonometric functions. For example, for two integers k and n,

$$\int_0^\pi d\theta \sin k\theta \sin n\theta = 0 \quad (k \neq n)$$

If a set of functions has the property that any pair is orthogonal, as for example the set $\{\sin n\theta; n = 1, 2, 3, \ldots\}$, then the functions form an orthogonal set. If all the functions belonging to an orthogonal set are also normalized then the set is termed **orthonormal**.

2.3 Measurements on a quantum system

In Postulate 1 the significance of the wave function is defined in operational terms, as the means of predicting the results of a position measurement. Although there may be intuitive and philosophical difficulties in the concept of a probability wave, the predictions of quantum mechanics concerning the results of real physical measurements are perfectly clear and unambiguous. The nature of quantum mechanical predictions can be illustrated, and compared with corresponding classical predictions, by considering a position measurement.

Suppose a quantum particle is in a state known to be represented by the wave function $\Psi(x, y, z, t)$, which is non-zero over a particular plane area A at time t. The size of this area is of the order of magnitude of the uncertainty in the position of the particle at this instant, which may be considerably larger than the actual size of the particle. For example, even though any experiment designed to measure the radius of a neutron will give a result of the order of 10^{-15} m, in a neutron diffraction experiment each neutron may be effectively spread over an area the size of a postage stamp. To measure the position of the particle at time t, an area containing all of A is covered by an array of suitable detectors. If the particle behaved classically (and the size of each detector were properly chosen) its arrival would trigger just one of the detectors, and Newtonian mechanics would predict precisely which one. If it behaved like a classical wave with the same intensity distribution, $|\Psi(x, y, z, t)|^2$, as the quantum wave, then all the detectors over area A would respond, since the energy in a classical wave is spread continuously over a region of space. The magnitude of the response of a particular detector would be proportional to the total energy of the wave over the area of that detector, and the variation of response from detector to detector would be predicted from the intensity of the wave. In the case of a quantum particle only one detector would respond, and record the arrival of the whole particle at the position of that detector, just as in the case of a particle that behaves classically. However, this time it is impossible to predict in advance which detector will be triggered. The quantum wave intensity predicts not the magnitude of the response of each detector but rather the probability, for each detector, that it will record the arrival of the particle.

In the classical case the accuracy of the predicted response of the detectors could, in principle, be checked by a single measurement. But this is not true for the quantum prediction. To compare the results of experiment with the predicted probability distribution, we need to examine the distribution of arrival points for a large number of quantum particles all in the same dynamical state, and therefore described by the same wave function. The statistical distribution of the results is an intrinsic characteristic of the state of the quantum particle, whereas in classical mechanics the spread of results obtained when any measurement is repeated many times is ascribed to unavoidable experimental error.

Similar remarks apply to the measurement of any dynamical variable: in general, it is impossible to predict a unique value for the result of a measurement on a quantum system – not because of any inadequacy in the experimental arrangement but because of the inherent indeterminacy in the value of the variable. But knowledge of the wave function describing the state of the particle will enable us to predict the probability distribution of the set of possible results. We shall find out how to predict the set of all possible results for any measurement in Chapter 3, and how to predict their relative probabilities in Chapter 9.

To summarize, quantum mechanical predictions are expressed in terms of probabilities, and can only be tested experimentally by carrying out a large number of identical measurements. By identical measurements we mean not only that the experimental procedure must not differ in any significant way from one measurement to the next, but that for each measurement the particle must be in the same quantum state initially. According to Section 1.5, a measurement will, in general, change the state of the particle on which it is performed, so that instead of repeating the measurement on one particle, we must perform the measurement on each of a large number of identical particles, all in the same quantum state. This means that the particles on which a particular measurement is to be made must first be prepared in the same state. This preparation is achieved by the experimental procedure through which particles are produced by a source and conveyed to the measuring apparatus. For example, in an electron diffraction experiment all the electrons should be in the same momentum state, so that the beam is parallel and has a unique de Broglie wavelength. Electrons with the same momentum to a sufficiently good degree of accuracy are selected by passing the beam through an electrostatic separator to pick out momenta of the same magnitude, and then through a system of electromagnetic lenses to ensure that the momenta are all in the same direction. (Of course any real beam will not conform exactly to the ideal, any more than a real beam of light can be exactly parallel and monochromatic.)

Suppose that a set of identical quantum particles are prepared in the same state, represented by the single-particle wave function $\Psi(x, y, z, t)$. Then when a measurement of the value of a particular variable is made on each particle, the set of results must be analysed statistically. Consider a set of N identical measurements that yield the following results:

numerical result	number of times obtained
a_1	n_1
a_2	n_2
a_3	n_3
$\vdots$	$\vdots$
a_k	n_k

That is, k different results are obtained, and (for each $i = 1, 2, \ldots, k$) n_i is the number of measurements that give the result a_i. Since the total number of measurements is N, we must have

$$\sum_{i=1}^{k} n_i = N \tag{2.6}$$

The statistical mean of the set of results [2] is defined to be

$$\langle a \rangle = \frac{1}{N} \sum_{i=1}^{k} n_i a_i = \sum_{i=1}^{k} \mathbb{P}_i a_i \tag{2.7}$$

where $\mathbb{P}_i = n_i/N$ is the proportion of measurements that result in the particular value a_i. In classical statistics the proportion $\mathbb{P}_i$ can be identi- fied as the probability of obtaining this particular value, provided the total number of measurements, N, is large enough. We shall assume that the number of measurements performed always satisfies this criterion, so that a probability predicted from a knowledge of the wave function (by methods to be discussed in Chapter 9) can be compared with the experi- mentally determined $\mathbb{P}_i$.

The spread of results about the mean value is conveniently described by the standard deviation [2], defined as

$$\Delta a = \sqrt{(\langle a^2 \rangle - \langle a \rangle^2)} \tag{2.8}$$

The first term is the mean value of the square of the result, which will be identical to the square of the mean value only if the set of measurements all give the same, unique, result. In this special case the standard devi- ation is zero.

In general, then, a set of identical measurements on quantum particles in identical states will give a distribution of results that can be character- ized by the statistical mean and standard deviation. Although quantum mechanics cannot, in general, predict the precise result of any individual measurement it *can* predict this statistical distribution. In quantum mechanics the predicted mean value of a variable is called its **expectation value**, and the predicted standard deviation is the **uncertainty** in the value of the variable. For example, the uncertainty Δx appearing in the uncer- tainty relation (1.12) is the predicted standard deviation of a set of position measurements.

2.4 The wave function for a free particle

The incident beam in an electron or neutron diffraction experiment consists, ideally, of free particles all with the same energy E and momentum p, distributed with constant uniform density throughout the volume occupied by the beam. Section 2.2 implies that each particle in the (non-relativistic) beam should have the same wave function, which must be a plane monochromatic wave in order to predict the correct probability distribution at the final, detector, screen. De Broglie's expressions (1.9) relate the energy of the particle and the magnitude of its momentum to the frequency and wavelength of the wave. Introducing $\omega = 2\pi\nu$ and the wave vector k, where $k = 2\pi/\lambda$, we have $E = \hbar\omega$ and $p = \hbar k$. We shall define the beam direction to be the z axis, so that the momentum p of each particle is in the z direction.

If the wave were a classical one, its amplitude would be proportional to $\cos(kz - \omega t + \gamma)$, where γ is an arbitrary constant phase. In classical electromagnetic theory we often use a complex wave amplitude, proportional to $e^{i(kz-\omega t+\gamma)}$, for mathematical convenience, with the understanding that the amplitude of the physical wave is the real part of the complex expression, and the intensity must be calculated from the real part of the amplitude only. In quantum mechanics, as we mentioned in Section 2.2, we must allow the wave function to be complex. We write the wave function for a single beam particle as

$$\Psi(z, t) = Ce^{i(kz-\omega t)} = Ce^{i(pz-Et)/\hbar} \qquad (2.9)$$

where C is an arbitrary constant, which can be complex. This form for the plane wave amplitude allows us to predict the intensity of the two-slit diffraction pattern in the same way as in classical wave theory, except that both the real and imaginary parts of the amplitude in (2.9) contribute to the intensity $|\Psi|^2$. One immediate advantage of the complex form (2.9) is that the intensity does not vary with z and t:

$$|\Psi(z, t)|^2 = |C|^2 \qquad (2.10)$$

This represents a probability distribution that has the same constant value at each point along the beam. The intensity of a classical, real, plane wave would oscillate as $\cos^2(kz - \omega t + \gamma)$, and if this were interpreted as a probability distribution, it would imply that the probability of finding a particle at any instant oscillates as a function of z, the distance along the beam. The constant distribution (2.10) is more appropriate for a beam containing a uniform density of particles, and from now on the complex wave function (2.9) will be used to represent the state of a free particle with energy E and momentum in the z direction of magnitude p.

The idea of a completely free particle is, of course, an idealization, in just the same way as a plane monochromatic wave. The wave extends from $-\infty$ to $+\infty$ along the z axis, and, since the amplitude of a plane wave is constant over any plane perpendicular to the propagation direction at any instant, the wave also extends over an infinite area in the (x, y) plane. In practice, of course, the beam used in a diffraction experiment is of finite width and length, but, provided these are both very large compared with the wavelength, and the spread in wavelengths in the beam is small enough, a plane monochromatic wave is a good approximation to the actual beam. Similarly, the probability density of (2.10) is independent of x and y as well as of z and t, and so the probability density for finding the particle extends through an infinite volume, and takes the same constant value at each point in space. In fact this is required by the uncertainty principle, since the momentum of the particle is assumed to be precisely known both in magnitude and direction: the z component of momentum is p, with uncertainty $\Delta p_z = 0$, and the x and y components of momentum are both zero, with uncertainties $\Delta p_x = 0$ and $\Delta p_y = 0$. According to the inequality (1.12) and the corresponding expressions for the y and z components, the uncertainties Δx, Δy and Δz in the position of the particle must be infinite, so the position is completely indeterminate.

As in the case of the classical plane monochromatic wave, the free particle state represented by the wave function (2.9) is a good approximation in situations such as those pertaining in a diffraction experiment. The actual finite width of the beam sets an upper limit on the uncertainties Δx and Δy, which means that the x and y components of momentum have corresponding minimum uncertainties, determined by (1.12). So the momentum cannot be precisely directed along the z axis, and the beam will actually diverge slightly. In practice, this divergence is usually very much smaller than that due to experimental limitations, or to the mutual repulsion of the particles in cases where they are electrically charged. Similarly, the finite length of the beam determines Δz, and the corresponding uncertainty in the z component of momentum represents a small spread in wavelength. Provided these departures from the ideal behaviour are small, a wave function of the form (2.9) can be used to analyse the experiment.

Although the wave function (2.9) is continuous and finite everywhere, we run into difficulty if we try to normalize it using (2.2). If we integrate the wave intensity over any volume $\mathbb{V}$, the result is proportional to $\mathbb{V}$:

$$\int_{\mathbb{V}} d^3 r \, |Ce^{i(pz - Et)/\hbar}|^2 = |C|^2 \mathbb{V} \qquad (2.11)$$

Consequently the integral over all space is infinite, and to satisfy the normalization condition (2.2), C would have to be infinitesimally small. This problem emphasizes that the completely free particle state, in which the momentum is precisely defined and the position completely undetermined, is an idealization. In order to be able to use the free particle wave function (2.9) as a probability wave density, we have to modify the normalization condition (2.2). In practice, the particle is one of a beam with a uniform density, N particles per unit volume, say. There are therefore N chances of detecting a particle in a unit volume, and the volume in which there is probability one of finding a single particle is $V = 1/N$. So we normalize the wave function (2.9) over the volume $1/N$ instead of over all space, and require the right-hand side of (2.11) to be equal to 1 when $V = 1/N$. This gives $|C|^2 = N$, or

$$C = \sqrt{N} \tag{2.12}$$

where we have chosen C to be real, since, according to Section 2.2, the phase of the normalization constant has no physical significance.

Except when much experimental ingenuity has been used to prepare particles in a state in which the momentum of each has a negligibly small uncertainty, the wave function (2.9) is not suitable for representing the state of a free particle. Consider a free particle that is known to exist at a time t within some finite region that is small enough for the particle to have a significant uncertainty (determined by the size of the region) in its momentum. The position of such a particle is **localized**, and its state is not well approximated by one with precise values of either momentum or energy. We shall discover in Chapter 9 that a localized free particle state can be represented by a **wave packet**, which is a Fourier superposition of a large number of wave functions of the form (2.9) with different values of p and $E = p^2/2m$.

2.5 Free particle beams and scattering experiments

Free particle beams are very important in experiments that probe the structure of quantum systems. Just as the shape and structure of macroscopic objects is observed by scattering light beams from them, the structure of quantum systems can be investigated in scattering experiments using light or any type of quantum particle that interacts with the system under investigation. We have already met the example of electron and neutron diffraction experiments, which can provide information about the atomic structure of a solid. Similarly, experiments in which electrons are scattered from atoms probe atomic structure, while the characteristics of nuclei can be investigated by neutron or proton scattering experiments. High-energy scattering experiments use large accelerators to produce very high-energy beams of electrons, protons or some of the more exotic subnuclear particles (such as pions and kaons), and probe the fundamental structure and interactions of matter.

An ideal scattering experiment is illustrated in Figure 2.2. Just as in an

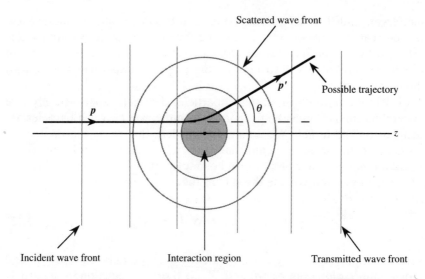

Figure 2.2 A schematic representation of an ideal scattering experiment, showing the plane wave fronts of the incident beam and the beam that is transmitted without being scattered, together with the spherical scattered wave fronts. A possible trajectory for a scattered particle is also shown, when θ is the scattering angle.

optical diffraction experiment the incident beam is ideally a plane mono-chromatic wave, the incident particles in an ideal quantum scattering experiment all have the same energy and momentum, and can be de-scribed by a wave function of the form (2.9). The target scatters the beam, and produces a spherical wave spreading out in all directions. (Think of the circular ripples produced in calm water when a small wave travelling towards the shore meets a sharp rock.) The term 'spherical wave' is not meant to imply that the scattered wave necessarily has the same intensity in all directions, and in the case of quantum particles the scattered wave intensity in a particular direction determines the probabil-ity that an incident particle will be scattered in that direction. The way in which this probability varies with direction contains important informa-tion about the interaction between an incident particle and the target.

In Figure 2.2 the incident beam is parallel to the z axis, and it is assumed that the interaction that scatters an incident particle depends on its distance from a single point within the target – the **scattering centre** – which we choose to be the origin. The scattered particles are detected only when they are once again free, and far from the region in which they interact with the target. The measurable quantities are [3] the **total cross-section** σ and the **differential cross-section** $d\sigma/d\Omega$. The total cross-section is found by counting the rate at which particles are scattered by the target, and this will be proportional to the probability that a particle will be scattered as opposed to passing through the target without interacting with it. The differential cross-section is a measure of the way in which the probability distribution for a scattered particle varies with direction.

Clearly the number of scattered particles detected experimentally in a given time interval must be proportional to the rate at which particles are incident on the target. We define the **incident flux** to be the number of particles in the incident beam that cross unit area perpendicular to the beam direction in unit time. For a beam of particles with uniform density N particles per unit volume this means that

$$\text{incident flux} = Nv = N\frac{p}{m} = |\Psi_{\text{inc}}|^2 \frac{p}{m} \qquad (2.13)$$

where $v = p/m$ is the velocity of an incident particle of mass m and momentum p along the z axis, and Ψ_{inc} is the wave function for a particle in the incident beam, normalized according to (2.12). The total cross-section is then defined as

$$\sigma = \frac{\text{total number of particles scattered per unit time}}{\text{incident flux}} \qquad (2.14)$$

Thus σ has the dimensions of area.

A billiard ball travelling towards a solid target will be scattered if it hits the target but not if it misses it. So in this classical case the total cross-section is the area presented by the target in the plane perpendicular to the motion of the billiard ball (the (x, y) plane in our case). This is why σ is termed the cross-section. In the scattering processes that provide information about quantum particles the target is usually a collection of scattering centres, like the nuclei of the atoms in the target in Rutherford's α-particle scattering experiment, and of course the scattering interaction is not a simple billiard-ball-like collision, but an electromagnetic or nuclear process that varies in strength with the separation of the two particles. So σ is not the geometrical cross-section of the target, but can be thought of as the effective cross-sectional area that it presents.

The differential cross-section $d\sigma/d\Omega$ is a measure of the spatial distribution of the scattered beam, and is found by counting the rate at which particles are incident on detectors placed at different positions around the target. An element of solid angle $d\Omega$ is defined as shown in Figure 2.3: a small section of the surface of a sphere of radius r, subtending a small element of solid angle $d\Omega$ at the centre, is perpendicular to the direction (θ, ϕ), and its area is defined to be $r^2 d\Omega$. It can be seen from the diagram that the area of the surface element is $r\,d\theta\,r\sin\theta\,d\phi$, so that

$$d\Omega = d\phi \sin\theta\,d\theta = -d\phi\,d(\cos\theta) \qquad (2.15)$$

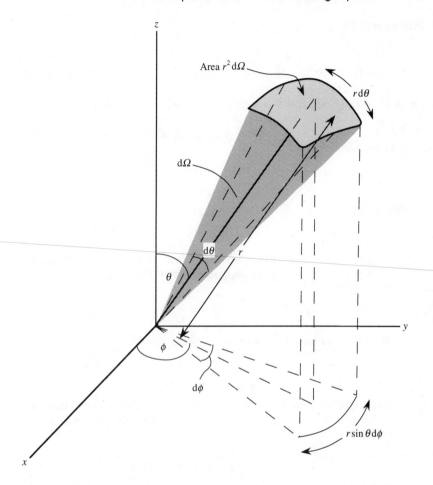

Figure 2.3 A small element of the surface of a sphere of radius r subtending an element of solid angle $d\Omega = \sin\theta \, d\theta \, d\phi$ at the centre.

The differential cross-section is defined through

$$d\sigma = \frac{\text{number of particles scattered per unit time into } d\Omega}{\text{incident flux}} \qquad (2.16)$$

Clearly the differential cross-section is a function of the direction (θ, ϕ). If the scattering interaction is symmetric about the z axis, which is the beam direction, then we expect the differential cross-section to be independent of ϕ, and to depend only on θ, which is termed the **scattering angle**. The total cross-section can be calculated from the differential cross-section by integrating over all directions:

$$\sigma = \int d\Omega \, \frac{d\sigma}{d\Omega} = \int_0^{2\pi} d\phi \int_{-1}^{1} d(\cos\theta) \, \frac{d\sigma}{d\Omega} \qquad (2.17)$$

References

[1] Young, H.D. (1992). *University Physics* 8th edn, Section 13-9. Reading MA: Addison-Wesley.

[2] Boas M.L. (1983). *Mathematical Methods in the Physical Sciences* 2nd edn, Chap. 16, Section 5. New York: Wiley.

[3] Blin-Stoyle R.J. (1991). *Nuclear and Particle Physics*, Section 5.2. London: Chapman & Hall.

Problems

2.1   At time $t = 0$ a quantum particle constrained to move one-dimensionally in the region $0 \leqslant z \leqslant \mathbb{L}$ is in a state described by the wave function

$$\Psi(z, 0) = \begin{cases} Cz(\mathbb{L} - z) & (0 \leqslant z \leqslant \mathbb{L}) \\ 0 & (z < 0, z > \mathbb{L}) \end{cases}$$

Show that this function is normalized by the choice

$$C = \sqrt{\left(\frac{30}{\mathbb{L}^5}\right)}$$

and sketch the wave function and the probability distribution as functions of z. Show that exactly the same probability distribution results from the choice

$$C = \sqrt{\left(\frac{30}{\mathbb{L}^5}\right)}(\cos \tfrac{1}{5}\pi + i \sin \tfrac{1}{5}\pi)$$

Find the position where the probability density is a maximum, and calculate the probability of finding the particle within each of the regions $0 \leqslant z \leqslant \tfrac{1}{2}\mathbb{L}$ and $0 \leqslant z \leqslant \tfrac{1}{4}\mathbb{L}$.

2.2   Show that the wave function

$$\Phi(z, 0) = \begin{cases} C'z(z - \tfrac{1}{2}\mathbb{L})(\mathbb{L} - z) & (0 \leqslant z \leqslant \mathbb{L}) \\ 0 & (z < 0, z > \mathbb{L}) \end{cases}$$

is orthogonal to the wave function $\Psi(z, 0)$ of Problem 2.1. Sketch the wave function and the probability distribution as functions of z. Compare this probability distribution with that found from $\Psi(z, 0)$.

2.3   Show that, if a particle is in the state represented by

$$\Psi(z, 0) = C \exp(-\tfrac{1}{2}z^2)$$

then at $t = 0$ the probability of detecting it between $z = 0$ and $z = \mathbb{L}$ is equal to the probability of finding it between $z = -L$ and $z = 0$. Calculate the probability of finding the particle

(a) between $z = 0$ and $z = 0.5$
(b) within a distance one unit from the origin
(c) between $z = 0.5$ and $z = 1$

(You will need to know the following integrals:

$$\int_{-\infty}^{\infty} dz \exp(-z^2) = \sqrt{\pi},$$

$$\int_{0}^{0.5} dz \exp(-z^2) \approx 0.5205 \frac{\sqrt{\pi}}{2}$$

$$\int_{0}^{1} dz \exp(-z^2) \approx 0.8427 \frac{\sqrt{\pi}}{2}$$

2.4 At $t = 0$ a particle is in a state described by a wave function $\Psi(z, 0)$. Show that if the wave function is either symmetric or antisymmetric about $z = 0$ (that is, $\Psi(-z, 0) = \pm \Psi(z, 0)$), the probability of finding the particle between $z = L_1$ and $z = L_2$ is equal to that of finding it between $z = -L_2$ and $z = -L_1$. Explain why in this case the mean of a set of position measurements made on particles in the state $\Psi(z, 0)$ is $\langle z \rangle = 0$. Would you expect to find $\langle z^2 \rangle = 0$?

2.5 A set of position measurements is made on 100 quantum particles, all in the same state. The numbers of particles found in each of 10 intervals of width 10 nm are as follows

distance from origin/nm	number of times obtained
0–10	1
10–20	5
20–30	15
30–40	25
40–50	30
50–60	15
60–70	5
70–80	3
80–90	0
90–100	1

Calculate the mean, $\langle x \rangle$, and standard deviation, Δx, of the results, using the approximation that all particles in a particular interval are at the centre of that interval. What is the minimum standard deviation you would expect to find from a set of momentum measurements on particles in the same state?

2.6 A beam of electrons with energy 300 eV travelling in the z direction has a circular cross-section of radius 10^{-2} m. How long would the beam have to be in order to observe an increase of 1% in its radius due to the uncertainty principle? What other effects can produce a spreading of the beam?

2.7   A scattering process is found to have a differential cross-section of the form

$$\frac{d\sigma}{d\Omega} = C(1 + 3\cos^2\theta)$$

where θ is the scattering angle. Plot the number of particles scattered per unit time as a function of θ. In which direction, relative to the incident beam direction, are most particles scattered? In which direction are fewest scattered? Find the total cross-section for the process if $C = 0.01$ barn. How does this compare with the geometrical cross-section of a proton, which has a radius of the order of 10^{-15} m? (Nuclear and elementary particle scattering cross-sections are measured in barns: 1 barn = 10^{-28} m^2.)

CHAPTER 3

The Representation of Dynamical Variables

3.1 Eigenvalue equations

According to Postulate 1, a characteristic wave function is associated with each dynamical state of a non-relativistic quantum particle. For example, we described the state of a free particle moving along the z axis with precise values of momentum p and energy E by the wave function (2.9):

$$\Psi(z, t) = C e^{i(pz - Et)/\hbar} \tag{3.1}$$

We need some way of choosing the correct wave function to describe any other dynamical state, such as that of a particle moving with a particular constant total energy under forces that can be represented by a potential energy function. We know that the shape of the wave function provides a statistical prediction for the results of a position measurement, but we also need to be able to predict the results of measurements of other variables such as the energy and momentum. To solve these problems, we need a new, quantum mechanical, way of representing any dynamical variable that might be measured experimentally, and the free particle wave function can give us a clue as to how we should proceed.

Although we introduced the wave function to describe the *spatial* probability distribution of the particle, the wave function (3.1) also contains the values E and p of the particle's energy and momentum. In this particular state the energy and momentum have precise values, and we should expect any experiment that measures them to give precisely these results. The values can be extracted from the wave function mathematically by differentiation with respect to the appropriate variable:

$$i\hbar \frac{\partial}{\partial t} \Psi(z, t) = E \Psi(z, t) \tag{3.2}$$

$$-i\hbar \frac{\partial}{\partial z} \Psi(z, t) = p \Psi(z, t) \qquad (3.3)$$

Operating with $i\hbar\,\partial/\partial t$ or $-i\hbar\,\partial/\partial z$ on the wave function (3.1) simply multiplies it by a constant, without altering its shape. In a sense, then, the differential operator $i\hbar\,\partial/\partial t$ represents the energy variable, since its effect on the wave function is equivalent to multiplication by E, the particular value of the energy in the state described by that wave function. In the same sense, $-i\hbar\,\partial/\partial z$ represents the z component of momentum.

Now, instead of assuming the form (3.1) of the free particle wave function, let us *start* from the assumption that the energy and momentum are represented, in the sense just defined, by the differential operators $i\hbar\,\partial/\partial t$ and $-i\hbar\,\partial/\partial z$ respectively. We then write (3.2) and (3.3) for unknown wave functions and unknown constants E and p, and solve them to determine respectively the t and z dependences of the wave function that describes a state in which the energy and momentum take particular precise values. Only functions of the exponential form (3.1) can satisfy these differential equations. So the operator representation of the energy and momentum variables determines the form given by (3.1) for the wave function representing a free particle with a precise value for its energy and the z component of its momentum.

Equations like (3.2) and (3.3) are **eigenvalue equations**, of the general form

$$\hat{A}\Psi = a\Psi \qquad (3.4)$$

where $\hat{A}$ is a linear operator, a is a constant **eigenvalue** of this operator, and Ψ is the corresponding **eigenfunction**. (A **linear operator** is one that satisfies

$$\hat{A}(c_1\Psi_1 + c_2\Psi_2) = c_1\hat{A}\Psi_1 + c_2\hat{A}\Psi_2$$

where Ψ_1 and Ψ_2 are functions on which $\hat{A}$ acts, and c_1 and c_2 are arbitrary complex numbers.) If a particular form is chosen for the linear operator, its eigenvalue equation can be solved to give a set of eigenfunctions, each corresponding to a particular eigenvalue. The set of all the eigenvalues of an operator is the **eigenvalue spectrum** of that operator. Equations (3.2) and (3.3) are respectively eigenvalue equations for the total energy and the z component of momentum, and the eigenvalues are the possible values of these variables. The energy eigenfunction corresponding to the eigenvalue E is of the form

$$\Psi(z, t) = e^{-iEt/\hbar}\psi(z) \qquad (3.5)$$

where $\psi(z)$ can be any function of z (and also of x and y when the motion is not restricted to one dimension). Similarly, the eigenfunction of the z component of momentum corresponding to the eigenvalue p has z dependence given by

$$\Psi(z, t) = e^{ipz/\hbar}\psi(t) \qquad (3.6)$$

where the time dependence $\psi(t)$ is not determined by the momentum eigenvalue equation (3.3).

Clearly the effect of the differential operator $i\hbar\,\partial/\partial t$ on any function of t other than the exponential form (3.5) is to change the shape of the function. Only the eigenfunctions of an operator retain their shape under the action of that operator. Conversely, any function that changes its shape when acted on by an operator cannot be an eigenfunction of that operator.

We shall assume that in completely general circumstances the total energy of a quantum system is represented by the operator

$$\hat{E}_{tot} = i\hbar\,\frac{\partial}{\partial t} \tag{3.7}$$

Similarly, we shall assume that the z component of the momentum can always be represented by the operator

$$\hat{P}_z = -i\hbar\,\frac{\partial}{\partial z} \tag{3.8}$$

All other dynamical variables will be represented by suitable linear operators whose eigenvalues are the possible values of the corresponding variable. We shall usually use a capital letter with a 'hat' to denote an operator, and the corresponding lower case letter for one of its eigenvalues. However, we retain the conventional notation E for an eigenvalue of the total energy operator $\hat{E}_{tot}$, and use $\hat{\mu}$ for the magnetic moment operator.

The eigenfunction corresponding to a particular eigenvalue a represents an **eigenstate** of the system, in which it is certain that the variable represented by the operator $\hat{A}$ takes precisely this value, a. This, in essence, is the content of our second postulate. The 'possible values' of a dynamical variable are the values that might be obtained in any experimental measurement of that variable. We assume that the result of a measurement is always a real number (so that it takes two measurements to determine a complex quantity). Then the linear operators we choose to represent physical variables must only have real eigenvalues. Operators that satisfy this requirement are called **Hermitian** operators; we shall define more precisely what is meant by this term in Chapter 10.

Postulate 2

A dynamical variable is represented by a linear Hermitian operator $\hat{A}$ whose (real) eigenvalue spectrum is the set of all possible results of a measurement of that variable. The eigenfunction corresponding to a particular eigenvalue a describes an eigenstate of the quantum system, on which a measurement of the variable represented by $\hat{A}$ yields the value a with probability 1.

The x and y components of the momentum can be represented by differential operators of similar form to that in (3.8) for the z component, so the three-dimensional momentum operator may be defined as

$$\hat{P} = -i\hbar\nabla \qquad (3.9)$$

The eigenfunctions of this operator are of the form

$$\psi(x, y, z) = Ce^{ip \cdot r/\hbar} \qquad (3.10)$$

where p is the vector eigenvalue of the momentum operator $\hat{P}$, and x, y and z are the components of the vector r. (Since the momentum eigenvalue equation does not determine the time dependence of the wave function, we have written the eigenfunction as a function of the spatial variables only.)

The operators used to represent dynamical variables need not be differential ones, but can be of any type, provided they are linear and Hermitian. For example, the algebraic variables x, y and z can be thought of as linear operators on functions of the spatial variables, and we shall use them to represent the components of the position operator $\hat{R}$. The effect of the operator $\hat{R}$ on a wave function is then identical to multiplication by the normal algebraic variable r. Matrices are yet another type of linear operator, and in Heisenberg's formulation of quantum mechanics all dynamical variables are represented by matrices. We shall discover the relationship between the matrix formulation and the Schrödinger formulation, which we are using, in Chapter 10.

3.2 Energy eigenstates

Postulate 1 (Section 2.2) requires the wave function to be non-singular (so that it can be normalized). The momentum eigenfunction (3.10) satisfies this for any real vector p, since in that case it is finite for all x, y and z. So, provided p is real, no restrictions need be imposed on the eigenfunction, and it describes the state of a free particle with a constant probability density throughout space. The momentum spectrum for a free particle is therefore a continuous range of real values, each component taking any value in the range $(-\infty, +\infty)$.

Since the free particle wave function (3.1) is an energy eigenfunction that is finite for any t provided E is real, it appears that the energy can also take any real value from $-\infty$ to $+\infty$. But in classical mechanics the total energy of a free particle is restricted to positive values or zero, because of the relationship between the energy and momentum variables, expressed by $E = p^2/2m$. No such relationship is implicit in the energy and momentum eigenvalue equations (3.2) and (3.3), since the energy and momentum are represented by completely independent operators. The energy eigenvalues for a free quantum particle can be restricted to the range $(0, \infty)$ by imposing a relationship between the energy and momentum operators, corresponding to that between the classical variables. Replacing the variables E and p in the classical relation by the operators $\hat{E}_{\text{tot}}$ and $\hat{P}_z$ respectively, acting on a wave function $\Psi(x, y, z, t)$, we obtain

$$i\hbar \frac{\partial}{\partial t} \Psi(x, y, z, t) = -\frac{\hbar^2}{2m} \frac{\partial^2}{\partial z^2} \Psi(x, y, z, t) \tag{3.11}$$

It is easy to check that the free particle wave function (3.1) is an eigenfunction of both of the operators in (3.11), and that the equality of the operators implies the classical relationship between the eigenvalues. Equation (3.11) therefore limits the free particle energy spectrum to the continuous range of values $(0, \infty)$.

In general, as in classical mechanics, we expect there to be an expression for the total energy of any particle, whether free or not, in terms of its momentum. For example, for a non-relativistic conservative system the total energy is expressed classically as the sum of the kinetic and potential energies; similarly, a free relativistic particle has an energy satisfying $E^2 = p^2 c^2 + m^2 c^4$. Corresponding equations in quantum mechanics are relations between operators, obtained by using the **correspondence principle**: dynamical variables in the appropriate classical equation are replaced by the corresponding quantum mechanical operators, as we did to obtain (3.11). This procedure gives a second representation for the total energy, which is called the **Hamiltonian** operator $\hat{H}$. The case of the non-relativistic conservative motion of a particle is of particular interest:

$$\hat{H} = -\frac{\hbar^2}{2m} \nabla^2 + \hat{V}(x, y, z) \tag{3.12}$$

Very often the operator $\hat{V}(x, y, z)$ is obtained from a classical potential function, by replacing x, y and z by components of the position operator. But since we have chosen to represent the position operator by ordinary algebraic variables, the potential operator in this case is just the ordinary classical potential energy function: $\hat{V}(x, y, z) = V(x, y, z)$. When this is the case we shall often omit the hat on the potential energy operator. The operator on the right-hand side of (3.11) is the special case of the Hamiltonian (3.12) that applies to a free particle moving in one dimension.

If we equate the effect of the operator $\hat{E}_{\text{tot}}$ of (3.7) on an arbitrary wave function $\Psi(x, y, z, t)$ with that of the Hamiltonian operator, we obtain the **time-dependent Schrödinger equation**

$$i\hbar \frac{\partial}{\partial t} \Psi(x, y, z, t) = \hat{H} \Psi(x, y, z, t) \tag{3.13}$$

(Equation (3.11) is a particular example of this.) This is the quantum mechanical equation of motion, which we mentioned in Section 2.1, and it determines the way in which the wave function describing a particular system evolves with time. However, we defer a discussion of the time dependence of states that are not energy eigenstates until Chapter 12.

States in which the total energy is conserved are important in both classical and quantum mechanics. In quantum mechanics such constant-energy states are energy eigenstates, which are described by eigenfunctions with the time dependence given in (3.5), and characterized by a probability distribution $|\Psi|^2$ that is constant in time. This clearly guarantees that if the wave function has been normalized at one time then it is also normalized at any other time, so that the probability of finding the particle somewhere in space remains unity for all t. Of course we expect this conservation of probability to apply even when the particle is not in an energy eigenstate, and this is the case if the total energy is represented by the operator (3.7). This can be demonstrated using the time-dependent Schrödinger equation (3.13) and a Hamiltonian of the form (3.12). Consider the variation with time of the probability $\mathbb{P}_V$ defined in (2.1):

$$\frac{d}{dt} \mathbb{P}_V = \int_V d^3r \frac{\partial}{\partial t} |\Psi|^2 = \int_V d^3r \left[\Psi^* \frac{\partial}{\partial t} \Psi + \left(\frac{\partial}{\partial t} \Psi^* \right) \Psi \right] \quad \textbf{(3.14)}$$

To write the right-hand side in terms of the Hamiltonian, we use (3.13) and its complex conjugate:

$$-i\hbar \frac{\partial}{\partial t} \Psi^* = -\frac{\hbar^2}{2m} \nabla^2 \Psi^* + V(x, y, z) \Psi^* \quad \textbf{(3.15)}$$

Then (3.14) gives

$$i\hbar \frac{d}{dt} \mathbb{P}_V = -\frac{\hbar^2}{2m} \int_V d^3r \left[\Psi^*(\nabla^2 \Psi) - (\nabla^2 \Psi^*) \Psi \right]$$

$$+ \int_V d^3r (\Psi^* V \Psi - V \Psi^* \Psi)$$

The two terms in the second integral cancel because they are both just products of the same three functions. However, the first integral is non-zero, because the differential operator acts on a different function in each term. The integrand is equal to $\nabla \cdot [\Psi^*(\nabla \Psi) - (\nabla \Psi^*)\Psi]$, and the equation can be written as

$$\int_V d^3r \frac{\partial}{\partial t} |\Psi|^2 = \frac{i\hbar}{2m} \int_V d^3r \nabla \cdot [\Psi^*(\nabla \Psi) - (\nabla \Psi^*) \Psi]$$

Since V is an arbitrary volume, the integrands on either side of this equation can be equated, giving

$$\frac{\partial}{\partial t} |\Psi|^2 + \nabla \cdot j = 0 \quad \textbf{(3.16)}$$

where

$$j = -\frac{i\hbar}{2m} [\Psi^*(\nabla \Psi) - (\nabla \Psi^*) \Psi] \qquad (3.17)$$

Equation (3.16) is a continuity equation just like that representing conservation of electric charge [1], with $|\Psi|^2$ taking the place of the charge density and j, defined through (3.17), representing a probability current density. Accordingly, the only way the total probability of finding the particle within a particular volume V can change is through a flow of probability current across the surface of this volume, and so the probability is a conserved quantity.

The conservation of probability depends on the validity of the time-dependent Schrödinger equation, which in turn depends on the representation (3.7) for the total energy operator. But the eigenfunctions of this operator are complex, and so we cannot avoid the use of complex wave functions if we are to have a theory in which the probability of finding a particle somewhere in space is to remain constant

If we write an energy eigenvalue equation using the Hamiltonian (3.12) as the total energy operator, we obtain a second-order differential equation

$$\left[-\frac{\hbar^2}{2m} \nabla^2 + V(x, y, z) \right] \Psi(x, y, z, t) = E \Psi(x, y, z, t) \qquad (3.18)$$

Clearly this equation will determine the spatial dependence of an energy eigenfunction. The time dependence is determined by solving the energy eigenvalue equation with $\hat{E}_{tot}$ of (3.7) as the total energy operator, and is always of the form (3.5). So the energy eigenfunction may be factorized into time- and space-dependent parts:

$$\Psi(x, y, z, t) = \psi(x, y, z)e^{-iEt/\hbar} \qquad (3.19)$$

Substitution of this into (3.18) and cancellation of the exponentials on either side gives the **time-independent Schrödinger equation**

$$\left[-\frac{\hbar^2}{2m} \nabla^2 + V(x, y, z) \right] \psi(x, y, z) = E \psi(x, y, z) \qquad (3.20)$$

This is the equation that is used to determine the energy spectrum and energy eigenfunctions of any conservative non-relativistic quantum particle, such as an electron in an atom, or an atom in a crystal lattice (provided each is treated as a single particle in a static potential due to all the other particles in the system).

For example, for a free particle in one-dimensional motion along the z axis (3.20) becomes

$$\frac{d^2}{dz^2} \psi(z) = -k^2 \psi(z) \qquad (3.21)$$

where we have defined

$$k^2 = \frac{2mE}{\hbar^2} \geqslant 0 \qquad (3.22)$$

A second-order differential equation has two independent solutions, and the general solution involves two arbitrary constants. The general solution [2] of (3.21) is

$$\psi(z) = C_1 e^{ikz} + C_2 e^{-ikz} \qquad (3.23)$$

This is a linear superposition of two eigenfunctions of momentum, of the form (3.10) with p in the z direction, one corresponding to the eigenvalue $p = +\hbar k$, and the other to $p = -\hbar k$. Equation (3.22) guarantees that the square of the momentum eigenvalue is correctly related to the energy eigenvalue. However, the energy eigenfunction (3.23) is not, in general, an eigenfunction of $\hat{P}_z$, since operating on $\psi(z)$ with the operator (3.8) changes the wave function unless $C_1 = 0$ or $C_2 = 0$. This means that in general (3.23) describes a state in which the energy and the magnitude of the momentum take precise values, but in which the sign of the z component of momentum is indeterminate. When an energy eigenfunction can be written as some linear combination of two or more independent eigenfunctions of some other operator, the energy eigenstate is said to be **degenerate**. The energy eigenstate of a free particle moving in one dimension is doubly degenerate, since there are two different momentum eigenstates corresponding to the same energy.

3.3 Bound states of a particle in a one-dimensional square potential well

Consider a particle moving along the z axis, with the potential energy (Figure 3.1)

$$V(z) = \begin{cases} V_0 & (z < 0, z > L) \\ 0 & (0 \leqslant z \leqslant L) \end{cases} \qquad (3.24)$$

If the motion were classical with total energy $E < V_0$, the particle would be bound in Region II, where $0 \leqslant z \leqslant L$. This means that there would be no chance of finding it in Region I, $z < 0$, or Region III, $z > L$, and the particle would move backwards and forwards between the turning points $z = 0$ and $z = L$. The corresponding quantum behaviour is found by solving the Schrödinger equation (3.20) for the potential (3.24).

In Region II the Schrödinger equation is of the same form as the free particle equation (3.21), with wavenumber k given by (3.22), and the

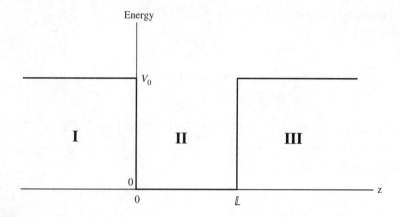

Figure 3.1 The one-dimensional square potential well (3.24).

general solution (3.23). The wave function in Region II can therefore be written as

$$\psi_2(z) = A_2 e^{ikz} + B_2 e^{-ikz} \quad (0 \leqslant z \leqslant L) \qquad (3.25)$$

In Regions I and III, with $E \leqslant V_0$, the Schrödinger equation becomes

$$\frac{d^2\psi}{dz^2} = \gamma^2 \psi, \qquad \gamma^2 = \frac{2m}{\hbar^2}(V_0 - E) \geqslant 0 \qquad (3.26)$$

with the solutions

$$\psi_1(z) = A_1 e^{\gamma z} + B_1 e^{-\gamma z} \quad (z < 0) \qquad (3.27)$$

$$\psi_3(z) = A_3 e^{\gamma z} + B_3 e^{-\gamma z} \quad (z > L) \qquad (3.28)$$

According to Postulate 1 (Section 2.2), the wave function should be continuous everywhere. If the slope $d\psi/dz$ were discontinuous at some point, the second derivative $d^2\psi/dz^2$, would be infinite there. But the Schrödinger equation implies that the value of $d^2\psi/dz^2$ is finite everywhere (as long as the potential remains finite), though it has discontinuities at the same points $z = 0$ and $z = L$ as the potential. So both $d\psi/dz$ and the wave function itself must be continuous at $z = 0$ and $z = L$:

$$\psi_1(0) = \psi_2(0), \qquad \left.\frac{d\psi_1}{dz}\right|_{z=0} = \left.\frac{d\psi_2}{dz}\right|_{z=0} \qquad (3.29)$$

$$\psi_2(L) = \psi_3(L), \qquad \left.\frac{d\psi_2}{dz}\right|_{z=L} = \left.\frac{d\psi_3}{dz}\right|_{z=L} \qquad (3.30)$$

These restrictions relate the six arbitrary constants A_1, B_1, A_2, B_2, A_3 and B_3:

$$\left.\begin{array}{c} A_1 + B_1 = A_2 + B_2 \\ \gamma(A_1 - B_1) = ik(A_2 - B_2) \\ A_3e^{\gamma L} + B_3e^{-\gamma L} = A_2e^{ikL} + B_2e^{-ikL} \\ \gamma(A_3e^{\gamma L} - B_3e^{-\gamma L}) = ik(A_2e^{ikL} - B_2e^{-ikL}) \end{array}\right\} \quad \text{(3.31)}$$

A fifth relationship between the six constants is provided by the normalization condition (2.2):

$$\int_{-\infty}^{0} dz \, |\psi_1|^2 + \int_{0}^{L} dz \, |\psi_2|^2 + \int_{L}^{\infty} dz \, |\psi_3|^2 = 1 \quad \text{(3.32)}$$

Equations (3.31) and (3.32) allow five of the constants to be determined in terms of the sixth.

The computer program *Finite square well* generates the solution to the Schrödinger equation for particular values of m, L, V_0 and E. Some solutions are plotted in Figure 3.2 for a particular set of values m, L and V_0 and eight different values of E. The wave function becomes infinite as $z \to \pm\infty$ for six of these values, and these solutions are not physically acceptable since they cannot be normalized. Inspection of (3.27) and (3.28) shows that the problem is caused by the terms that increase exponentially as $z \to -\infty$ in Region I and $z \to +\infty$ in Region III. These must vanish in an acceptable energy eigenfunction, so we must put $B_1 = 0$ and $A_3 = 0$:

$$\psi_1(z) = A_1e^{\gamma z} \quad (z < 0) \quad \text{(3.33)}$$

$$\psi_3(z) = B_3e^{-\gamma z} \quad (z > L) \quad \text{(3.34)}$$

This is equivalent to imposing the boundary condition

$$\psi(z) \to 0 \quad (z \to \pm\infty) \quad \text{(3.35)}$$

But it is only possible to implement this condition for a particular set of discrete energy values $\{E_n; \, n = 1, 2, 3, \ldots\}$, as can be discovered by trial and error using the computer program. These are the energy eigenvalues for the bound particle.

An equation to determine the energy eigenvalues can be found by solving (3.31) with $B_1 = 0$ and $A_3 = 0$ (see Problem 3.7), but there is no analytical expression for the energy eigenvalues for a square well of finite depth. The eigenvalues and corresponding eigenfunctions can be found numerically. However, in the limit where the well depth becomes infinite a complete analytical solution to the energy eigenvalue problem is possible. As $V_0 \to \infty$, $\gamma \to \infty$, and the wave functions of (3.33) and (3.34) become identically zero:

$$\psi_1(z) \equiv 0, \quad \psi_3(z) \equiv 0 \quad \text{(3.36)}$$

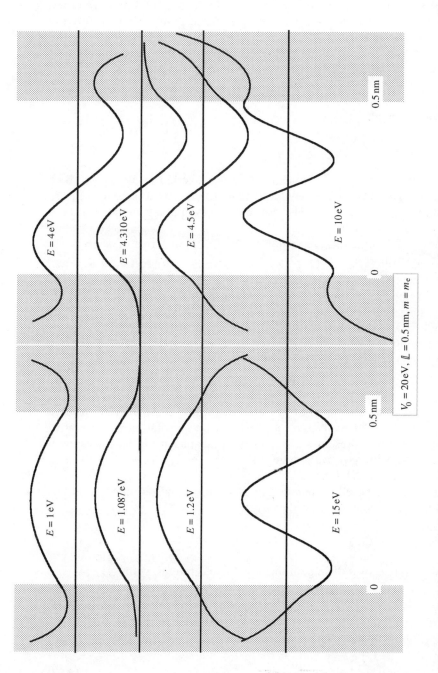

$V_0 = 20\,\mathrm{eV}$, $\mathbb{L} = 0.5\,\mathrm{nm}$, $m = m_e$

Figure 3.2 Some solutions of the Schrödinger equation for an electron in a square well of depth 20 eV and width 0.5 nm. These functions have been chosen to be symmetric or antisymmetric about the centre of the well, but the restriction to functions that are finite at all z has not been applied. Only the functions corresponding to $E = 1.087\,\mathrm{eV}$ and $E = 4.310\,\mathrm{eV}$ remain finite as $z \to \pm\infty$, and so are acceptable energy eigenfunctions. The shading indicates regions that are not classically accessible.

59

The wave function in Region II is still of the form (3.25). Since the potential becomes infinite at $z = 0$ and $z = L$, the second derivative of the wave function need no longer be finite at these points, so that only the first condition in each of (3.29) and (3.30) can be applied, requiring the wave function but not its derivative to be continuous. The condition at $z = 0$ gives $A_2 = -B_2$, so that the wave function is of the form

$$\psi(z) = C \sin kz \quad (0 \leqslant z \leqslant L) \tag{3.37}$$

The condition at $z = L$ gives

$$\sin kL = 0, \quad \text{that is,} \quad k = n\frac{\pi}{L} \quad (n = 1, 2, \dots) \tag{3.38}$$

(Note that this is the condition for the wave function in the well to be a standing wave. The energy eigenfunctions for the finite square well are also standing waves, but in this case the wave amplitude cannot vanish at the edges of the well.)

Using (3.38) with the expression (3.22) for E in terms of k gives the energy spectrum

$$E_n = \frac{h^2}{8mL^2}n^2 \quad (n = 1, 2, \dots) \tag{3.39}$$

with the corresponding eigenfunctions

$$\psi_n(z) = \begin{cases} C \sin \dfrac{n\pi z}{L} & (0 \leqslant z \leqslant L) \\ 0 & (z < 0, z > L) \end{cases} \tag{3.40}$$

(The solution with $n = 0$ is of no interest because it would yield a wave function that is identically zero everywhere, and a solution with negative integer does not represent a different physical state from the corresponding solution with positive integer, since the eigenfunctions $\psi_n(z)$ and $\psi_{-n}(z)$ differ only by a constant factor -1.)

For the infinite well, remembering that the normalization constant can be chosen to be real, (3.32) reduces to

$$\int_0^L dz\, C^2 \sin^2 \frac{n\pi z}{L} = 1$$

So for all values of n the eigenfunctions (3.40) are real, with

$$C = \sqrt{\left(\frac{2}{L}\right)} \tag{3.41}$$

The eigenfunctions form an orthogonal set, as defined by (2.5), because

$$\int_0^L dz \sin \frac{n\pi z}{L} \sin \frac{j\pi z}{L} = 0 \quad (n \neq j) \tag{3.42}$$

The forms of the energy eigenfunctions for four eigenstates of a finite and an infinite square well are compared in Figure 3.3. They are qualitatively similar, except that the wave function penetrates into the classically forbidden Regions I and III in the case of the finite well. This is an example of barrier penetration, and means that for a finite well there is a non-zero probability of finding the particle outside the classically allowed Region II, though the probability density for this decreases exponentially with distance from the region. For both the finite and infinite wells the wave function is symmetric about the centre of the well for states with odd n, and antisymmetric if n is even. The eigenfunctions are oscillatory within Region II, with the number of zeros (other than those at $z = 0$ and $z = L$ in the infinite case) equal to $n - 1$ for the nth eigenfunction. According to the classical picture, the particle moves back and forth in Region II with a constant velocity (except at $z = 0$ and $z = L$), so that the particle passes through every point in the region. But in the quantum case the zeros of the energy eigenfunctions are points where there is zero probability of finding the particle, so their existence implies that the classical picture of a continuous trajectory is not appropriate.

The energy spectrum for the finite square well is qualitatively similar to that for the infinite well: the eigenvalues increase approximately as n^2, and the lowest eigenvalue is lower the greater the width of the well. However, only a finite number of discrete levels can be accommodated in a finite well. An estimate of the number of bound states can be obtained by inserting $E_n = V_0$ in (3.39). This gives a maximum value of n close to $w = \sqrt{(8mV_0L^2/h^2)}$; this quantity is called the **well parameter**. A more accurate assessment shows that the number of levels in the finite well is the next integer greater than the well parameter. (See Problem 3.7.)

The energy eigenstates for a particle in the square well that we have just discussed show certain characteristics that are common to any quantum bound state, irrespective of the shape of the binding potential or of whether the motion is in one, two or three dimensions.

(1) The energy spectrum of a bound particle is discrete. Mathematically this is a result of the boundary conditions that must be imposed on the energy eigenfunctions to prevent them diverging to infinity at large distances from the region occupied by the potential well. It is in contrast to the continuous spectrum of a free particle, whose energy eigenfunctions remain finite at all points in space for any positive value of the energy.

(2) The lowest energy eigenvalue for a bound particle is always greater than the minimum that would be expected classically. For the square well the minimum energy is $E_1 > 0$, while the classical minimum energy is zero. This zero-point energy can be accounted for by Heisenberg's uncertainty principle, as explained in Section 1.5.

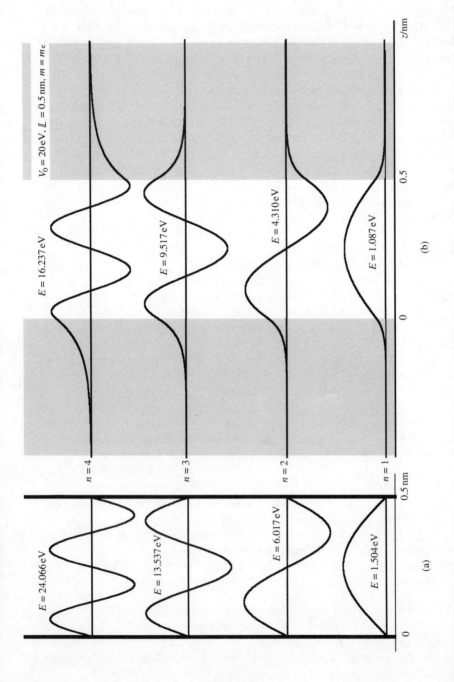

Figure 3.3 Energy eigenfunctions for an electron bound in (a) an infinite square well and (b) a finite square well of depth 20 eV, both of width 0.5 nm. In (b) the shading indicates regions that are not classically accessible.

(3) Within the classically allowed region (where the total energy E is greater than the potential energy V) the energy eigenfunctions of the bound particle have zeros, or **nodes**, like the nodes of a standing wave on a string. The ground state, like the fundamental mode of the string, has no node, and in general the number of nodes increases with the order n of the energy level. In this region, therefore, an energy eigenfunction oscillates.

(4) The energy eigenfunctions for a bound particle extend into the region of space that would be inaccessible to a classical particle of the same energy, bound by the same potential. This is an example of barrier penetration, mentioned in Section 1.5, and implies that there is a probability of finding the quantum particle in a classically forbidden region. Energy eigenfunctions in a barrier region are not oscillatory, and fall off rapidly as the distance from the classicaly allowed region increases.

(5) A pair of eigenfunctions corresponding to different eigenvalues is always orthogonal, in the sense defined by (2.5). This result is true for the eigenfunctions of any Hermitian operator. We shall prove this in Chapter 10.

Quantum well structures

It might seem that the potential we have been discussing is too artificial to be useful in any real situation. However, recent technical developments have made it possible to construct semiconducting devices in which electrons behave as if they were bound in a one-dimensional square potential well [3, 4]. It is possible to lay down a very uniform and thin layer of one semiconducting material on another. For example a layer of gallium arsenide (GaAs) only 5 nm thick can be laid on a gallium aluminium arsenide (GaAlAs) substrate. So-called **quantum well structures** are composed of thin layers of one semiconductor sandwiched between layers of another.

In a bulk semiconductor the energy levels of the valence electrons form an almost continuous band – the **valence band** (this will be discussed in Chapter 7). The highest energy in this band is separated by a forbidden energy gap E_G from the lowest energy in a higher band of continuous levels known as the **conduction band**. To excite an electron from the valence band to the conduction band therefore requires an energy at least as great as the band gap energy E_G. The excited electron leaves a **hole** in the filled levels of the valence band, which behaves like a positively charged particle (with an effective mass that is not, in general, the same as that of a valence electron).

In quantum well structures the layers are composed of semiconductors with different band gap energies. For example, GaAs has a lower band gap energy than GaAlAs. A layer of GaAs sandwiched between two layers of GaAlAs effectively creates a one-dimensional square potential well in the conduction band, of width equal to the width of the layer, as shown in Figure 3.4. An electron in the GaAs region may be excited from the valence band to one of the quantized energy levels in the well in the conduction band. The hole left in the valence band effectively moves in an inverted square well, the top of which is separated by the GaAs band gap energy from the bottom of the well in the conduction band. Such an electron–hole pair is confined in the region of the GaAs layer, but in the other two dimensions it is essentially free. So the total energy of the electron is the combination of contributions from the continuous spectrum of kinetic energies associated with motion in the plane of the GaAs layer, and from the set of discrete one-dimensional square well levels. Similarly, the hole energy is the sum of a continuous part and a discrete square well energy eigenvalue. Energies predicted by a one-dimensional finite square well calculation are in good agreement with experimental results.

It is also possible to confine electrons two-dimensionally, in **quantum wires**, or even three-dimensionally in **quantum dots** [4]. Quantum well structures have great potential as the basis of new devices, because of their interesting nonlinear optoelectronic properties, and because their electronic energy levels can be tailored to suit particular applications.

3.4 Scattering by a one-dimensional potential step

Suppose that the potential energy of a particle moving in one dimension changes abruptly from a constant value V_0 to another constant value V_1 at the point $z = 0$, as shown in Figure 3.5. Classically, if a particle moves along the z axis with total energy E greater than both V_0 and V_1, its momentum changes abruptly at $z = 0$. That is, the potential step scatters the particle, though if the motion is restricted to one dimension, only the magnitude (and not the direction) of the momentum is changed. But if the particle behaves quantum mechanically, its wave-like characteristics manifest themselves in the possible reflection of the particle at $z = 0$.

The energy eigenfunction of a quantum particle moving with energy $E > V_0$ in the potential of Figure 3.5 is of the form (3.23), since the Schrödinger equation is of the form (3.21) in each of the regions $z < 0$ and $z \geqslant 0$. The appropriate expression for the wave vector in each of

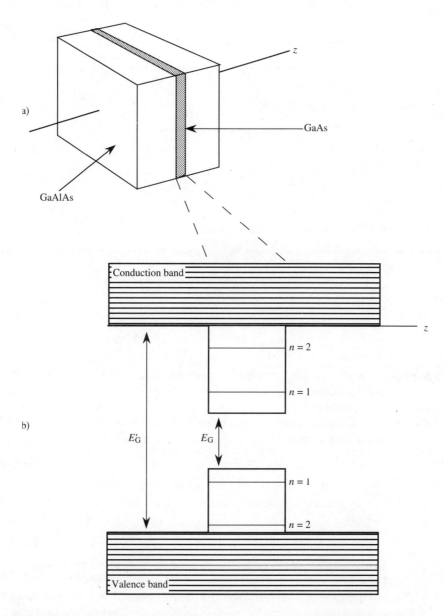

Figure 3.4 A single GaAs/GaAlAs quantum well: (a) the 'sandwich' of GaAs in GaAlAs represented in real space; (b) the energy diagram for the quantum well, where E_G is the GaAs band gap energy and E'_G that of GaAlAs.

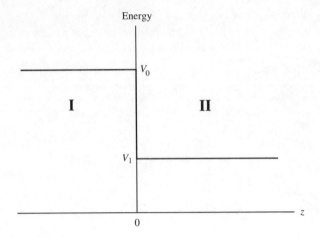

Figure 3.5 A one-dimensional potential barrier.

these regions is

$$k_0^2 = \frac{2m}{\hbar^2}(E - V_0) \qquad (z < 0)$$

$$k_1^2 = \frac{2m}{\hbar^2}(E - V_1) \qquad (z \geqslant 0)$$

(3.43)

Suppose that the particle is incident from the left in Figure 3.5, so that there is no probability of finding it travelling in the $-z$ direction in Region II, where $z \geqslant 0$. The energy eigenfunction in that region must therefore be simultaneously an eigenfunction of momentum corresponding to the eigenvalue $p = +\hbar k_1$, and can be written as

$$\psi_1(z) = A_1 e^{ik_1 z} \qquad (z \geqslant 0)$$

(3.44)

In the region $z < 0$ it is not possible to assume the absence of a reflected component of the wave function, and the eigenfunction must be written as

$$\psi_0(z) = A_0 e^{ik_0 z} + B_0 e^{-ik_0 z} \qquad (z < 0)$$

(3.45)

This is like a superposition of an incident wave and a reflected wave in classical wave theory (except that the quantum wave is complex).

The coefficients A_0 and B_0 are the amplitudes of the incident and reflected waves (since $|e^{\pm ik_0 z}| = 1$). We shall see in Chapter 9 that these constants can be identified as probability amplitudes for finding the particle in Region I ($z < 0$) travelling to the right and to the left respectively, if a measurement of the momentum direction is made. A scattering experiment involves a set of such measurements, as we discussed in Section 2.5, though in the present one-dimensional case the momentum can only be along the positive or negative z direction, so that the wave vectors of both the incident and scattered waves are parallel to the z axis. The probability of finding the particle incident on the step at $z = 0$ from the left is $|A_0|^2$, and the probabilities of finding it reflected back into Region I and transmitted into Region II are $|B_0|^2$ and $|A_1|^2$ respectively.

The wave function refers to a *single* particle, so that $\psi_0(z)$ does *not* represent a mixture of particles, some travelling to the left and others to the right. However, according to Section 2.3, if identical measurements are carried out on a large number of particles in identical quantum states, the proportion of measurements that yield a particular result is equal to the probability of obtaining that result in a single measurement. Thus if a beam of particles, all with the same momentum, is scattered by a target equivalent to the potential step, the numbers of particles found to be incident and reflected in the region $z < 0$ is proportional to $|A_0|^2$ and $|B_0|^2$ respectively, while the number transmitted into the region $z \geqslant 0$ is proportional to $|A_1|^2$. The proportion of incident particles that is reflected is therefore equal to the single-particle reflection probability $|B_0|^2/|A_0|^2$, while the proportion that are transmitted is $|A_1|^2/|A_0|^2$.

In the scattering of classical waves the reflection and transmission coefficients, $\mathbb{R}$ and $\mathbb{T}$ respectively, are defined in terms of the rate of flow of energy in the reflected and transmitted beams relative to that in the incident beam [5]. In quantum scattering $\mathbb{R}$ and $\mathbb{T}$ are defined in a similar way, but the rate of flow of energy is replaced by the rate of flow of probability. For a free particle the rate of flow of probability per unit area is the **probability current density**, equal to the probability density multiplied by the velocity $\hbar k/m$ (Problem 3.5a). For the incident wave the probability current density $(\hbar k_0/m)|A_0|^2$ multiplied by the number of particles per unit volume in the incident beam is identical to the incident flux defined by (2.13). Similar probability current densities determine the rate of flow of particles in the reflected beam $(k = k_0)$ and transmitted beam $(k = k_1)$. The **reflection** and **transmission coefficients** are therefore defined as

$$\mathbb{R} = \frac{|B_0|^2}{|A_0|^2}, \qquad \mathbb{T} = \frac{k_1 |A_1|^2}{k_0 |A_0|^2} \qquad (3.46)$$

The wave function and its derivative must be continuous at $z = 0$, so that equations like (3.29) must be satisfied:

$$\psi_0(0) = \psi_1(0), \qquad \frac{\mathrm{d}\psi_0}{\mathrm{d}z}\bigg|_{z=0} = \frac{\mathrm{d}\psi_1}{\mathrm{d}z}\bigg|_{z=0} \qquad (3.47)$$

Applying these conditions to the wave functions (3.44) and (3.45), we obtain

$$\frac{A_1}{A_0} = \frac{2k_0}{k_0 + k_1}, \qquad \frac{B_0}{A_0} = \frac{k_0 - k_1}{k_0 + k_1} \qquad (3.48)$$

These expressions are real, so that if A_0 is chosen to be real, A_1 and B_0 are real too. There is no restriction on the value of k_0 or k_1, since ψ_0 and ψ_1 are finite everywhere. The reflection and transmission coefficients are

$$R = \left(\frac{k_0 - k_1}{k_0 + k_1}\right)^2, \qquad T = \frac{4k_0k_1}{(k_0 + k_1)^2} \tag{3.49}$$

Note that $R + T = 1$, which must be true if each incident beam particle is either reflected or transmitted.

R and T are independent of Planck's constant, whether the wave vectors are expressed in terms of the wavelengths in the two regions or in terms of the energy E and well depth V_0, so that, according to the rule of thumb given in Section 1.6, the phenomenon is a classical one. Indeed it is, though it is the behaviour of a classical wave rather than of a classical particle. Phenomena described by an equation involving h cannot be derived from either classical wave theory or classical particle theory.

Figure 3.6 shows the probability density for the particle. It has the same value, A_1^2, at each point in the region $z \geqslant 0$. However, in the region $z < 0$ it is

$$|\psi_0|^2 = A_0^2 + B_0^2 + 2A_0B_0 \cos 2k_0 z \tag{3.50}$$

The oscillations in this probability density are caused by interference between the incident and reflected components of the wave function.

3.5 Scattering by a one-dimensional square well

A particle in the potential of Figure 3.1 with $E < V_0$ is bound; but if $E > V_0$, it is free to move anywhere along the z axis. According to classical mechanics, the free particle can travel from $z = -\infty$ to $z = +\infty$ without any chance of its direction being reversed, though it will suffer abrupt changes in kinetic energy at $z = 0$ and $z = L$. On the other hand, according to quantum mechanics, there is a·non-zero probability that the free particle will be reflected at $z = 0$ or $z = L$, where the potential is discontinuous. Reflection and transmission coefficients R and T can be defined and calculated as in the case of scattering by a potential step, and in this case interference effects produce oscillations in the probability densities in both Regions I and II of Figure 3.1, and also in R and T.

The potential is given by (3.24), and for $E > V_0$ the Schrödinger equation is of a similar form to (3.21) in each of the three regions. In Region II the wave vector is k, given by (3.22), while in Regions I and III it is k_0, (3.43). Assuming that there is no probability of finding the particle travelling in the $-z$ direction in Region III, the wave function in that region is of the same form as (3.44):

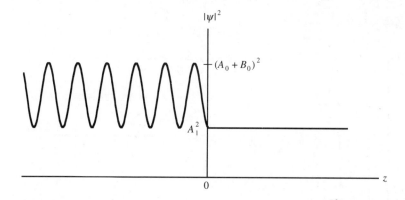

Figure 3.6 The probability density $|\psi|^2$ for an electron in an energy eigenstate corresponding to the eigenvalue $E = 30\,\text{eV}$, in the potential of Figure 3.5 with $V_0 = 20\,\text{eV}$ and $V_1 = 0\,\text{eV}$. A_0^2 is the square of the amplitude (assumed real) of the incident component of the wave function.

$$\psi_3(z) = A_3 e^{ik_0 z} \quad (z > L) \tag{3.51}$$

In the other two regions it is not possible to assume the absence of reflected components, and the energy eigenfunctions must be written as

$$\psi_1(z) = A_1 e^{ik_0 z} + B_1 e^{-ik_0 z} \quad (z < 0) \tag{3.52}$$

$$\psi_2(z) = A_2 e^{ikz} + B_2 e^{-ikz} \quad (0 \leqslant z \leqslant L) \tag{3.53}$$

The wave function and its derivative must be continuous at $z = 0$ and $z = L$, so that (3.29) and (3.30) must be satisfied. These conditions give four relationships between the five constants A_1, A_2, A_3, B_1 and B_2:

$$\left. \begin{array}{l} A_1 + B_1 - (A_2 + B_2) = 0 \\[4pt] k_0(A_1 - B_1) - k(A_2 - B_2) = 0 \\[4pt] A_2 e^{ikL} + B_2 e^{-ikL} - A_3 e^{ik_0 L} = 0 \\[4pt] k(A_2 e^{ikL} - B_2 e^{-ikL}) - k_0 A_3 e^{ik_0 L} = 0 \end{array} \right\} \tag{3.54}$$

This set of equations can be solved to determine four of the constants in terms of the fifth, while the fifth can be determined by normalization. In general, therefore, neither B_1 nor B_2 will vanish, and there will be reflected waves in Regions I and II. Acceptable solutions exist for any value of $E > V_0$, so that the particle has a continuous energy spectrum.

The coefficients $\mathbb{R}$, for reflection into Region I, and $\mathbb{T}$, for transmission into Region III, are defined in a corresponding way to those of (3.46) (but remember that the wave vector is the same in Regions I and III):

$$\mathbb{R} = \left| \frac{B_1}{A_1} \right|^2, \quad \mathbb{T} = \left| \frac{A_3}{A_1} \right|^2 \tag{3.55}$$

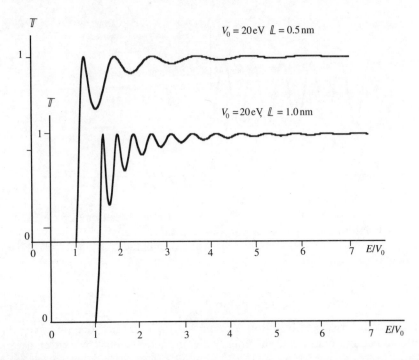

Figure 3.7 The transmission coefficient T for an electron moving in the potential of Figure 3.1 with energy E greater than V_0, shown for two different sets of well parameters.

When the simultaneous equations (3.54) are solved, the following expressions are found (Problem 3.12):

$$R = \left[1 + \frac{4k_0^2 k^2}{(k_0^2 - k^2)^2 \sin^2 kL} \right]^{-1} \tag{3.56}$$

$$T = \left[1 + \frac{(k_0^2 - k^2)^2 \sin^2 kL}{4k_0^2 k^2} \right]^{-1} \tag{3.57}$$

The form of the transmission coefficient as a function of the energy eigenvalue E is shown in Figure 3.7. The oscillations in its magnitude are due to the $\sin^2 kL$ term in (3.57). Their height and width depend on the size of the well parameter $\sqrt{(8mV_0L^2/h^2)}$, and they are damped out as the energy becomes very large compared with the depth of the potential well. So for energies such that $E/V_0 \gg 1$ the transmission probability is 1, just as in the classical case. At lower energies **transmission resonances**, where $T = 1$, occur for certain discrete values of E determined by $\sin kL = 0$. This is identical to the condition (3.38) that determines the energy eigenvalues for an infinite square well. In both cases it is the condition that Region II contains a standing wave. Although the particle behaves classically at these energies in that it has zero probability of being reflected back into Region I, it is easy to check that its behaviour in Region II is not classical – there is a non-vanishing reflected component to the wave function $\psi_2(z)$, which interferes with the component corresponding to a particle travelling in the incident direction.

The scanning tunnelling microscope

A square potential barrier scatters free particles with energy greater than the height of the barrier in a similar way to the potential well we have just discussed (Problem 3.13a). In the case of a barrier free particles may be incident at an energy less than the height of the barrier, and can penetrate through to the far side only through quantum tunnelling (Problem 3.13b). The probability of tunnelling decreases rapidly as the width of the barrier is increased, and the **scanning tunnelling microscope (STM)** exploits this fact to measure variations in the height of a conducting surface on an atomic scale. In an STM [6] a potential difference is applied between a sharp tungsten tip and the surface of the (conducting) sample under examination. If the applied voltage is small, there is a potential barrier between tip and sample of the form shown in Figure 3.8(a), and, provided the barrier is narrow enough, electrons pass from tip to sample through quantum tunnelling, producing an electric current. An increase in the barrier width by only 0.1 nm decreases the tunnelling current by a factor of about 10, so that as the surface is scanned the 'hills' due to individual atoms and the 'valleys' between them can be detected!

If the STM is operated in **field emission mode**, with a relatively large potential difference between tip and sample (of the order of volts), transmission resonances can be observed in the current flowing from tip to sample. In the field emission mode there is a region between tip and sample where the energy of an electron from the tip is greater than the potential energy due to the applied potential difference, as shown in Figure 3.8(b). So electrons can escape from the tip by tunnelling into this classically allowed region, and are then accelerated towards the sample, where they are absorbed and the resulting current measured.

In Figure 3.8(b) Region I is a classically forbidden region, $E < V$, through which an electron from the tip tunnels into a classically allowed Region II. The potential decreases linearly in this region, because of the applied potential difference, and there is a potential discontinuity at the surface of the sample, $z = L$. The energy of the electron is typically the Fermi energy E_F of the tip – that is, the energy of the highest occupied electronic states in the tip. The applied voltage raises the Fermi level of the tip above the Fermi level E_F' of the sample. So changing the applied

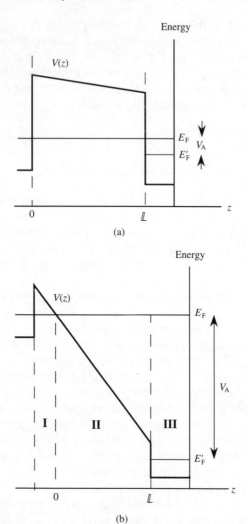

Figure 3.8 The approximate form for the potential energy of an electron in an STM: (a) for a small applied potential difference; (b) in field emission mode. The electron has total energy E_F, which is the Fermi energy of the tip, E'_F is the Fermi energy of the sample, and $V_A = -e\mathcal{V}_A$ is the potential energy difference between tip and sample due to the applied potential difference $\mathcal{V}_A$.

voltage $\mathcal{V}_A$ effectively changes the total energy E of the electron, making it greater than the potential energy V in Region II.

Just as in the case of scattering by a square well, interference between the components of the wave function that are incident and reflected at $z = \mathbb{L}$ in Figure 3.8(b) produces periodic variations in the coefficient of transmission into Region III. This results in transmission resonances at

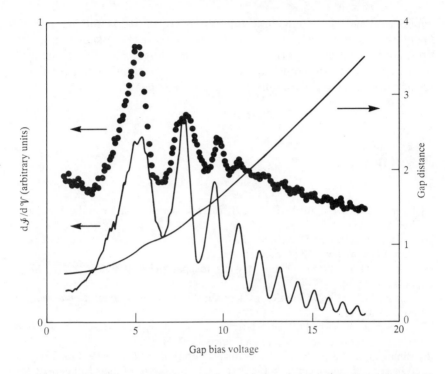

Figure 3.9 Experimental curves of effective conductance $d\mathcal{I}/d\mathcal{V}$ (circles) and of tip–sample separation (smooth curve) as functions of the applied potential difference $\mathcal{V}_A$, in the case of a tungsten tip and a gold sample. (Adapted from Figure 2 of [7].) The feedback mechanism maintained the average current at $1\,\text{nA}$. The oscillatory solid curve is the theoretical prediction, based on a potential energy of the form shown in Figure 3.8(b).

energies for which there are standing waves in Region II. These energies are precisely those at which bound states would occur in a triangular potential well with the same slope as the potential in Region II, but extending (with the same slope) throughout the region $z < 0$ and terminated at $z = \mathcal{l}$ by an infinitely high potential step.

Experimentally, when an STM is run in field emission mode, small oscillations in the current are observed as the applied voltage and/or the separation of the tip from the sample is varied. The energies at which the transmission resonances occur agree well with the predicted energy eigenvalues of a particle in an infinite triangular potential well. In practice, an STM includes a feedback mechanism that can alter the tip–sample separation to keep the current constant as the voltage is varied. A small additional potential difference, oscillating too rapidly for the feedback mechanism to follow it, allows the effective conductance $d\mathcal{I}/d\mathcal{V}$ to be measured. The data in Figure 3.9 show resonances in $d\mathcal{I}/d\mathcal{V}$ as the applied voltage and tip–sample separation are varied simultaneously to maintain a constant average current. The smooth curve shows how the tip–sample separation varies with applied voltage, and the oscillatory solid curve is the theoretical prediction. The positions of the predicted resonances agree well with those of the observed peaks.

References

[1] Grant I.S. and Phillips W.R. (1990). *Electromagnetism* 2nd edn, Section 10.1. London: Wiley.
[2] Boas M.L. (1983). *Mathematical Methods in the Physical Sciences* 2nd edn. New York: Wiley, Chapter 8, Section 5.
[3] Chemla D.S. (1985). Quantum wells for photonics. *Physics Today*, May, p. 57.
[4] Challis L.J. (1992). Physics in less than three dimensions. *Contemporary Physics* **33**, 111.
[5] Lorrain P., Corson D.P. and Lorrain F. (1988). *Electromagnetic Fields and Waves* 3rd edn, Section 30.6. San Francisco: W.H. Freeman.
[6] Binnig G. and Rohrer H. (1987). *Reviews of Modern Physics* **59**, 615. This is a transcript of their lecture on the occasion of the award of their Nobel prize for the invention. A more detailed review of the theory and applications of the STM can be found in van den Leemput L.E.C. and van Kempen H. (1992). *Reports on Progress in Physics* **55**, 1165.
[7] Becker R.S., Golovchenko J.A. and Swartzentruber B.S. (1985). *Physical Review Letters* **55**, 987.

Problems

3.1   Show that the functions

$$\psi_1(z) = C_1 \exp\left(-\tfrac{1}{2}z^2\right), \qquad \psi_2(z) = C_2 z \exp\left(-\tfrac{1}{2}z^2\right)$$

are eigenfunctions of the operator $\tfrac{1}{2}(z^2 - \mathrm{d}^2/\mathrm{d}z^2)$, and find the corresponding eigenvalues. Is the function $\psi = \psi_1 + \psi_2$ an eigenfunction of the operator? Compare the properties of ψ_1 and ψ_2 with the general properties (iii) and (v) of energy eigenfunctions given in Section 3.3.

3.2   Solve the eigenvalue equation

$$-\mathrm{i}\,\frac{\mathrm{d}}{\mathrm{d}\phi}\,\psi(\phi) = \beta\psi(\phi)$$

subject to the boundary condition $\psi(\phi) = \psi(\phi + 2\pi)$. Normalize the wave function over the range $(0, 2\pi)$ of ϕ, and sketch the probability distribution of a particle in one of the eigenstates $\psi(\phi)$. The operator $-i\hbar\, d/d\phi$ represents orbital angular momentum about the z axis, and ϕ is the angle of rotation. A set of measurements of the z component of the orbital angular momentum of particles in a particular quantum state gives the mean result $1.25\hbar$. How many of the individual measurements could yield exactly this result? What can you deduce about the state of the particles?

3.3   Are the following functions eigenfunctions of the given operator?

    (a) $\psi(t) = \sin \omega t$; $\hat{E}_{tot}$ of (3.7)

    (b) $\psi(z) = C(1 + z^2)$; $\hat{P}_z$ of (3.8)

    (c) $\psi(z) = C_1 e^{ikz} + C_2 e^{-ikz}$, (3.23); $\hat{P}_z$

    (d) $\psi(z) = Ce^{-3z}$; the Hamiltonian $\hat{H}$ for a free particle moving along the z axis

3.4   Check that the function (3.1) is an eigenfunction of the operator on each side of (3.11), and prove that the corresponding energy and momentum eigenvalues are related by the usual classical expression.

3.5   (a) Show that for a free particle in a state described by the wave function (3.1) the probability current density defined through (3.17) is exactly what we should expect for a flow of probability density $|\psi|^2$ at the classical velocity of the particle. Use this result to obtain the expressions for the reflection and transmission coefficients (3.46) and (3.55).

    (b) Show that the probability current associated with any energy eigenstate must be divergenceless.

3.6   Prove that the constant (3.41) normalizes any energy eigenfunction for a particle in an infinitely deep one-dimensional square potential well.

3.7   Show that for a particle bound in the finite one-dimensional square well (3.24) the energy eigenfunctions that are symmetric about the centre of the well are of the form

$$\psi(z) = C\cos\left[k\left(z - \tfrac{1}{2}L\right)\right] \quad (0 \leqslant z \leqslant L)$$

and the corresponding eigenvalues are determined by

$$\tan\tfrac{1}{2}kL = \gamma/k$$

Show that this condition is equivalent to

$$\cos\tfrac{1}{2}kL = \pm\, \frac{k}{\sqrt{(k^2 + \gamma^2)}} = \pm\, \sqrt{\left(\frac{E}{V_0}\right)}$$

$$((n-1)\pi \leqslant \tfrac{1}{2}kL \leqslant (n - \tfrac{1}{2})\pi;\ n = 1, 2, \ldots)$$

Demonstrate graphically that there is always at least one such bound state.

Show that eigenfunctions that are antisymmetric about the centre of the well are of the form

$$\psi(z) = C \sin[k(z - \tfrac{1}{2}L)] \quad (0 \leqslant z \leqslant L)$$

where $\cot\tfrac{1}{2}kL = -\gamma/k$, or equivalently

$$\sin\tfrac{1}{2}kL = \pm \frac{k}{\sqrt{(k^2 + \gamma^2)}} = \pm \sqrt{\left(\frac{E}{V_0}\right)}$$

$$((n - \tfrac{1}{2})\pi \leqslant \tfrac{1}{2}kL \leqslant n\pi; \; n = 1, 2, \ldots)$$

Prove that, for the highest bound state in the well, n is the next integer greater than

$$\frac{L}{\pi} \sqrt{(k^2 + \gamma^2)} = \sqrt{\left(\frac{8mV_0L^2}{h^2}\right)}$$

*3.8   Use the programs on the computer disk to investigate the solutions of the Schrödinger equation in an infinite and a finite one-dimensional square well. You may choose the mass of the bound particle in units of the free electron mass, the width of the well in nm, and, in the finite case, the well depth in eV.

(a) Choose the parity of the solution you wish to investigate ('even' for solutions that are symmetric about the centre of the well and 'odd' for those that are antisymmetric). Enter an energy value E, and select *Draw psi*. The program will display the solution to the Schrödinger equation (3.20) for one-dimensional motion in the potential (3.24) and the chosen value of E. To see the corresponding probability density, select *Draw psi²*. Unless your choice of E is close to an energy eigenvalue, the wave function will not satisfy the boundary conditions (that it should vanish at the well edges in the case of infinite depth, or that it should approach zero far outside the well in the case of finite depth). Change the value E by a small amount and try again. By a process of trial and error, you should be able to find an eigenvalue correct to several decimal places. (For example, for the default values of the parameters in the case of the infinite square well $E = 0.3$ eV gives a wave function that is greater than zero at the well edges. Changing to $E = 0.4$ eV makes the wave function less than zero at the well edges. At some intermediate energy the wave function will vanish at the well edges.)

(b) Now change the parity, and enter a higher value of E. Trial and error should give you a new eigenvalue, corresponding to a state with the opposite symmetry to the first you examined. In this way you can find successive energy eigenvalues, and verify that

(i) there is only a discrete set of values of E for which the boundary conditions are satisfied;

(ii) energy eigenfunctions are alternately symmetric and antisymmetric about the centre of the well;

(iii) The number of zeros of the probability distribution within the well is $n - 1$ for the nth energy eigenstate;

(iv) the number of bound states in the finite well is the next integer greater than the well parameter $\sqrt{(8mV_0L^2/h^2)}$.

(c) Examine the dependence of the energy eigenvalues on particle mass, and the width and depth of the well. Note that the shape of the nth eigenfunction is the same for different parameters that correspond to the same well parameter. In the case of the infinitely deep well (3.39) indicates that the energy varies inversely as the mass and the square of the well width. So plot an eigenvalue of a finite well of fixed depth against $1/m$ and $1/L^2$. (The option *Find E* will calculate energy eigenvalues to five significant figures, using the equations derived in Problem 3.7. The eigenfunction that is displayed in this case satisfies the boundary conditions exactly.)

Suggested ranges of m, L and V_0:

(i) free electron mass m_e, L between 0.5 and 50 nm, V_0 between 5 and 50 eV;
(ii) proton mass, $1836m_e$, L between 10^{-5} and 10^{-4} nm (nuclear dimensions), V_0 ranging from 1 keV to 1 MeV.

*3.9 (a) The energy levels in a quantum well material are often calculated using the approximation that the well is infinitely deep. Using the computer program, investigate the accuracy of this approximation for an electron in the conduction band in a GaAs/GaAlAs quantum well of width $L = 10$ nm. The electron has an effective mass $m^* = 0.067m_e$, and the actual depth of the GaAs well is $V_0 = 0.26$ eV.

(b) Compare the amount of spreading of the probability density into the GaAlAs for different energy levels of the electron of part (a), and for the same level in wells of different widths, ranging from 5 to 50 nm.

*3.10 In bulk GaAs the band gap energy is 1.515 eV. In a GaAs/GaAlAs quantum well the effective band gap in the GaAs is the energy difference between the highest level in the inverted well in the valence band and the lowest level in the well in the conduction band (Figure 3.4). Show that this increases as the width of the GaAs layer is decreased. Find (using the computer program) the effective band gap for a well of width 10 nm, if the effective masses of the electron and hole are respectively $0.07m_e$ and $0.48m_e$, and the electron and hole wells are of depths 0.268 eV and 0.112 eV respectively. (The depth of the energy level of a hole below the top of the inverted well is equal to the height of the corresponding level above the bottom of a normal square well, for a particle with the same mass.)

3.11 Solve the Schrödinger equation for a particle incident from the left on a potential step

$$V = \begin{cases} 0 & (z < 0) \\ V_0 & (z \geqslant 0) \end{cases}$$

in the case where $E < V_0$. Show that the reflection coefficient is 1, and explain this with reference to the behaviour of the wave function in the two regions.

3.12 Solve the set of simultaneous equations (3.54) to find B_1/A_1 and A_3/A_1, and verify (3.56) and (3.57) for the reflection and transmission coefficients.

3.13 A particle is incident from $z < 0$ on a square potential barrier

$$V = \begin{cases} 0 & (z < 0, z > L) \\ V_0 & (0 \leqslant z \leqslant L) \end{cases}$$

(a) Show that if it has energy $E > V_0$, the transmission coefficient can be written in the same form as that for a square well, (3.57), but with the values of k^2 and k_0^2 redefined as

$$k^2 = \frac{2m(E - V_0)}{\hbar^2}, \qquad k_0^2 = \frac{2mE}{\hbar^2}$$

(b) Show that if $E < V_0$, the transmission coefficient is

$$T = \left\{ 1 + \frac{\sinh^2\left[\pi w \sqrt{(1 - E/V_0)}\right]}{(4E/V_0)(1 - E/V_0)} \right\}^{-1}$$

where $w = \sqrt{(8mV_0 L^2/\hbar^2)}$.

3.14 When the applied voltage is small, the potential barrier presented to an electron in an STM, shown in Figure 3.8(a), can be approximated by a square barrier. Use the expression calculated for T in Problem 3.13(b), with $V_0 = 6\,\mathrm{eV}$, $L = 1\,\mathrm{nm}$ and $E = 1\,\mathrm{eV}$, to show that the STM current will be reduced by a factor of 10 if the spacing between tip and sample changes by $0.1\,\mathrm{nm}$. (You may use the free electron mass.)

*3.15 Using the computer program *Transmission: plane wave, well*, plot the transmission coefficient for the square well as a function of E/V_0 for different choices of particle mass, and width and depth of the well. Investigate how this graph varies with m, L and V_0 and well parameter $\sqrt{(8mV_0L^2/\hbar^2)}$. (Suitable ranges of values are given in Problem 3.8c.)

   Similarly investigate the transmission coefficient for a square barrier, using the program *Transmission: plane wave, barrier*. Explain why the shape of T near $E/V_0 = 1$ is different for the barrier and the well.

*3.16 Use the program *Triangular Well* to find the seven lowest energy levels of a particle in the infinitely deep triangular potential well

$$V = \begin{cases} -Cx & (x \leqslant 0) \\ \infty & (x > 0) \end{cases}$$

for several different choices of the parameter $C > 0$, which determines the slope of the potential. Show (analytically) that the classically allowed region for the particle when its energy is E is $-E/C \leqslant x \leqslant 0$. Note that the energy eigenfunctions satisfy the general characteristics (iii) and (iv) given in Section 3.3. That is, they oscillate within $-E/C \leqslant x \leqslant 0$, and decrease towards zero when $x < -E/C$.

CHAPTER 4

More about Dynamical Variables

4.1 Compatible and incompatible variables

We saw in Section 1.5 that a quantum particle cannot be in a state where both its position and momentum have precisely defined values – that is, a quantum particle cannot be simultaneously in an eigenstate of position and of momentum. Indeed, the position of a particle described by the momentum eigenfunction (3.10) is completely indeterminate, in accordance with the Heisenberg uncertainty relation. Corresponding components of position and momentum are said to be **incompatible variables**. On the other hand, all three components of the momentum can simultaneously have precisely defined values, as can all three position coordinates, or any pair of different components of position and momentum such as x and p_y. These are sets of **compatible variables**, and an eigenstate of one can simultaneously be an eigenstate of another. For example, the wave function (3.10) is simultaneously an eigenfunction of all three components of momentum.

If any two variables are compatible, there exist states that are simultaneously eigenstates of both, and a precise measurement of one will not introduce any indeterminacy into the value of the other. For example, the wave function (3.10) is simultaneously an eigenfunction of $\hat{P}_x$, $\hat{P}_y$ and $\hat{P}_z$, and the eigenvalues of the three components of momentum can be determined precisely by measuring each in turn. The order in which the measurements are made will have no effect on the result. In contrast, if a particle is initially in an eigenstate of one variable represented by operator $\hat{A}$, and a measurement of an incompatible variable is made, the state is changed to one that is not an eigenstate of $\hat{A}$, and the value of the variable represented by $\hat{A}$ is no longer precisely defined. For example, a measurement of the x component of the position of a particle that is in an eigenstate of $\hat{P}_x$ reduces the complete uncertainty in the particle's position, and thus, according to the uncertainty principle, the uncertainty in $\hat{P}_x$ must increase. So, after the position measurement, the particle's momentum can no longer have a precise value, and the particle cannot be in a momentum eigenstate.

The difference between compatible and incompatible variables is reflected in a mathematical property of the operators that represent them: if the operators $\hat{A}$ and $\hat{B}$ represent a pair of incompatible variables then the product of the two operators depends on the order in which they are multiplied, and $\hat{A}\hat{B} \neq \hat{B}\hat{A}$. The operators $\hat{A}$ and $\hat{B}$ are **non-commuting** operators. Compatible variables, on the other hand, are represented by operators that **commute** with one another – that is, they satisfy $\hat{A}\hat{B} = \hat{B}\hat{A}$ (see Problem 4.2). We define the **commutator** of two operators $\hat{A}$ and $\hat{B}$ as

$$[\hat{A}, \hat{B}] \equiv \hat{A}\hat{B} - \hat{B}\hat{A} = -[\hat{B}, \hat{A}] \qquad (4.1)$$

To evaluate the commutator of, for example, the operators $\hat{X} = x$ and $\hat{P}_x = -i\hbar\, \partial/\partial x$, we consider its effect on an arbitrary wave function $\Psi(x, y, z, t)$:

$$[\hat{X}, \hat{P}_x]\Psi(x, y, z, t) = -i\hbar\left(x\,\frac{\partial}{\partial x} - \frac{\partial}{\partial x}\,x\right)\Psi(x, y, z, t)$$

$$= -i\hbar\left(x\,\frac{\partial\Psi}{\partial x} - \frac{\partial(x\Psi)}{\partial x}\right)$$

Note that each term in the commutator is an operator acting on a wave function to its right, so that in the second term the derivative acts on the function formed by operating on Ψ with x. Differentiation of this product gives

$$[\hat{X}, \hat{P}_x]\Psi(x, y, z, t) = -i\hbar\left(x\,\frac{\partial\Psi}{dx} - \Psi - \frac{\partial\Psi}{\partial x}\,x\right) = i\hbar\Psi(x, y, z, t)$$

$$(4.2)$$

Thus the commutator is equivalent to multiplication by $i\hbar$ when acting on an arbitrary wave function, and we can write $[\hat{X}, \hat{P}_x] = i\hbar$. The fact that this is not zero is the mathematical representation of the incompatibility of the x components of the position and momentum. The y and z components satisfy similar commutation relations:

$$[\hat{X}, \hat{P}_x] = i\hbar, \qquad [\hat{Y}, \hat{P}_y] = i\hbar, \qquad [\hat{Z}, \hat{P}_z] = i\hbar \qquad (4.3)$$

Any two components of the momentum are compatible, since

$$[\hat{P}_x, \hat{P}_y]\psi(x, y, z) = -\hbar^2\left(\frac{\partial^2}{\partial x\,\partial y} - \frac{\partial^2}{\partial y\,\partial x}\right)\psi(x, y, z) = 0$$

and similarly for the other pairs of components, so that

$$[\hat{P}_x, \hat{P}_y] = 0, \qquad [\hat{P}_y, \hat{P}_z] = 0, \qquad [\hat{P}_z, \hat{P}_x] = 0 \qquad (4.4)$$

Note also that different components of the position and momentum operators commute. For example,

$$[\hat{X}, \hat{P}_y] = 0, \qquad [\hat{X}, \hat{P}_z] = 0 \tag{4.5}$$

The commutation relations (4.3) and (4.4) can be used to investigate the compatibility of the total energy and the momentum of a particle. The Hamiltonian for a particle moving in a potential $V(x, y, z)$ with conserved total energy is given by (3.12). Since $\hat{P}_x$ commutes with $\hat{P}_x^2$, $\hat{P}_y^2$ and $\hat{P}_z^2$, it is easy to show that

$$[\hat{H}, \hat{P}_x] = -i\hbar\left[V(x, y, z), \frac{\partial}{\partial x}\right] = i\hbar\frac{\partial V}{\partial x} \tag{4.6}$$

(To prove this, calculate the effect of the commutator on a wave function.) Clearly this commutator will vanish only if V is independent of x, and for all three components of $\hat{P}$ to commute with $\hat{H}$ the potential must be constant. So it is only for a free particle that the total energy and the momentum are compatible variables. For any other form of potential function an energy eigenstate of the particle is not a momentum eigenstate: a momentum measurement will inevitably change the state of the particle to one in which the energy is indeterminate. Note that this also applies to the particle in a square well, since the slope of the potential is not zero (in fact it is infinite!) at the edges of the well: even when $E > V_0$ the energy eigenstate is not a momentum eigenstate, since, although E is constant, the momentum is different in regions such as Regions I and II of Figure 3.1, where the potential takes different values.

4.2 The angular momentum operators

So far we have discussed the representation of energy, momentum and position in quantum mechanics. In classical mechanics other dynamical variables can be expressed in terms of the momentum and position variables. For example, the angular momentum is $M = r \times p$. The quantum mechanical operator representing such a variable is obtained, according to the correspondence principle that we invoked in writing the Hamiltonian (3.12), by substituting the appropriate operators in the classical expression. Of course we must be careful about the order in which we write the operators when the classical expression involves a product of variables. In the case of the angular momentum there is no problem: each component of M is a product of *different* components of r and p, and, according to (4.5), the corresponding operators commute: $\hat{R} \times \hat{P} = \hat{P} \times \hat{R}$. (In Section 4.3 we shall see an example where the order of the operators does make a difference.) Thus the angular momentum M is represented unambiguously by $\hbar \hat{L}$, where $\hat{L}$ is the dimensionless operator

$$\hat{L} = \frac{1}{\hbar} \hat{R} \times \hat{P} = -\mathrm{i} r \times \nabla \qquad (4.7)$$

Expression (4.7) defines a linear Hermitian operator with Cartesian components

$$
\left.
\begin{aligned}
\hat{L}_x &= -\mathrm{i}\left(y \frac{\partial}{\partial z} - z \frac{\partial}{\partial y} \right) = \mathrm{i}\left(\sin\phi \frac{\partial}{\partial \theta} + \cot\theta \cos\phi \frac{\partial}{\partial \phi} \right) \\
\hat{L}_y &= -\mathrm{i}\left(z \frac{\partial}{\partial x} - x \frac{\partial}{\partial z} \right) = \mathrm{i}\left(-\cos\phi \frac{\partial}{\partial \theta} + \cot\theta \sin\phi \frac{\partial}{\partial \phi} \right) \\
\hat{L}_z &= -\mathrm{i}\left(x \frac{\partial}{\partial y} - y \frac{\partial}{\partial x} \right) = -\mathrm{i}\frac{\partial}{\partial \phi}
\end{aligned}
\right\} \qquad (4.8)
$$

For convenience, we have included the forms for these operators in terms of spherical polar coordinates, since we shall need these later on. The square of the angular momentum is represented (apart from a factor $\hbar^2$) by

$$\hat{L}^2 = \hat{L}_x^2 + \hat{L}_y^2 + \hat{L}_z^2 = -\left(\frac{\partial^2}{\partial \theta^2} + \cot\theta \frac{\partial}{\partial \theta} + \frac{1}{\sin^2\theta} \frac{\partial^2}{\partial \phi^2} \right) \qquad (4.9)$$

Unlike the three components of the momentum operator or the position operator, the three components of the angular momentum operator do not commute with one another. Their commutation relations can be found using the expressions for the components of $\hat{L}$ in terms of the Cartesian variables, (4.8), and the commutation relations (4.3):

$$[\hat{L}_x, \hat{L}_y] = \mathrm{i}\hat{L}_z \qquad \text{and cyclic permutations} \qquad (4.10)$$

This means that a precise measurement of one component of angular momentum leaves the values of the other two indeterminate. A particle cannot be simultaneously in an eigenstate of more than one component of the angular momentum. However, it is possible for the square of the angular momentum and one of the components to take precise values simultaneously, since

$$[\hat{L}_x^2 + \hat{L}_y^2 + \hat{L}_z^2, \hat{L}_i] = 0 \quad (i = x, y, z) \qquad (4.11)$$

This can be proved using the commutation relations (4.10), and the identities

$$
\left.
\begin{aligned}
[\hat{L}_x^2, \hat{L}_x] &= \hat{L}_x^3 - \hat{L}_x^3 = 0 \\
[\hat{L}_x^2, \hat{L}_y] &= \hat{L}_x[\hat{L}_x, \hat{L}_y] + [\hat{L}_x, \hat{L}_y]\hat{L}_x
\end{aligned}
\right\} \qquad (4.12)
$$

with similar relations between the other components.

A particle may therefore be in a simultaneous eigenstate of $\hat{L}^2$ and one of the components of $\hat{L}$, usually chosen to be $\hat{L}_z$. To find the possible results of simultaneous measurements of the square of the angular momentum and of its z component, we must, according to Postulate 2 (Section 3.1), solve simultaneously the appropriate eigenvalue equations. Since the operators $\hat{L}^2$ and $\hat{L}_z$ given by (4.8) and (4.9) do not contain the radial variable r or derivatives with respect to r, the eigenfunctions of these operators depend only on the angles θ and ϕ. So the eigenvalue equations can be written as

$$\hat{L}^2 \psi(\theta, \phi) = \alpha^2 \psi(\theta, \phi) \qquad (4.13)$$

$$\hat{L}_z \psi(\theta, \phi) = \beta \psi(\theta, \phi) \qquad (4.14)$$

where α^2 and β are respectively the eigenvalues of $\hat{L}^2$ and $\hat{L}_z$. We shall solve these equations in Section 4.4. If we operate on each side of (4.14) with $\hat{L}_z$, we can see that the eigenvalue of $\hat{L}_z^2$ is β^2. So, from the expression (4.9) for $\hat{L}^2$ in terms of the Cartesian components of $\hat{L}$, we see that $\psi(\theta, \phi)$ must also be an eigenfunction of

$$\hat{L}_x^2 + \hat{L}_y^2 = \hat{L}^2 - \hat{L}_z^2 \qquad (4.15)$$

corresponding to the eigenvalue $\alpha^2 - \beta^2$. If this were zero, $\hat{L}_x^2$ and $\hat{L}_y^2$, and thus also $\hat{L}_x$ and $\hat{L}_y$, would each have to take the precise value zero. However, $\hat{L}_x$ and $\hat{L}_y$ individually cannot have definite values at the same time as $\hat{L}_z$, so that the eigenvalue β can never be exactly equal to α. The only exception to this is the zero-angular-momentum state, in which $\beta = \alpha = 0$.

A comparison with the classical picture is useful for understanding the nature of the quantum mechanical angular momentum eigenstate. $\psi(\theta, \phi)$ represents the state of a particle in which the total length of the angular momentum vector is $\alpha\hbar$ and its z component is $\beta\hbar$. Classically this angular momentum vector would make an angle $\theta = \cos^{-1}(\beta/\alpha)$ with the z axis, as shown in Figure 4.1. The x and y components of the classical angular momentum then have the precise values

$$M_x = \alpha\hbar \sin\theta \cos\phi, \qquad M_y = \alpha\hbar \sin\theta \sin\phi$$

Quantum mechanically, $\hat{L}_x$ and $\hat{L}_y$ do not have precise values, and this corresponds to complete uncertainty in the orientation ϕ of the vector in the plane perpendicular to the z axis: if the z component of angular momentum is precisely determined, the angular position of the particle in its rotation about this axis is completely unknown. This is reminiscent of the complete uncertainty in the z component of the position vector when the z component of momentum is precisely determined. Indeed, the polar form of the expression for $\hat{L}_z$, (4.8), shows that

$$[\hat{\phi}, \hbar\hat{L}_z] = \left[\phi, -i\hbar \frac{\partial}{\partial \phi}\right] = i\hbar \qquad (4.16)$$

which is similar in form to the position–momentum commutation relations (4.3). The important difference is that the range of physically

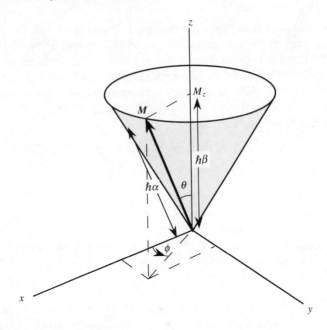

Figure 4.1 A semiclassical representation of an angular momentum vector $\boldsymbol{M}$ of length $\hbar\alpha$ and z component $\hbar\beta$. The angle made by the vector with the z axis is θ, where $\cos\theta = \beta/\alpha$. As ϕ is varied from 0 to 2π, the vector $\boldsymbol{M}$ is rotated about the z axis, defining the surface of a cone of semi-angle θ.

significant eigenvalues for the operator $\hat{\phi}$ (represented by the algebraic variable ϕ) is the finite range $(0, 2\pi)$, instead of the infinite range of eigenvalues for the Cartesian components of the position vector.

The operators $\hat{L}_z$ and $\hat{L}^2$ given by (4.8) and (4.9) refer to **orbital angular momentum**, associated with rotational motion of a particle in space. As we shall see in detail in Chapter 8, the intrinsic spin of a particle is also represented by angular momentum operators that satisfy the commutation relations (4.10) and (4.11). However, spin, as we mentioned in Section 1.6, is a non-classical variable, which cannot be expressed in terms of the spatial position and momentum variables, and so the differential operators of (4.8) and (4.9) do not represent spin: spin is *not* a physical rotation in space.

4.3 The radial momentum operator

In classical mechanics the square of the momentum can be separated into two parts (Appendix A), one involving the angular momentum $\boldsymbol{M}$ and the other the radial momentum p_r:

$$p^2 = p_r^2 + \frac{M^2}{r^2} \tag{4.17}$$

$$p_r = \frac{r}{r} \cdot p \tag{4.18}$$

As before, we may use the correspondence principle, and determine the operator representing p_r^2 by substituting the operators (3.9) and (4.9) into (4.17). If ∇^2 is expressed in spherical polar coordinates, we find [1]

$$\hat{P}_r^2 = -\hbar^2\left(\nabla^2 + \frac{\hat{L}^2}{r^2}\right) = -\hbar^2\left(\frac{\partial^2}{\partial r^2} + \frac{2}{r}\frac{\partial}{\partial r}\right) \tag{4.19}$$

Equation (4.19) is a perfectly satisfactory form for $\hat{P}_r^2$, but there is a pitfall in attempting to use the correspondence principle with (4.18) to derive the operator $\hat{P}_r$. Simply taking the radial component of the momentum operator, we should expect $\hat{P}_r = -i\hbar \partial/\partial r$, but squaring this operator does not give the operator (4.19). The problem is that, although classically there is no difference between $r \cdot p$ and $p \cdot r$, the operators $\hat{P}$ and $\hat{R}$ do not commute. To find the form of the operator $\hat{P} \cdot (\widehat{R/R})$, note that the effect of $\widehat{R/R}$ on an arbitrary function $\psi(r)$ is to produce the new function $r\psi(r)/r$, so that

$$\hat{P} \cdot \left(\frac{R}{R}\right)\psi(r) = -i\hbar\nabla \cdot \left[\frac{r\psi(r)}{r}\right] = -i\hbar\left\{r \cdot \nabla\left[\frac{\psi(r)}{r}\right] + \frac{\psi(r)}{r}\nabla \cdot r\right\}$$

$$= -i\hbar\left(\frac{\partial}{\partial r} + \frac{2}{r}\right)\psi(r) \tag{4.20}$$

(where we have used $r \cdot \nabla = r\partial/\partial r$ and $\nabla \cdot r = 3$). On the other hand,

$$\left(\frac{R}{R}\right) \cdot \hat{P}\psi(r) = -i\hbar\frac{r}{r} \cdot \nabla\psi(r) = -i\hbar\frac{\partial}{\partial r}\psi(r) \tag{4.21}$$

The correct operator for the radial momentum is obtained by using the correspondence principle with the symmetric expression

$$p_r = \frac{1}{2}\left(\frac{r}{r} \cdot p + p \cdot \frac{r}{r}\right)$$

This gives

$$\hat{P}_r = -i\hbar\left(\frac{\partial}{\partial r} + \frac{1}{r}\right) \tag{4.22}$$

and the square of this operator gives (4.19).

This example shows that we must be cautious in applying the correspondence principle when products of variables are involved. If the classical expression contains a product of incompatible variables, there will be more than one corresponding operator, depending on the order in which the product is written. In such circumstances the correct, Hermitian, operator is obtained by applying the correspondence principle to a symmetrized product, as we did for the radial momentum. There was no difficulty in the case of the angular momentum because each component involves products of different, compatible, components of position and momentum. This is evident from the explicit expressions for the components of $\hat{L}$ in (4.8).

4.4 The parity operator

The energy eigenfunctions for a particle in a square well, shown in Figure 3.3, are alternately symmetric and antisymmetric about the centre of the well. If we label the centre of the well $z = 0$ (instead of $z = \frac{1}{2}\mathcal{L}$), the eigenfunctions are either even or odd (symmetric or antisymmetric) under the transformation $z \to -z$. This is the one-dimensional form of the **parity transformation**, which is an operation on the coordinate system that inverts all three space axes:

$$x \to -x, \qquad y \to -y, \qquad z \to -z \tag{4.23}$$

Any variable or function that is unchanged by this tranformation is said to possess **even parity**, while one that changes sign has **odd parity**. For example, the function $x^2 - y^2$ has even parity, while $x + y + z$ has odd parity. On the other hand $\cos x + \sin x$ does not have a definite parity, but can be expressed as the sum of an even-parity term $(\cos x)$ and an odd-parity term $(\sin x)$.

In classical mechanics the vectors $p = m\,dr/dt$ and r clearly have odd parity, while the kinetic energy is a scalar with even parity since it is unchanged by the transformation (4.23). The angular momentum, although it is a vector, has even parity, since it is defined as a product of the position and momentum vectors. It is called an **axial vector**, to distinguish it from a **polar vector**, which has odd parity. The quantum mechanical operators have the same parity as the corresponding classical variables. For example, it is easy to check that the form (3.9) for the momentum operator is odd under the transformation (4.23).

In classical mechanics we do not usually think of parity as a dynamical variable. However, in quantum theory a **parity operator** $\hat{\Pi}$ can be defined through its effect on any wave function:

$$\hat{\Pi}\Psi(x, y, z, t) = \Psi(-x, -y, -z, t) \tag{4.24}$$

If the wave function is either even or odd under this transformation then

$$\hat{\Pi}\Psi(x, y, z, t) = \pm\Psi(x, y, z, t) \tag{4.25}$$

This is of exactly the same form as the general eigenvalue equation (3.4), so that $\Psi(x, y, z, t)$ is an eigenfunction of the parity operator $\hat{\Pi}$ corresponding to the eigenvalue $+1$ if Ψ is an even function, and -1 if it is an odd function. $\hat{\Pi}$ is a linear operator with two real eigenvalues. Using the operator a second time is equivalent to restoring the original orientation of the coordinate axes, so that

$$\hat{\Pi}^2 \Psi(x, y, z, t) = \hat{\Pi}\Psi(-x, -y, -z, t) = \Psi(x, y, z, t) \quad (4.26)$$

This means that $\Psi(x, y, z, t)$ is an eigenfunction of $\hat{\Pi}^2$ corresponding to the eigenvalue $+1$, and this is true not only for eigenfunctions of $\hat{\Pi}$ but for any function that can be written as the sum of such eigenfunctions (like $\cos x + \sin x$). Equation (4.26) can also be obtained by operating with $\hat{\Pi}$ on (4.25).

We can establish whether or not an eigenfunction of some operator is simultaneously a parity eigenfunction by examining the commutator of the operator with $\hat{\Pi}$. In the case of the z component of the momentum operator note that

$$\hat{\Pi}\frac{\partial}{\partial z}\, \psi(x, y, z) = \frac{\partial}{\partial(-z)}\,\hat{\Pi}\psi(x, y, z) = -\frac{\partial}{\partial z}\,\hat{\Pi}\psi(x, y, z)$$

where $\psi(x, y, z)$ is an arbitrary wave function. Similar expressions apply in the case of the x and y components, so that

$$\hat{\Pi}\hat{P} = -\hat{P}\hat{\Pi}$$

and

$$[\hat{\Pi}, \hat{P}] = 2\hat{P}\hat{\Pi} \quad (4.27)$$

This means that the momentum and parity operators do not commute, so that a momentum eigenfunction cannot also be a parity eigenfunction. Indeed, we can see that the parity transformation (4.24) applied to the momentum eigenfunction (3.10) converts it into a different momentum eigenfunction, corresponding to the opposite momentum eigenvalue:

$$\hat{\Pi}Ce^{-i p \cdot r/\hbar} = Ce^{i p \cdot (-r)/\hbar} = Ce^{-i p \cdot r/\hbar} \quad (4.28)$$

However, the square of the momentum operator does commute with $\hat{\Pi}$, since

$$\hat{\Pi}\frac{d^2}{dz^2} = \frac{d^2}{d(-z)^2}\,\hat{\Pi} = \left(-\frac{d}{dz}\right)^2\hat{\Pi} = \frac{d^2}{dz^2}\,\hat{\Pi} \quad (4.29)$$

and similarly $\hat{\Pi}$ commutes with d^2/dx^2 and d^2/dy^2. So $\hat{\Pi}$ commutes with the kinetic energy term in the Hamiltonian (3.12), and whether or not an energy eigenstate can also be a parity eigenstate depends on whether or not $\hat{\Pi}$ commutes with the potential energy operator $V(x, y, z)$. Clearly it will do so if V is an even function, like the square well potential when $z = 0$ is at the centre of the well. If V is not an even function of x, y and

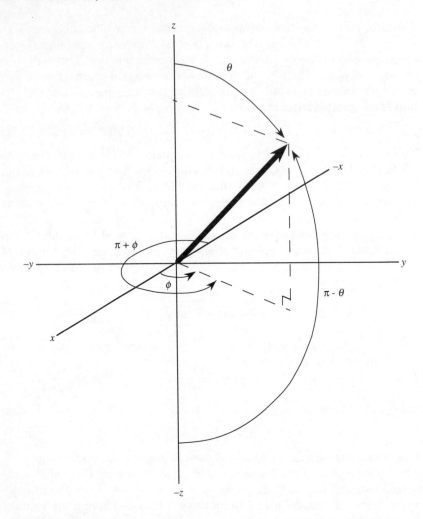

Figure 4.2 The parity transformation in terms of spherical polar coordinates.

z, the energy eigenstates cannot have a definite parity. This is the case, for example, for a triangular potential well like that in Region II of Figure 3.8.

The classical angular momentum vector M has even parity, and correspondingly the operator $\hbar\hat{L}$ commutes with $\hat{\Pi}$. Eigenstates of $\hat{L}^2$ and $\hat{L}_z$ can therefore be simultaneously eigenstates of parity. It is useful, in dealing with angular momentum eigenfunctions, to use spherical polar coordinates, and accordingly, using Figure 4.2, we rewrite the parity transformation (4.23) as

$$r \to r, \qquad \theta \to \pi - \theta, \qquad \phi \to \phi + \pi \tag{4.30}$$

Using the parity transformation a second time gives

$$r \rightarrow r, \qquad \theta \rightarrow \theta, \qquad \phi \rightarrow \phi + 2\pi \qquad (4.31)$$

That is, the angle θ returns to its original value, but the angle ϕ is increased by 2π. Although $\phi + 2n\pi$ specifies the same point as ϕ if n is an integer, not all functions automatically take the same value at angles differing by $2n\pi$ (for example, $\sin\frac{1}{2}(\phi + 2\pi) = -\sin\frac{1}{2}\phi$). The requirement that a function be unchanged by the operation $\hat{\Pi}^2$, as in (4.26), is equivalent to requiring that the function be periodic in ϕ, with period 2π:

$$\hat{\Pi}^2 \psi(r, \theta, \phi) = \psi(r, \theta, \phi + 2\pi) = \psi(r, \theta, \phi) \qquad (4.32)$$

This ensures that the wave function has a unique value at each point in space – that is, the wave function is **single-valued**. We shall impose this as a boundary condition on the orbital angular momentum eigenfunctions that we derive in Section 4.5.

4.5 Orbital angular momentum eigenfunctions and eigenvalues

Consider first a particle rotating in the (x, y) plane, which, according to classical mechanics, would mean that the angular momentum vector is along the z axis. The possible results of a measurement of the angular momentum in the quantum case are found by solving the eigenvalue equation for $\hat{L}_z$, (4.14), which takes the simplest form in terms of polar coordinates:

$$-i \frac{\partial}{\partial \phi} \psi(\phi) = \beta \psi(\phi) \qquad (4.33)$$

This is of the same form as (3.2) and (3.3), so the eigenfunction is an exponential:

$$\psi(\phi) = C e^{i\beta\phi} \qquad (4.34)$$

The corresponding eigenvalue is β, and the probability density is the same, $|C|^2$, at each angle ϕ (Problem 3.2). Since the physical range of ϕ is 2π, the normalization condition (2.2) becomes

$$\int_0^{2\pi} d\phi \, |\psi(\phi)|^2 = 1 \qquad (4.35)$$

giving $C = \sqrt{(1/2\pi)}$.

As a boundary condition we impose (4.32), requiring the $\hat{L}_z$ eigenfunction to be single-valued:

$$\psi(\phi + 2\pi) = \psi(\phi) \qquad (4.36)$$

Applying this to the wave function (4.34) gives $e^{2\pi i \beta} = 1$, so that the eigenvalues of $\hat{L}_z$ are restricted to the discrete set

$$\beta = 0, \pm 1, \pm 2, \pm 3, \ldots \tag{4.37}$$

With this eigenvalue spectrum the $\hat{L}_z$ eigenfunctions (4.34) are simultaneously eigenfunctions of parity, as can be checked by using the parity transformation in polar form, (4.30). The parity eigenvalue is $(-1)^\beta$.

In the general case of a three-dimensional system we look for simultaneous eigenfunctions of $\hat{L}_z$ and $\hat{L}^2$, which are solutions to both (4.13) and (4.14). If the form (4.9) is substituted for $\hat{L}^2$ in (4.13), we obtain the differential equation

$$-\left(\frac{\partial^2}{\partial \theta^2} + \cot \theta \, \frac{\partial}{\partial \theta} + \frac{1}{\sin^2 \theta} \frac{\partial^2}{\partial \phi^2} \right) \psi(\theta, \phi) = \alpha^2 \psi(\theta, \phi) \tag{4.38}$$

Since the solutions for which we are looking are eigenfunctions of $\hat{L}_z$, their ϕ dependence is given by (4.34), and we can write the wave function as the product

$$\psi(\theta, \phi) = \chi(\theta) e^{i\beta \phi} \tag{4.39}$$

Then $\partial^2 \psi / \partial \phi^2 = -\beta^2 \psi$, and when this is inserted in (4.38) and the exponential factor $e^{i\beta\phi}$ is cancelled, we find that $\chi(\theta)$ satisfies the differential equation

$$\left(\frac{d^2}{d\theta^2} + \cot \theta \, \frac{d}{d\theta} - \frac{1}{\sin^2 \theta} \beta^2 + \alpha^2 \right) \chi(\theta) = 0 \tag{4.40}$$

With the substitution $x = \cos \theta$ this becomes

$$\left[(1 - x^2) \frac{d^2}{dx^2} - 2x \frac{d}{dx} - \frac{\beta^2}{1 - x^2} + \alpha^2 \right] \chi(x) = 0 \tag{4.41}$$

When $\beta = 0$ this is Legendre's equation [2]. We need solutions that are finite everywhere in the range $-1 \leqslant x \leqslant 1$, and, as shown in Appendix B1, these only exist for discrete sets of values for α and β:

$$\alpha^2 = l(l + 1) \quad (l = 0, 1, 2, 3, \ldots) \tag{4.42}$$

and, for a particular value of l, β can take any of the $2l + 1$ values

$$\beta = m_l = l, l - 1, l - 2, \ldots, 0, \ldots, -(l - 2), -(l - 1), -l \tag{4.43}$$

(Note that these values for β belong to the spectrum (4.37).) We shall refer to the numbers l and m_l as the **angular momentum quantum numbers** characterizing an eigenstate of orbital angular momentum.

The solutions to (4.41) with $\beta = 0$ that are finite in the range $-1 \leqslant x \leqslant 1$ are the well-known **Legendre polynomials** $P_l(x)$. Table 4.1 gives these for the first few values of l. When $\beta = m_l \neq 0$ the solutions are

Table 4.1 Legendre polynomials and associated Legendre functions.

Legendre polynomials

$$P_0(\cos\theta) = 1$$
$$P_1(\cos\theta) = \cos\theta$$
$$P_2(\cos\theta) = \tfrac{1}{2}(3\cos^2\theta - 1)$$
$$P_3(\cos\theta) = \tfrac{1}{2}(5\cos^3\theta - 3\cos\theta)$$
$$P_4(\cos\theta) = \tfrac{1}{8}(35\cos^4\theta - 30\cos^2\theta + 3)$$

Associated Legendre functions

$$P_l^0(\cos\theta) \equiv P_l(\cos\theta)$$
$$P_1^1(\cos\theta) = \sin\theta$$
$$P_2^1(\cos\theta) = 3\sin\theta\cos\theta$$
$$P_2^2(\cos\theta) = 3\sin^2\theta$$
$$P_3^1(\cos\theta) = \tfrac{3}{2}\sin\theta\,(5\cos^2\theta - 1)$$
$$P_3^2(\cos\theta) = 15\sin^2\theta\cos\theta$$
$$P_3^3(\cos\theta) = 15\sin^3\theta$$

The associated Legendre functions $P_l^{m_l}$ form an orthogonal set:

$$\int_{-1}^{1} d(\cos\theta)\, P_{l'}^{m_l} P_l^{m_l} = \delta_{l'l}\, \frac{2}{2l+1}\, \frac{(l+|m_l|)!}{(l-|m_l|)!}$$

The following recursion relations are satisfied:

$$(2l+1)\cos\theta\, P_l^{m_l} = (l - |m_l| + 1)P_{l+1}^{m_l} + (l + |m_l|)P_{l-1}^{m_l}$$
$$(2l+1)\sin\theta\, P_l^{m_l-1} = P_{l+1}^{m_l} - P_{l-1}^{m_l}$$

(Note that we have defined $P_l^{m_l}$ through (4.44), which implies that $P_l^{-m_l} = P_l^{m_l}$. However, the associated Legendre functions for $m_l < 0$ are sometimes normalized differently.)

the **associated Legendre functions** $P_l^{m_l}(x)$, and in Appendix B1 these are shown to be related to the usual Legendre polynomials by

$$P_l^{m_l}(x) = (1 - x^2)^{|m_l|/2}\, \frac{d^{|m_l|}}{dx^{|m_l|}}\, P_l(x) \qquad (4.44)$$

Some of these functions are also given in Table 4.1. The complete angular momentum eigenfunctions of (4.39) are called the **spherical harmonics** $\psi(\theta, \phi) = Y_{lm_l}(\theta, \phi)$, defined for $m_l \geqslant 0$ by

$$Y_{lm_l}(\theta, \phi) = (-1)^{m_l}\left[\frac{(2l+1)(l-m_l)!}{4\pi(l+m_l)!}\right]^{1/2} P_l^{m_l}(\cos\theta)e^{im_l\phi} \qquad (4.45)$$

For negative m_l they are defined by

$$Y_{l,-|m_l|}(\theta, \phi) = (-1)^{m_l} Y^*_{l|m_l|}(\theta, \phi)$$

The spherical harmonics are normalized, and different functions in the set are orthogonal to one another:

$$\int_0^{2\pi} d\phi \int_{-1}^1 d(\cos \theta) Y^*_{lm_l}(\theta, \phi) Y_{l'm_l'}(\theta, \phi) = \delta_{ll'} \delta_{m_l m_l'} \qquad (4.46)$$

Each $Y_{lm_l}(\theta, \phi)$ is an eigenfunction of parity, with eigenvalue $(-1)^l$, independent of m_l.

The angle between a classical angular momentum vector and the z axis can take any value in the continuous range $(0, \pi)$. In contrast, for a quantum angular momentum vector this angle is quantized! This can be seen from Figure 4.1, which shows an angular momentum vector of length $\hbar\alpha$ and z component $\hbar\beta$: if β is restricted to the discrete set of values given by (4.43) then only $2l + 1$ discrete values of $\cos \theta = \beta/\alpha$ are allowed for a particular value of $\alpha = \sqrt{[l(l + 1)]}$. As mentioned in Section 4.2, the angle ϕ is indeterminate in an eigenstate of $\hat{L}_z$, so that in such a state the orientation of the quantum mechanical angular momentum is restricted to the surface of one of a discrete set of cones of semi-angle θ. Furthermore, in the quantum case the vector is never aligned parallel or antiparallel to the z axis, since the maximum and minimum values of $\cos \theta$ are $\pm l/\sqrt{[l(l + 1)]}$. This last characteristic is a consequence of the incompatibility of different components of the angular momentum in quantum mechanics: if the angular momentum could lie parallel or antiparallel to the z axis, so that $\beta = \pm\alpha$, then both the x and y components of angular momentum would have values precisely equal to zero, and the state would be a simultaneous eigenstate of three incompatible variables. Note, however, that if l is very large, the number of allowed values of θ also becomes very large, and the maximum and minimum values of $\cos \theta$ approach $+1$ and -1 respectively. So in this limit it becomes more and more difficult to distinguish the quantum features of the rotator, and its behaviour appears more and more classical.

4.6 Angular distributions in orbital angular momentum eigenstates

The probability distribution for a particle in an orbital angular momentum eigenstate is determined by the appropriate spherical harmonic (4.45). For the eigenstate with quantum numbers (l, m_l) the probability density at angular position (θ, ϕ) is $|Y_{lm_l}(\theta, \phi)|^2$. Because the spherical harmonics depend on angle ϕ only through the exponential factor $e^{im_l\phi}$, the probability density is independent of ϕ, and the probability distribution is symmetric about the z axis. The square root of the probability distribution, $|Y_{lm_l}(\theta, \phi)|$, can be calculated and displayed as a function of θ for a large range of different angular momentum eigenstates using the computer program *Orbital angular momentum*, and Figure 4.3 shows these distributions for $l = 0$, 1 and 2. In these diagrams the value of $|Y_{lm_l}(\theta, \phi)|$ at angle θ is represented by the distance of the surface from the origin at that angular position. Since the spherical harmonic $Y_{00}(\theta, \phi)$ is constant, the $l = 0$ distribution is completely spherically symmetric. (The square root of the probability distribution rather than the probability distribution itself is plotted so that the subsidiary lobes are easier to see.)

The probability distribution predicted for a particle in an orbital angular momentum eigenstate is not as easy to check experimentally as quantization of the angular momentum spectrum. However, in a scattering experiment the differential cross-section, defined by (2.16), is determined by the angular probability distribution of the scattered particles. Suppose that in a two-particle scattering process each scattered particle is produced in an eigenstate of orbital angular momentum (l, m_l) about the target particle. Then the differential cross-section, measured in a frame of reference in which the centre of mass of the two particles is at rest (the **centre-of-mass frame**), is proportional to the angular distribution $|Y_{lm_l}(\theta, \phi)|^2$.

For example, evidence for the angular distribution expected in an $l = 1$ state is dramatically evident in the elastic scattering of pions from protons at a certain energy (as shown in Figure 4.4c). A pion is a subnuclear particle with mass approximately one-seventh that of the proton [3], and it can be produced in high-energy scattering processes. There are three different pions, π^+, π^0 and π^-, with different electric charges equal to +1, 0 and −1 in units of the electron charge. A beam of charged pions can be accelerated, and scattered from protons, for example by passing the pion beam through liquid hydrogen in a bubble chamber. The proportion of the beam of incident particles that is scattered determines the total scattering cross-section (2.14), while the angular distribution of the scattered particles gives the differential cross-section. The protons interact with the pions electromagnetically through their electric charge, and also through the strong nuclear force, which has only a very short range, of the order of 10^{-15} m. The few pions that come close enough to a proton for this latter interaction to be effective are scattered by the nuclear force, and the details of this non-Coulomb scattering provide information about the nature of the interaction. The experimental data (with the Coulomb scattering effects subtracted) show **scattering resonances** – sharp peaks in the total cross-section at certain incident pion energies (Figure 4.4b). At an energy where such a resonance peak occurs there is a particularly strong interaction between the pion and the proton, and a metastable

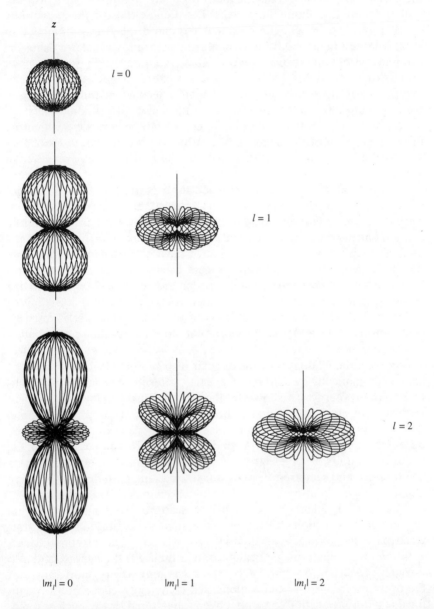

Figure 4.3 The square root of the probability distribution, $|Y_{lm}|$, for a particle in each of the possible orbital angular momentum eigenstates with $l = 0$, 1 and 2. The magnitude of $|Y_{lm}|$ in the direction represented by the angles (θ, ϕ) is proportional to the length of the radius vector in this direction. The distributions are all symmetric about the z axis.

bound state is formed, as shown schematically in Figure 4.4(a). This resonance state decays with a characteristic lifetime of the order of 10^{-23} s, which is too short for it to be observed directly, but its properties can be deduced by examining the decay products – a pion and a proton, or a pion and a neutron. (Electric charge must be conserved in the decay.)

The lowest-energy resonance of this kind is formed when a pion beam with kinetic energy approximately 190 MeV is incident on a stationary proton target. At this energy nearly all the scattering is due to the formation of the metastable state $\Delta(1260)$, where 1260 is the mass of the state in MeV/c^2 (somewhat larger than the mass of the proton, which is 939 MeV/c^2). The $\Delta(1260)$ is an eigenstate of orbital angular momentum with $l = 1$, which means that when it decays the pion and proton that are produced must be in a state with relative orbital angular momentum $l = 1$. We should expect the distribution of the pions relative to the protons in the centre-of-mass frame to be proportional to $|Y_{10}(\theta, \phi)|^2 = (3/4\pi)\cos^2\theta$, where θ is the angle made by a scattered pion with the incident direction in this frame of reference. The differential cross-section shown in Figure 4.4(c) appears qualitatively to have this type of behaviour, with a minimum at $\theta = 90°$ and approximate symmetry about this point. However, there are quantitative differences, and in particular the minimum value is not zero. These departures from a $\cos^2\theta$ distribution are largely accounted for by the fact that the proton, like the electron, has an intrinsic spin. This complicates matters and has the effect of adding a term $\frac{1}{4}(|Y_{11}(\theta, \phi)|^2 + |Y_{1-1}(\theta, \phi)|^2) = (3/16\pi)\sin^2\theta$. The resulting $1 + 3\cos^2\theta$ shape of the distribution is a reasonably good fit to the observed data.

4.7 Rotational energy levels in nuclei and molecules

The quantization of the orbital angular momentum spectrum explains why a discrete spectrum of light is emitted or absorbed when the rotational state of a particle, or system of particles, changes. We shall treat these rotating particles as freely rotating rigid bodies. In general, a system of quantum particles can never behave even approximately like a rigid body, but non-spherical nuclei and molecules are exceptions, and have certain rotational energy levels that are well described as states of a rigid rotator.

A collection of classical particles bound together in a rigid structure may be set rotating as a whole. If the rotation is about a principal axis of the system the total energy is $E = M^2/2\mathcal{I}$, where $\mathcal{I}$ is the moment of inertia and M the angular momentum about the principal axis [6]. This

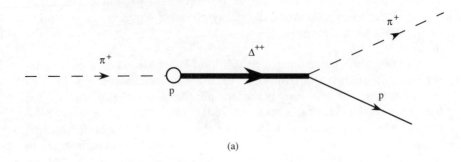

(a)

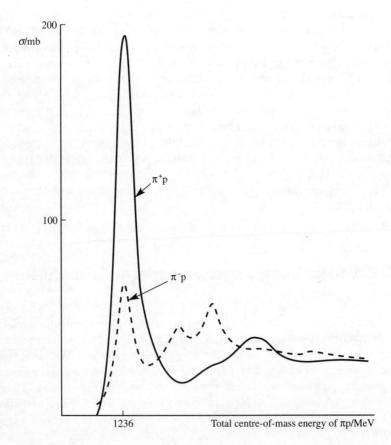

(b)

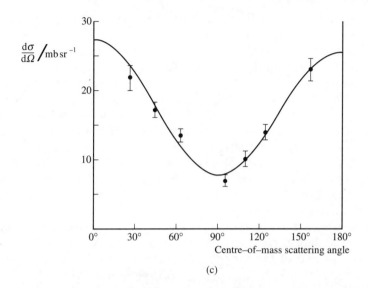

(c)

Figure 4.4 The scattering of charged pions by protons through the formation of a metastable intermediate state Δ. (a) A schematic view of the scattering of a positively charged pion by a proton (which is initially at rest) through the production of a Δ. (b) Resonance peaks in the total cross sections for π^+p and π^-p scattering mark the creation of metastable particles (adapted from Figure 19.3, p. 294, of [4]). (c) The angular distribution of π^+ relative to the proton in π^+p scattering at an incident pion kinetic energy of 200 MeV; the solid curve is a best fit to the experimental data (adapted from Figure 13 of [5]).

energy is purely kinetic, and depends on the internal structure of the body through the constant $\mathcal{J}$. In practice, no body can be perfectly rigid, and most tend to stretch as the angular momentum increases, producing a corresponding increase in the moment of inertia.

If the correspondence principle is applied to the classical expression $E = M^2/2\mathcal{J}$, the Hamiltonian for a quantum rigid rotator is found to be

$$\hat{H} = \frac{\hbar^2}{2\mathcal{J}} \hat{L}^2 \tag{4.47}$$

The eigenvalue equation is the same as the angular momentum eigenvalue equation (4.38), apart from the constant factor $\hbar^2/2\mathcal{J}$, and the eigenvalues are given by

$$E_l = \frac{\hbar^2}{2\mathcal{J}} l(l + 1) \quad (l = 0, 1, 2, \ldots) \tag{4.48}$$

The corresponding eigenfunctions are the spherical harmonics (4.45), which are simultaneously eigenfunctions of parity with eigenvalue $(-1)^l$. Note that the energy of the rotator is independent of the eigenvalue m_l of the z component of angular momentum, so that for any particular value of l there are $2l + 1$ distinct L_z eigenstates each with the same energy: the energy level is $(2l + 1)$-fold degenerate.

Nuclei that are non-spherical even in the ground state are called **permanently deformed nuclei**. They are ellipsoidal in shape, and occur only with mass numbers between $A = 150$ and 190 (elements belonging to the rare earths) and for A above 220 (the actinides). The mass number of a nucleus is the total number of protons and neutrons it contains, so that if the number of protons is Z (the atomic number) then the number of neutrons is $A - Z$ [7]. **Even–even nuclei**, which contain even numbers of both protons and neutrons, always have a ground state with total angular momentum quantum number 0^+, where the superscript refers to the parity of the state. This state therefore has the correct eigenvalues for the ground state of a rigid rotator. Deformed even–even nuclei have excited states corresponding to the rotational levels in (4.48), but a nucleus possesses additional symmetry that restricts the parity of these levels to be the same as that of the ground state. This means that only the even-l states occur: $l = 0, 2, 4, 6, \ldots$.

Two sets of rotational levels, based on the ground states of the isotopes $^{160}_{66}\text{Dy}$ of dysprosium and $^{238}_{92}\text{U}$ of uranium, are shown in Figure 4.5. (The subscript gives the atomic number Z and the superscript the mass number A.) Each level is labelled by the appropriate angular momentum quantum number l and the parity, which is $+$ in each case. These are not the only excited states of these isotopes in the energy range up to 300 keV, since other types of excitation also occur. However, the levels that are not purely rotational have been omitted for clarity.

The qualitative pattern of energy levels in each set shown in Figure 4.5 is characteristic of a quantum rigid rotator. The spacing between the levels increases with l, as we should expect from (4.48). Substituting the energy values from the diagram into this equation shows that the moment of inertia $\mathcal{I}$ is not quite constant: in the case of $^{160}_{66}\text{Dy}$ the value calculated from the difference in energy between the 0^+ and 2^+ levels is approximately $0.035\hbar^2\,\text{keV}^{-1}$, but the value increases when adjacent levels with higher l are used, reaching $0.043\hbar^2\,\text{keV}^{-1}$ in the case of the 12^+ and 14^+ levels. Similarly, the moment of inertia for $^{238}_{92}\text{U}$ increases from $0.067\hbar^2\,\text{keV}^{-1}$, calculated from the energy difference between the lowest two levels, to $0.073\hbar^2\,\text{keV}^{-1}$, calculated from the energy difference between the 10^+ and 12^+ levels. This shows that the nucleus is not completely rigid, but tends to stretch as the angular momentum increases, as would be expected for a classical body that is elastic rather than completely rigid.

As the angular momentum of a large nucleus is increased, it stretches more and more, until finally fission occurs. But a few rare earth nuclei can exist in states with very high angular momenta, reaching $l = 60$ in the case of $^{152}_{66}\text{Dy}$. These exceptional high-l states are **superdeformed** states of the nucleus, in which the major axis of the ellipsoid is close to twice the

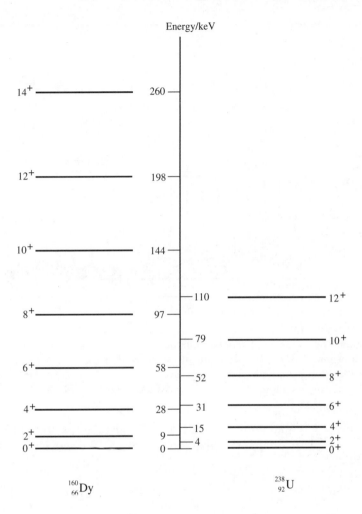

Figure 4.5 Rotational energy levels of the isotopes $^{160}_{66}$Dy of dysprosium and $^{238}_{92}$U of uranium. (For tables of nuclear energy level data see [8].)

length of the minor axes. They are produced in collisions between heavy ions: superdeformed states of dysprosium have been produced by bombarding a palladium-108 ($^{108}_{48}$Pd) target with calcium-48 ($^{48}_{20}$Ca) ions from a van der Graaff accelerator. The superdeformed states lose angular momentum in units of $2\hbar$, by emitting gamma rays – high-energy photons. The energy carried off by a gamma is equal to the difference between the energies of adjacent rotational levels (apart from a very small amount needed to balance the recoil kinetic energy of the nucleus), and again only even l values are allowed, as for normal deformed nuclei. The

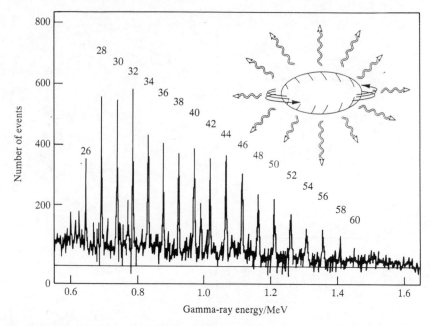

Figure 4.6 The spectrum of gamma radiation emitted by a superdeformed nucleus of dysprosium, $^{152}_{66}$Dy. The peaks are labelled by the spin of the nucleus from which they were emitted. Below the level $l = 22$ the superdeformed nucleus decays abruptly into more normally deformed states. (Adapted from [9].)

gamma energy is therefore $h\nu_{l+2} = E_{l+2} - E_l$, and the spectrum of Figure 4.6 shows a series of almost equally spaced peaks. We should expect this from (4.48):

$$h\nu_{l+4} - h\nu_{l+2} = (E_{l+4} - E_{l+2}) - (E_{l+2} - E_l) = 8\frac{\hbar^2}{2\mathcal{J}} \qquad \textbf{(4.49)}$$

The almost-constant spacing of about 47 keV between the observed peaks implies that this superdeformed nucleus is behaving like a rigid rotator with moment of inertia approximately $0.085\hbar^2$ keV^{-1}. The $^{152}_{66}$Dy nucleus is exceptional in the constancy of its moment of inertia. In other superdeformed nuclei $\mathcal{J}$ varies more rapidly with l, though the range of angular momentum over which it is approximately constant is still much larger than is the case for more normal deformed nuclei. (In Figure 4.6 it can be seen that the spacing increases slightly with increasing l, implying that the moment of inertia decreases, contrary to what we normally expect. This is typical of superdeformed nuclei, and indicates that more subtle changes than simple stretching occur in their structure as their angular momentum is increased.)

Purely rotational molecular spectra can also be observed [10]. The energy levels of a rigidly rotating molecule are given by (4.48), and in

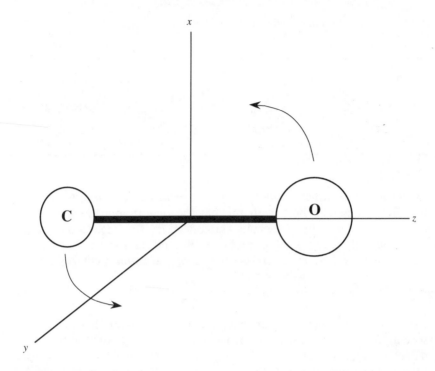

Figure 4.7 A schematic drawing of a carbon monoxide molecule with its axis along the z direction. The molecule can rotate end-over-end about the y axis (as shown) or the x axis.

general both odd and even values of l are allowed. The simplest type of molecule to consider is a diatomic one, such as CO (carbon monoxide), shown schematically in Figure 4.7. In an approximation that treats the atoms as point particles there are no rotations about the axis of the molecule (the corresponding moment of inertia is zero), and in practice a molecule only behaves like a rigid rotator in the case of end-over-end rotations. The energies of the end-over-end rotational levels of diatomic molecules are typically of the order of 10^{-4} or 10^{-5} eV, and the separations between adjacent levels correspond to radiation of wavelength in the millimetre and submillimetre regions. Multi-atom molecules have larger moments of inertia, and their rotational spectra extend into the microwave region.

Electromagnetic radiation can be emitted or absorbed in transitions between adjacent rotational energy levels of a molecule if the molecule has a permanent electric dipole moment. In such a transition, called an **electric dipole transition**, the l value changes by ± 1, as we shall show in Chapter 12. In contrast, nuclei do not have permanent electric dipole moments, and the nuclear rotational spectrum shown in Figure 4.6 is due

Table 4.2 The rotational spectrum of the CO molecule [11].

Transition ($l + 1 \rightarrow l$)	$(1/\lambda)/\text{cm}^{-1}$	ν/GHz	$\Delta\nu/\text{GHz}$
$1 \rightarrow 0$	3.845 033 19	115.271 195	
$2 \rightarrow 1$	7.689 919 07	230.537 974	115.266 779
$3 \rightarrow 2$	11.534 509 6	345.795 900	115.257 926
$4 \rightarrow 3$	15.378 662	461.040 68	115.244 78
$5 \rightarrow 4$	19.222 223	576.267 75	115.227 07
$6 \rightarrow 5$	23.065 043	691.472 60	115.204 85

to transitions in which l changes by ± 2, produced by the interaction of the electromagnetic field with the electric quadrupole moment of the nucleus. For a diatomic molecule, and indeed for any linear molecule, the rotational energy levels are given by (4.48), and the spectrum of frequencies of the absorbed or emitted **electric dipole radiation** is determined by

$$h\nu = E_{l+1} - E_l = \frac{\hbar^2}{2\mathcal{J}}(2l + 2) \tag{4.50}$$

Clearly the frequency spectrum should consist of discrete lines with equal spacing $\hbar/2\pi\mathcal{J}$ if the molecule behaves like a rigid body. Carbon monoxide is a typical diatomic molecule, and the frequencies of six spectral lines in the rotational emission spectrum of CO are given in Table 4.2, together with the frequency separation of adjacent lines. In molecular spectroscopy a spectral line is usually characterized by its wavenumber rather than its frequency, so the wavenumbers $1/\lambda = \nu/c$, are also given in the table. The separation in frequency between adjacent spectral lines is constant throughout the series of six lines to approximately 1 part in 10 000, showing that at these energies CO behaves very much like a rigid body. However, there is a slight increase in moment of inertia with angular momentum, corresponding to a stretching of the molecule.

References

[1] Kreyszig E. (1988). *Advanced Engineering Mathematics* 6th edn, Section 8.11, Example 5. New York: Wiley.
[2] Boas M.L. (1983). *Mathematical Methods in the Physical Sciences* 2nd edn, Chap. 12, Sections 2–11. New York: Wiley.
[3] Blin-Stoyle R.J. (1991). *Nuclear and Particle Physics*, Section 3.4.1. London: Chapman & Hall.
[4] Gasiorowicz S. (1966). *Elementary Particle Physics*. New York: Wiley.
[5] Mukhin A.I., Ozerov E.B. and Pontecorvo B. (1957). *Soviet Physics JETP* **4**, 237.

[6] Feynman R.P., Leighton R.B. and Sands M. (1964). *The Feynman Lectures on Physics* Vol. I, Sections 19-3, 19-4 and 20-4. Reading MA: Addison-Wesley.
Synge J.L. and Griffith B.A. (1959). *Principles of Mechanics* 3rd edn, Sections 11.3–11.5. New York: McGraw-Hill.
[7] Ref. [3], Chap. 2.
[8] Lederer C. and Shirley V.S., eds (1978). *Tables of Isotopes* 7th edn. New York: Wiley.
[9] Goss Levi B. (1988). *Physics Today* **41**, 17.
[10] For further reading see Hollas J.M. (1992). *Modern Spectroscopy* 2nd edn, Chap. 5. Chichester: Wiley.
[11] Data from Ref. [10], Table 5.1.

Problems

4.1   (a) Prove the commutation relations (4.6), (4.10) and (4.11).

      (b) Using the forms for $\hat{L}_x$, $\hat{L}_y$ and $\hat{L}_z$ given in (4.8), show that each component of the angular momentum operator commutes with the parity operator.

      (c) Prove that for any pair of linear operators $\hat{A}$ and $\hat{B}$

$$[\hat{A}, \hat{B}^2] = [\hat{A}, \hat{B}]\hat{B} + \hat{B}[\hat{A}, \hat{B}]$$

Show that $[x, \hat{P}_x^2] = 2i\hbar\hat{P}_x$ and $[x, \hat{P}_x^3] = 3i\hbar\hat{P}_x^2$.

      (d) Show that $[\hat{L}_z, \hat{P}_x] = i\hat{P}_y$, $[\hat{L}_z, \hat{P}_y] = -i\hat{P}_x$, $[\hat{L}_z, \hat{P}_z] = 0$, $[\hat{L}_z, \hat{P}^2] = 0$.

4.2   If ψ is an eigenfunction of the linear operator $\hat{A}$, prove that it can simultaneously be an eigenfunction of another linear operator $\hat{B}$ that commutes with $\hat{A}$. (*Hint*: show that $\hat{B}\psi$ is an eigenfunction of $\hat{A}$, corresponding to the same eigenvalue of $\hat{A}$ as ψ.)

    In each of the following cases show that the two operators commute with one another, that the given wave function is an eigenfunction of both of them, and find the corresponding eigenvalue of each operator:

      (a) $\hat{P}_x$, $\hat{P}_y$; $\psi(x, y, z, t) = Ce^{i(ax+by+cz-\omega t)}$

      (b) $\hat{P}_r$, (4.22), $\hat{L}_z$ (4.8); $\psi(r, \phi) - Ce^{i(ar-3\phi)}/r$

      (c) the parity operator $\hat{\Pi}$, $\hat{H} = \hat{P}_x^2/2m + \frac{1}{2}m\omega^2 x^2$; $\psi(x) = Ce^{-m\omega x^2/2\hbar}$.

    (In all cases C, a, b, c, m and ω are constants.)

4.3   (a) Prove that (4.19) gives the correct expression for the square of the operator $\hat{P}_r$ defined by (4.22).

      (b) Show that the radial momentum operator $\hat{P}_r$ defined by (4.22) satisfies the commutation relation $[r, \hat{P}_r] = i\hbar$. Explain the physical significance of this.

4.4   Write down the energy eigenfunctions for a particle in an infinitely deep one-dimensional square well extending from $z = -\frac{1}{2}L$ to $z = \frac{1}{2}L$, and check that they are eigenfunctions of parity $(z \to -z)$ corresponding to the eigenvalue $(-1)^{n-1}$, where n labels the energy level. (To obtain the required eigenfunctions, rewrite the functions (3.40) as functions of $z' = z - \frac{1}{2}L$.)

4.5   (a) Show that the following functions are eigenfunctions of the parity operator, $\hat{\Pi}$, and find the corresponding eigenvalue in each case:

(i) $\psi_1(x) = C\left(\sin\dfrac{\pi x}{L} + \sin\dfrac{3\pi x}{L}\right)$

(ii) $\psi_2(x, y, z) = C\exp[-a\sqrt{(x^2 + y^2 + z^2)}]$, where a is a constant

(iii) $\psi_3(r, \theta, \phi) = Cf(r)(\cos\theta + \cos^3\theta)e^{i\phi}$, where $f(r)$ is an arbitrary function of r

(b) If $\psi_+(x, y, z)$ and $\psi_-(x, y, z)$ are eigenfunctions of $\hat{\Pi}$ corresponding to the eigenvalues $+1$ and -1 respectively, is the function $\Psi(x, y, z) = 2\psi_+(x, y, z) + 3\psi_-(x, y, z)$ an eigenfunction of parity? Show that it is an eigenfunction of $\hat{\Pi}^2$, and find the eigenvalue.

4.6   Find all possible orientations of the angular momentum vector relative to the z axis if (a) $l = 1$, (b) $l = 2$. In each case sketch the cones on which the angular momentum vector can lie.

4.7   (a) Prove that different eigenfunctions of $\hat{L}_z$ are orthogonal.
       (b) Prove explicitly (by performing the appropriate integrations) that the following pairs of functions, given in Table 4.1, are orthogonal:

$$P_1(\cos\theta) \text{ and } P_2(\cos\theta)$$

$$P_1(\cos\theta) \text{ and } P_3(\cos\theta)$$

$$P_1^1(\cos\theta) \text{ and } P_2^1(\cos\theta)$$

(The integration is over $\cos\theta$ in the range $(-1, +1)$.)

*4.8   Use the computer program *Orbital angular momentum* to study $|Y_{lm_l}(\theta, \phi)|$, the square root of the probability density, for a particle in various orbital angular momentum eigenstates. How many nodes are there in a distribution with quantum numbers (l, m_l)? Plot (by hand) the two-dimensional curve of $P_l^{|m_l|}$ as a function of θ for several sets of values (l, m_l), and interpret the three-dimensional representation produced by the computer program in terms of your graph. (In particular, identify the zeros and maxima of the distribution.)

4.9   In an elastic scattering process at a particular energy one spinless particle is scattered by another, and in the centre-of-mass frame the final state is an eigenstate of orbital angular momentum with quantum numbers $l = 2$, $m_l = 0$. Sketch the shape of the differential cross-section as a function of scattering angle in the centre-of-mass frame. In which directions is there least probability of the particle being scattered?

4.10 The observed rotational spectrum of the hydrogen chloride molecule consists of a set of equally spaced lines produced by the emission (or absorption) of electric dipole radiation. If the spacing is $20.68 \, \text{cm}^{-1}$, calculate the
moment of inertia of the HCl molecule. What are the energies of its four
lowest rotational energy levels? Assuming that the moment of inertia of a
diatomic molecule is of the form $\mathcal{I} = mr^2$, estimate the mean separation r
of the atoms in the molecule. (Here $m = m_{\text{H}} m_{\text{Cl}} / (m_{\text{H}} + m_{\text{Cl}})$ is the reduced
mass of the molecule, defined in (A1.7). Take $m_{\text{H}} = 1.66 \times 10^{-27} \, \text{kg}$, and
use the mass of the isotope ^{35}Cl: $m_{\text{Cl}} = 35 m_{\text{H}}$.)

4.11 Light can be scattered from molecules that have no permanent electric
dipole moment through first inducing such a moment. If there is no change
in the rotational state of the molecule, the light is scattered with no change
in its frequency (**Rayleigh scattering**). However, the rotational state of the
molecule may be changed from one characterized by the quantum number l
to one corresponding to $l' = l \pm 2$ (**Raman scattering**). Show that in this
case the frequency of the scattered light differs from that of the incident
light by $\Delta \nu = B(4l + 6)$, where B is a constant characteristic of the
molecule, and $l = 0, 1, 2, \ldots$.

CHAPTER 5

Ladder Operators: The One-Dimensional Simple Harmonic Oscillator

5.1 The energy spectrum of a one-dimensional simple harmonic oscillator

In classical mechanics small displacements from a state of stable equilibrium can always be treated, to a first approximation, as simple harmonic oscillations. This is true for the vibrations of individual particles and also of many-body systems. In a state of stable equilibrium interacting particles within a many-body system are arranged in a regular, well-defined way. If the system is disturbed, this arrangement is altered, and the individual particles vibrate about their equilibrium positions. But the particles do not move independently, because of the interactions between them, so the disturbance must be treated as a vibration of the overall distribution of the particles about the equilibrium configuration. This can be analysed into **normal modes**, each of which is a simple harmonic oscillation of the internal configuration of the system, at a unique frequency [1]. Simple harmonic oscillations have a similar importance in quantum physics. For example, the vibrational motion of molecules can be treated as simple harmonic to a first approximation, and vibrations of the atoms within a crystalline solid can be analysed into normal modes. As in classical physics, corrections to the simple harmonic approximation involve the addition of anharmonic terms to the potential energy.

If a classical particle of mass m performs simple harmonic motion in one dimension, its potential energy is $V = \frac{1}{2}m\omega^2 x^2$, where x is the displacement from the centre of the motion and ω is the constant angular frequency. The correspondence principle allows us to use this form in the time-independent Schrödinger equation (3.20), giving

$$\left(-\frac{\hbar^2}{2m}\frac{d^2}{dx^2} + \tfrac{1}{2}m\omega^2 x^2\right)\psi(x) = E\psi(x) \tag{5.1}$$

This equation can be reduced to its simplest form by introducing the dimensionless variable ξ and eigenvalue ϵ, where

$$\xi = \sqrt{\left(\frac{m\omega}{\hbar}\right)}x, \qquad \epsilon = \frac{E}{\hbar\omega} \tag{5.2}$$

If these substitutions are made, (5.1) becomes

$$\frac{1}{2}\left(\xi^2 - \frac{d^2}{d\xi^2}\right)\psi(\xi) = \epsilon\psi(\xi) \tag{5.3}$$

This differential equation may be solved analytically by a power series expansion method (Appendix B2), and there is a program on the computer disk for plotting the solutions. Physically acceptable solutions $\psi(\xi)$ must remain finite everywhere. As in the case of the square well, this restriction is only satisfied for a discrete set of eigenvalues, which means that the energy of the oscillator is quantized.

A different method of solving the eigenvalue equation (5.3) is the **ladder operator method**, which depends on exploiting the commutation properties of operators. The left-hand side of (5.3) can be written in terms of the product of two operators $\hat{A}$ and $\hat{A}^\dagger$, defined by

$$\hat{A} = \frac{1}{\sqrt{2}}\left(\xi + \frac{d}{d\xi}\right), \qquad \hat{A}^\dagger = \frac{1}{\sqrt{2}}\left(\xi - \frac{d}{d\xi}\right) \tag{5.4}$$

Then, since

$$\frac{d}{d\xi}\xi = \xi\frac{d}{d\xi} + 1$$

(to prove this, think of the operator acting on a function of ξ), we see that

$$\hat{A}\hat{A}^\dagger = \frac{1}{2}\left(\xi^2 - \frac{d^2}{d\xi^2} + 1\right), \qquad \hat{A}^\dagger\hat{A} = \frac{1}{2}\left(\xi^2 - \frac{d^2}{d\xi^2} - 1\right) \tag{5.5}$$

We can deduce the commutator of $\hat{A}$ and $\hat{A}^\dagger$:

$$[\hat{A}, \hat{A}^\dagger] = 1 \tag{5.6}$$

There are thus two different ways of writing the eigenvalue equation (5.3) in terms of $\hat{A}$ and $\hat{A}^\dagger$:

$$(\hat{A}^\dagger\hat{A} + \tfrac{1}{2})\psi(\xi) = \epsilon\psi(\xi) \tag{5.7}$$

$$(\hat{A}\hat{A}^\dagger - \tfrac{1}{2})\psi(\xi) = \epsilon\psi(\xi) \tag{5.8}$$

If we multiply (5.7) through from the left by $\hat{A}$, we get

$$\hat{A}\hat{A}^\dagger\hat{A}\psi(\xi) + \tfrac{1}{2}\hat{A}\psi(\xi) = \epsilon\hat{A}\psi(\xi) \tag{5.9}$$

(Since ϵ is just a number it can be placed to the left of the operator: $\hat{A}\epsilon = \epsilon\hat{A}$). This can be written in terms of $\psi'(\xi) = \hat{A}\psi(\xi)$:

$$(\hat{A}\hat{A}^+ + \tfrac{1}{2})\psi'(\xi) = \epsilon\psi'(\xi)$$

so that

$$(\hat{A}\hat{A}^\dagger - \tfrac{1}{2})\psi'(\xi) = (\epsilon - 1)\psi'(\xi) = \epsilon'\psi'(\xi) \qquad \textbf{(5.10)}$$

This is an eigenvalue equation of the form (5.8), with eigenfunction $\psi'(\xi)$ and eigenvalue $\epsilon' = \epsilon - 1$. The operator $\hat{A}$ transforms one energy eigenfunction $\psi(\xi)$, corresponding to eigenvalue ϵ, into another, $\psi'(\xi)$, which corresponds to an eigenvalue one unit lower.

Similarly, if we multiply (5.8) through from the left by $\hat{A}^\dagger$ and rearrange the terms, we obtain an eigenvalue equation of the form (5.7), in which the eigenfunction is $\hat{A}^\dagger\psi(\xi)$ and the eigenvalue is $\epsilon + 1$.

If each eigenfunction is labelled by the corresponding eigenvalue, the effect of the operators $\hat{A}$ and $\hat{A}^\dagger$ can be written as

$$\hat{A}\psi_\epsilon(\xi) = \frac{1}{\sqrt{2}}\left(\xi + \frac{\mathrm{d}}{\mathrm{d}\xi}\right)\psi_\epsilon(\xi) \propto \psi_{\epsilon-1}(\xi) \qquad \textbf{(5.11)}$$

$$\hat{A}^\dagger\psi_\epsilon(\xi) = \frac{1}{\sqrt{2}}\left(\xi - \frac{\mathrm{d}}{\mathrm{d}\xi}\right)\psi_\epsilon(\xi) \propto \psi_{\epsilon+1}(\xi) \qquad \textbf{(5.12)}$$

(where we have used the definitions (5.4)). Proportionality signs have been used, rather than equalities, so that the notation $\psi_\epsilon(\xi)$ can be reserved for the *normalized* eigenfunctions. The functions produced by operating on a normalized eigenfunction with $\hat{A}$ and $\hat{A}^\dagger$ are eigenfunctions, but they are not normalized. Starting from any particular state with eigenvalue ϵ, successive applications of $\hat{A}$ to ψ_ϵ produce eigenfunctions of states with eigenvalues $\epsilon - 1$, $\epsilon - 2$, $\epsilon - 3$, and so on. Similarly, successive applications of $\hat{A}^\dagger$ produce eigenfunctions of states with eigenvalues $\epsilon + 1$, $\epsilon + 2$, $\epsilon + 3$, and so on. So there is a ladder of discrete equally spaced energy eigenstates, as shown schematically in Figure 5.1. $\hat{A}$ and $\hat{A}^\dagger$ are **ladder operators**, connecting adjacent 'rungs' on the ladder, and effectively lowering or raising the energy by one unit.

It seems that after applying the lowering operator $\hat{A}$ a certain number of times we should generate eigenfunctions representing states with negative energy. This does not make physical sense – classically the total energy cannot be less than the minimum potential energy, and in fact the classical ground state has $E = V_{\min} = 0$. Similarly, in the quantum case there should be a ground state, in which the oscillator has its lowest possible energy. Indeed, it can be proved mathematically that the Hamiltonian in (5.1) does not have any negative eigenvalues, so there is a lowest energy eigenvalue $E_0 = \epsilon_0\hbar\omega \geq 0$, such that $E_0 - \hbar\omega < 0$.

Use of the ladder operator $\hat{A}$ on the ground state eigenfunction must give zero: if it did not, it would produce an eigenfunction representing a state with negative energy $E_0 - \hbar\omega$, and this is forbidden. So the ground state wave function $\psi_0(\xi)$ is defined to satisfy

$$\hat{A}\psi_0(\xi) = \frac{1}{\sqrt{2}}\left(\xi + \frac{\mathrm{d}}{\mathrm{d}\xi}\right)\psi_0(\xi) = 0 \qquad \textbf{(5.13)}$$

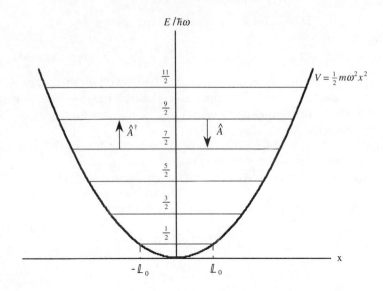

Figure 5.1 The potential energy curve and energy eigenvalues for a one-dimensional simple harmonic oscillator. $\mathbb{L}_0$ is the amplitude of a classical oscillator with energy equal to the quantum mechanical zero-point energy. The effect of the ladder operators $\hat{A}^\dagger$ and $\hat{A}$ is indicated.

Multiplying this equation from the left by $\hat{A}^\dagger$ and rearranging it, we get

$$(\hat{A}^\dagger\hat{A} + \tfrac{1}{2})\psi_0(\xi) = \tfrac{1}{2}\psi_0(\xi) \tag{5.14}$$

Comparing this with (5.7), we see that the eigenvalue corresponding to ψ_0 is $\epsilon_0 = \tfrac{1}{2}$. So there is a unique bottom rung to the ladder of energy levels shown in Figure 5.1, at $E_0 = \tfrac{1}{2}\hbar\omega$. Above this the rungs are equally spaced, at $E_1 = E_0 + \hbar\omega$, $E_2 = E_0 + 2\hbar\omega$, and so on. There is no physical restriction implying an upper limit to the energy of an oscillator, so there is no limit to the number of times we can use the operator $\hat{A}^\dagger$ that effectively raises the energy level. Therefore the energy spectrum of the one-dimensional simple harmonic oscillator consists of an infinite number of equally spaced levels:

$$E_n = (n + \tfrac{1}{2})\hbar\omega \quad (n = 0, 1, 2, 3 \ldots) \tag{5.15}$$

(You may wonder whether there are any other eigenvalues between these, which the ladder operators we have defined do not reveal. In fact, there are not, as you can check for yourself using the computer program – see Problem 5.1.)

The spectrum (5.15) shows two of the general characteristics of the energy eigenstates of bound particles, given in Section 3.3.

(1) The energy spectrum is discrete.
(2) The oscillator has a zero-point energy $E_0 > 0$ that is greater than the minimum energy of the corresponding classical oscillator.

5.2 The energy eigenfunctions of the one-dimensional simple harmonic oscillator

Equation (5.13) is a first-order differential equation for the ground state eigenfunction of the simple harmonic oscillator, with the solution

$$\psi_0(\xi) = C_0 e^{-\xi^2/2} \tag{5.16}$$

where C_0 is a normalization constant. The eigenfunction must be normalized in terms of the physical variable x rather than the dimensionless ξ defined in (5.2), so we require

$$|C_0|^2 \int_{-\infty}^{\infty} dx \exp\left(-\frac{m\omega}{\hbar}x^2\right) = 1 \tag{5.17}$$

The integral has the value $\sqrt{(\pi\hbar/m\omega)}$, and if C_0 is chosen to be real, we obtain

$$C_0 = \left(\frac{m\omega}{\pi\hbar}\right)^{1/4} \tag{5.18}$$

The eigenfunctions for the excited states can be found by successive applications of the operator $\hat{A}^{\dagger} = \sqrt{(\tfrac{1}{2})}\,(\xi - d/d\xi)$. For example,

$$\hat{A}^{\dagger}\psi_0(\xi) = \frac{C_0}{\sqrt{2}}\left(\xi - \frac{d}{d\xi}\right)e^{-\xi^2/2} = \frac{C_0}{\sqrt{2}}(2\xi)\,e^{-\xi^2/2} \tag{5.19}$$

This is an eigenfunction for the first excited state, proportional to the normalized eigenfunction $\psi_1(\xi)$. (We label the eigenfunctions by the integer n that determines the corresponding eigenvalue E_n.) If $\hat{A}^{\dagger}$ is applied a second time, we obtain

$$(\hat{A}^{\dagger})^2\psi_0(\xi) = C_0\left(\frac{1}{\sqrt{2}}\right)^2\left(\xi - \frac{d}{d\xi}\right)^2 e^{-\xi^2/2}$$
$$= C_0(\sqrt{\tfrac{1}{2}})^2(4\xi^2 - 2)\,e^{-\xi^2/2} \tag{5.20}$$

This must be proportional to $\psi_2(\xi)$. Continuing in this way, we find that the eigenfunction for the nth excited state, $\psi_n(\xi)$, is proportional to

$$(\hat{A}^{\dagger})^n\psi_0(\xi) = C_0\left(\frac{1}{\sqrt{2}}\right)^n\left(\xi - \frac{d}{d\xi}\right)^n e^{-\xi^2/2}$$
$$= C_0(\sqrt{\tfrac{1}{2}})^n\,H_n(\xi)\,e^{-\xi^2/2} \tag{5.21}$$

Table 5.1 Hermite polynomials

$$H_0(\xi) = 1$$
$$H_1(\xi) = 2\xi$$
$$H_2(\xi) = 4\xi^2 - 2$$
$$H_3(\xi) = 8\xi^3 - 12\xi$$
$$H_4(\xi) = 16\xi^4 - 48\xi^2 + 12$$

Higher polynomials can be generated using (5.25)

where $H_n(\xi)$ is a polynomial in ξ of order n, defined by

$$e^{-\xi^2/2} H_n(\xi) = \left(\xi - \frac{d}{d\xi} \right)^n e^{-\xi^2/2} \quad (n = 0, 1, 2, \ldots) \qquad (5.22)$$

The $H_n(\xi)$ are **Hermite polynomials** [2], some of which are listed in Table 5.1, and the functions $e^{-\xi^2/2} H_n(\xi)$ form an orthogonal set, normalized in such a way that

$$\int_{-\infty}^{\infty} d\xi\, e^{-\xi^2} H_n(\xi) H_m(\xi) = \delta_{nm} 2^n n! \sqrt{\pi} \qquad (5.23)$$

With this normalization the Hermite polynomials satisfy the recursion relations

$$\frac{d H_n}{d\xi} = 2n H_{n-1} \qquad (5.24)$$

$$2\xi H_n - \frac{d H_n}{d\xi} = H_{n+1} \qquad (5.25)$$

Equation (5.21) shows that the energy eigenfunctions for the one-dimensional simple harmonic oscillator are of the form

$$\psi_n(\xi) = C_n H_n(\xi)\, e^{-\xi^2/2} \qquad (5.26)$$

where C_n is a normalization constant. Using (5.23), we find

$$\int_{-\infty}^{\infty} dx\, \psi_n^*(x) \psi_m(x) = \delta_{nm} |C_n|^2\, 2^n n! \left(\frac{\pi\hbar}{m\omega} \right)^{1/2} \qquad (5.27)$$

Note that eigenfunctions corresponding to different energy eigenvalues (different integers n and m) are orthogonal, in accordance with the fifth general characteristic of energy eigenstates given in Section 3.3. If the eigenfunctions are normalized, the right-hand side of (5.27) must be equal to 1 when $n = m$, so that

$$C_n = \left(\frac{1}{2^n n!} \right)^{1/2} \left(\frac{m\omega}{\pi\hbar} \right)^{1/4} = \left(\frac{1}{2^n n!} \right)^{1/2} C_0 \qquad (5.28)$$

We can now evaluate the constants of proportionality in (5.11) and (5.12). First we use the operator $\hat{A}$ on the eigenfunction (5.26):

$$\hat{A}\psi_n(\xi) = \frac{1}{\sqrt{2}}\left(\xi + \frac{d}{d\xi}\right)C_n H_n(\xi)\, e^{-\xi^2/2}$$

$$= \frac{1}{\sqrt{2}}C_n\, e^{-\xi^2/2}\frac{d H_n(\xi)}{d\xi}$$

Use of the recursion relation (5.24) and the expression (5.28) for the normalization constants gives

$$\hat{A}\psi_n(\xi) = n^{1/2}\psi_{n-1}(\xi) \tag{5.29}$$

Next we apply $\hat{A}^\dagger$ to $\psi_n(\xi)$ and use the recursion relation (5.25):

$$\hat{A}^\dagger\psi_n(\xi) = \frac{1}{\sqrt{2}}\left(\xi - \frac{d}{d\xi}\right)C_n H_n(\xi)\, e^{-\xi^2/2} = \frac{1}{\sqrt{2}}C_n\, e^{-\xi^2/2}\left(2\xi - \frac{d}{d\xi}\right)H_n(\xi)$$

so that

$$\hat{A}^\dagger\psi_n(\xi) = (n + 1)^{1/2}\psi_{n+1}(\xi) \tag{5.30}$$

Equations (5.29) and (5.30) should be compared with (5.11) and (5.12). Note that (5.29) and (5.30) together imply

$$\hat{A}^\dagger\hat{A}\psi_n(\xi) = n\psi_n(\xi) \tag{5.31}$$

which is consistent with the eigenvalue equation (5.7) and the energy spectrum (5.15).

The energy eigenfunctions and probability distributions for a simple harmonic oscillator in the ground state and the first few excited states are plotted in Figure 5.2. Note that the eigenfunctions are alternately symmetric and antisymmetric about $x = 0$, so they are also eigenfunctions of parity, with eigenvalues $(-1)^n$. This is due to the symmetry of the potential, which ensures that the Hamiltonian in (5.1) commutes with the parity operator.

The behaviour of a quantum oscillator is very different from that of its classical counterpart. In its ground state a classical oscillator is at rest at the origin, with total energy zero. The origin is the most probable position for a quantum oscillator in its ground state, but it has a zero-point energy $E_0 = \frac{1}{2}\hbar\omega$, as we have already mentioned. At a higher energy E the classical particle would oscillate between $x = L$ and $x = -L$, where the amplitude L is determined by $E = \frac{1}{2}m\omega^2 L^2$. So a classical oscillator with energy equal to E_n (the energy of a quantum oscillator in its nth excited state) would be restricted to the region

$$|x| \leq L_n = \left[\frac{\hbar}{m\omega}(2n + 1)\right]^{1/2} \tag{5.32}$$

These limits to the classically allowed region are marked in Figure 5.2. As we should expect, the quantum oscillator can penetrate into the classically

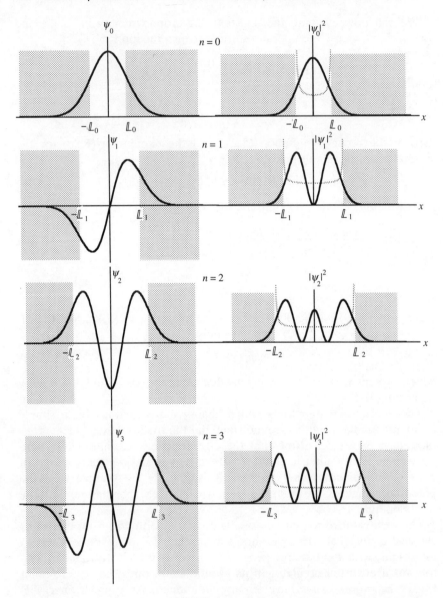

Figure 5.2 The energy eigenfunctions and corresponding probability distributions for a one-dimensional simple harmonic oscillator in its ground state and first three excited states. The classically forbidden regions are shaded, and the classical probability distribution is shown as a light line.

forbidden region, though the probability density decreases rapidly as we move beyond $|x| = L_n$. There are n nodes in the eigenfunction of the nth excited state, due to the n zeros of the polynomial H_n (which is of degree n), and these all occur within the classically allowed region. These two

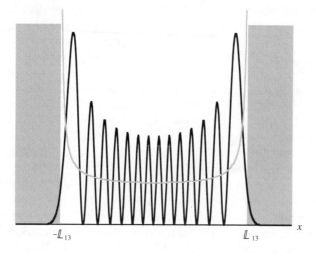

Figure 5.3 The classical (light line) and quantum (heavy line) probability distributions for a simple harmonic oscillator in the level $n = 13$.

properties of the eigenfunctions agree with the third and fourth general characteristics of bound states given in Section 3.3.

A classical oscillator can be observed continuously without its motion being affected, so its probability distribution is not a very useful concept, but it can be defined. Suppose we take a series of rapid glances at a classical oscillator, at random instants of time. At what point of its oscillation should we be most likely to see it? The oscillator spends most time furthest from the centre of its motion, since in these regions its velocity is small, reaching zero instantaneously at $x = \pm L$. The classical probability density therefore has maxima at $x = \pm L$, and the classical distribution is shown as a light line on the graphs of the quantum probability distribution in Figure 5.2. At low energies the quantum and classical distributions are completely different. However, when the level of excitation becomes sufficiently high, the average of the quantum distribution becomes similar to the classical distribution. This is evident in Figure 5.3, which shows the classical and quantum probability distributions for an oscillator with the energy appropriate to the level $n - 13$.

5.3 Vibrational spectra of molecules and nuclei

A diatomic molecule, such as that shown schematically in Figure 4.7, can not only rotate but can also vibrate linearly, parallel to its axis, like a one-dimensional simple harmonic oscillator, and the infrared spectrum of such a molecule shows evidence of this. A molecule vibrating like a one-dimensional quantum oscillator should have a set of equally spaced vibrational energy levels, corresponding to an energy spectrum of the form (5.15) and Figure 5.1. If the molecule has a permanent electric dipole moment, electric dipole radiation is emitted or absorbed in transitions between a pair of adjacent levels, and since the levels have equal spacing $\hbar\omega$ (where ω is characteristic of the particular molecule) this radiation should have a unique frequency $\nu = \omega/2\pi$.

In practice, if the spectrum of a diatomic molecule such as HCl (hydrogen chloride) or CO is measured in the near-infrared with low resolution, a single intense line is observed [3]. At slightly higher resolution this line is seen as a double peak, and at still higher resolution it becomes a band of closely spaced lines, as shown for HCl in Figure 5.4. For HCl and CO this **fundamental vibrational band** is centred on wavenumbers $1/\lambda = 2886\,\mathrm{cm}^{-1}$ and $1/\lambda = 2146\,\mathrm{cm}^{-1}$ respectively. These correspond to oscillator levels separated by $\hbar\omega = 0.36\,\mathrm{eV}$ for HCl and $0.27\,\mathrm{eV}$ for CO. The fine structure arises because when electric dipole radiation is emitted or absorbed the rotational state of the molecule is changed.

The energy associated with a transition from one vibrational state to the next is at least 100 times greater than that associated with transitions between the lower rotational levels – compare the wavenumber $1/\lambda = 2146\,\mathrm{cm}^{-1}$ for the vibrational spectrum of CO with those given in Table 4.2 for its rotational spectrum. So different rotational transitions accompanying the same vibrational transition produce small variations in the total energy absorbed or emitted. Although it is not possible to produce a purely vibrational electric dipole transition, a purely rotational spectrum can be observed in the far-infrared, where the frequency of the radiation is so low that there is insufficient energy to excite molecular vibrations.

Suppose that a diatomic molecule is initially in a state represented by the vibrational quantum number n_1 and the rotational quantum number l_1, and that it makes a transition to a state represented by the quantum numbers $n_2 = n_1 + 1$ and l_2, by absorbing a quantum of electric dipole radiation. As mentioned in Section 4.7, the rotational quantum number changes by one unit in an electric dipole transition. If $l_2 = l_1 + 1$, the change in rotational energy is, from (4.50), $\Delta E_{\mathrm{rot}} = (\hbar^2/2\mathcal{J})(2l_1 + 2)$, which (except for very high l_1 values) is very much less than the energy difference $\Delta E_{\mathrm{vib}} = h\nu_0$ between purely vibrational energy levels. The absorption spectrum therefore contains a set of closely spaced lines above the frequency ν_0 (which would correspond to a purely vibrational transition) at frequencies corresponding to the sum of the vibrational and rotational energy changes, $\Delta E_{\mathrm{vib}} + \Delta E_{\mathrm{rot}}$:

$$\nu = \nu_0 + \frac{\hbar}{4\pi\mathcal{J}}(2l_1 + 2) \quad (l_1 = 0, 1, 2, \dots) \tag{5.33}$$

This set of lines is called the **R branch** of the spectrum. But rotational

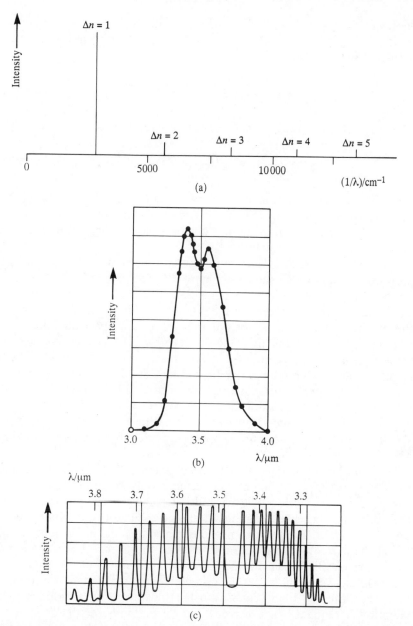

Figure 5.4 The infrared absorption spectrum of HCl (adapted from Figures 30–32, pp. 53–55, of [4]). (a) At very low resolution the intense line is the fundamental band, and the higher bands (which arise from anharmonicity) actually fall off in intensity about five times more rapidly than indicated. (b) At slightly higher resolution the fundamental band appears to be split into two. (c) At still higher resolution the rotational fine structure of the fundamental band is visible.

energy could also be lost in the transition, with $l_2 = l_1 - 1$, and there is a corresponding set of transition frequencies below v_0, referred to as the **P branch** of the spectrum:

$$v = v_0 - \frac{\hbar}{4\pi\mathcal{J}}2l_1 = v_0 - \frac{\hbar}{4\pi\mathcal{J}}(2l_2 + 2) \quad (l_2 = 0, 1, 2, \ldots) \quad (5.34)$$

The absorption band therefore consists of a set of lines at frequencies

$$v = v_0 \pm \frac{\hbar}{2\pi\mathcal{J}}, \quad v_0 \pm 2\frac{\hbar}{2\pi\mathcal{J}}, \quad v_0 + 3\frac{\hbar}{2\pi\mathcal{J}}, \quad \ldots \quad (5.35)$$

The lines in this spectrum have equal spacing $\hbar/2\pi\mathcal{J}$ (in terms of frequency), except for a gap of twice this width at the centre arising because the purely vibrational transition is not possible. The emission spectrum should show the same structure.

An example of such a **vibration–rotation spectrum** is the fundamental absorption band of HCl shown in Figure 5.4. The reason for the apparent double peak shown in Figure 5.4(b) is that at this particular resolution the gap in the centre is resolved but the separation between individual lines in the R and P branches is not. In the absorption spectrum, lines corresponding to transitions involving states with $l = 10$ have been observed, while lines have been seen in the emission spectrum involving states with angular momentum quantum numbers as high as 33. The spacing between the lines in the purely rotational spectrum of HCl is 20.68 cm^{-1}, and the separation of the lines associated with low l values in Figure 5.4(c) is approximately the same. However, the spacing is not exactly constant, showing departures from pure simple harmonic vibration and rigid rotation.

It should be emphasized that the expression (5.35) for the vibration–rotation spectrum of a diatomic molecule is only an approximation. It is based on the assumption that the molecule simultaneously vibrates along the molecular axis like a simple harmonic oscillator and rotates like a rigid body, and that these two motions are completely independent. This cannot be correct – the separation between the nuclei of the two atoms varies periodically because of the vibration, while it would be constant if the molecule were rigid! That the approximation works at all is a result of the very different time scales associated with the two motions. The vibrational frequency of a classical simple harmonic oscillator is $v_{vib} = \omega/2\pi$, and $\hbar\omega$ is the separation of the energy levels of the corresponding quantum oscillator. The value $\hbar\omega = 0.36 \text{ eV}$ found for the centre of the fundamental vibrational band of HCl corresponds to $v_{vib} = 8.65 \times 10^{13} \text{ s}^{-1}$, showing that in a semiclassical picture the HCl molecule performs 8.65×10^{13} vibrations per second. For a classical rigid rotator the energy E at angular frequency ω_{rot} is $E = \frac{1}{2}\mathcal{J}\omega_{rot}^2$. So the rotational frequency of a classical rigid rotator with the energy E_l given by (4.48) is

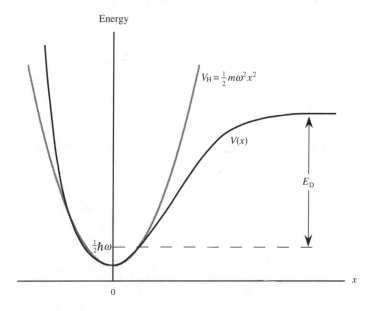

Figure 5.5 The form of the vibrational potential energy $V(x)$ of a diatomic molecule (heavy line), compared with a simple harmonic oscillator potential $V_H(x)$ (light line). The variable x is the difference between the actual separation of the two nuclei and their equilibrium separation. The amount of vibrational energy that will cause the molecule in its ground state to dissociate is the dissociation energy E_D.

$$\nu_{rot} = \frac{\omega_{rot}}{2\pi} = \frac{\hbar}{2\pi\mathcal{I}} \sqrt{[l(l+1)]} \qquad (5.36)$$

The value of $\hbar/2\pi\mathcal{I}$ can be found from the separation between the frequencies of adjacent lines in the rotational spectrum of the quantum oscillator, (5.35). In terms of wavenumber, the separation is in the region of $20.6\,\mathrm{cm}^{-1}$ for both the purely rotational spectrum of HCl and its fundamental vibration–rotation band, leading to a semiclassical picture of the molecule rotating approximately 8.7×10^{11} times per second in the $l = 1$ state. So the frequency of vibrations is around a hundred times greater than the rotational frequency – the molecule vibrates 100 times during each complete rotation. This means that it is the *average* nuclear separation that determines the moment of inertia $\mathcal{I}$ in the expression (5.35) for the vibration–rotation spectrum, and it is reasonable to treat this average value as a constant.

If a diatomic molecule behaved exactly like a simple harmonic oscillator, with a parabolic potential energy function $V(x) \propto x^2$, it would remain bound, however large its energy. But in practice the molecule can split apart if it is given sufficient vibrational energy, and a more realistic potential is shown in Figure 5.5. At low energies the parabolic behaviour is a reasonable approximation to this curve, but, as the energy increases, the two curves separate – the real oscillator becomes **anharmonic**, and terms involving higher powers of x must be added to the parabolic potential energy function. The energy levels of an anharmonic oscillator are not evenly spaced, and in practice, as the energy increases, the separation between the vibrational levels of the molecule decreases slightly. In addition, the selection rule that electric dipole transitions occur only between adjacent vibrational levels is no longer satisfied exactly, and higher vibrational bands occur, centred approximately on frequencies $2v_0$, $3v_0$ and so on (where v_0 is the central frequency of the fundamental band) as shown in Figure 5.4(a). However, these bands are very much less intense than the fundamental.

The vibrational motion of a nucleus is not as simple as that of a diatomic molecule, since most nuclei contain many constituent nucleons, and the oscillations in shape are three-dimensional rather than being restricted to linear vibrations along one particular axis [5]. However, it turns out that quantum mechanics predicts equally separated energy levels for a simple harmonic oscillator even if its vibration is not one-dimensional. Even–even nuclei have the simplest vibrational energy spectra, and show evidence of equally spaced levels with fine structure due to rotational effects. As in the case of purely rotational nuclear motion discussed in Section 4.7, only rotational states with the same parity are allowed in each vibrational band, so that states with odd l values are absent. In contrast to the purely rotational case, however, nuclei need not be deformed for vibrational levels to be seen – such levels can be associated with vibrations about a spherical configuration. Figure 5.6 shows vibrational energy levels in nickel and iron nuclei.

5.4 Thermal oscillations, phonons and photons

Because the energy levels of a one-dimensional simple harmonic oscillator are equally spaced, as shown in Figure 5.1, it can only emit or absorb energy in integer multiples of $\hbar\omega = hv$, where in a semiclassical picture v is the frequency of the oscillations. This type of discrete energy change is exactly what we assumed in Section 1.2 in order to explain the spectrum of black body radiation. In that case the oscillations were associated with the thermal motion of electric charge distributions in the material forming the walls of a black body cavity, and a change in oscillator energy was associated with the emission or absorption of electromagnetic radiation at the same frequency as the oscillation. But (5.15) refers to a single oscillating particle, while in a solid the thermal oscillations are not the vibrations of individual atoms. In a solid the atoms do not move independently.

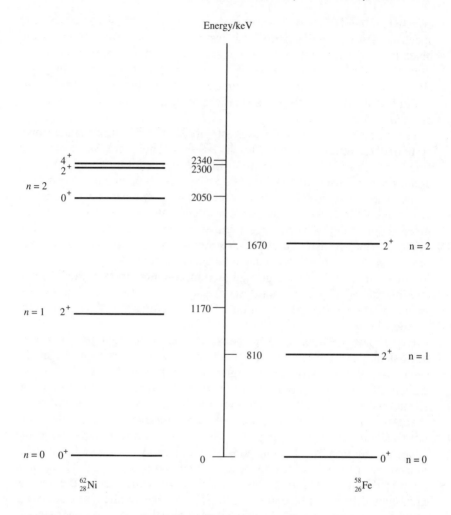

Figure 5.6 Low-lying vibrational energy levels of isotopes of nickel and iron. The $n = 2$ level in nickel is split into a triplet of states with different rotational properties. (For tables of nuclear energy level data see [6].)

Their positive ionic cores are bound together in a regular array, or **lattice**, and thermal energy is the vibrational energy of the lattice as a whole.

As mentioned in Section 5.1, small vibrations of a many-body system can be analysed into a linear combination of normal modes of vibration. In a solid each of these is a simple harmonic oscillation of the overall arrangement of the ions about their equilibrium configuration, with a characteristic frequency v. In a normal mode all the ions of which the lattice is made are vibrating coherently – that is, with a definite fixed relationship between the phases of vibration of individual ions – and effectively the lattice is behaving like a single unit. Since the energy spectrum of (5.15) and Figure 5.1 is just what we need to account for the black body spectrum, it appears that the normal modes of thermal vibration of the lattice should have the same type of energy spectrum as a single oscillator. Changes in thermal energy will then occur in integer multiples of hv, and electromagnetic energy can only be absorbed or emitted in corresponding quanta.

The number of normal modes of vibration of a three-dimensional system of n interacting particles is $3n - 6$. (There are $3n$ degrees of freedom, of which 3 are associated with the motion of the centre of mass, and a further 3 with rotation of the system as a whole about the centre of mass. The remaining degrees of freedom are associated with internal vibrations.) Since the number of constituent atoms is enormous in any macroscopic solid, a very large number of normal modes, each with a different characteristic frequency, combine to give the actual thermal motion of the lattice. Electromagnetic energy can be emitted or absorbed in integer multiples of each normal mode frequency. In fact, the frequencies of the normal modes are so closely spaced that the black body spectrum appears to span a continuous, rather than a discrete, range of frequencies.

Experimental evidence from many sources shows that the energy of normal modes of vibration in a many-body system is quantized in the same way as the energy of a single oscillating particle. As well as the evidence of the black body spectrum, there is evidence from the nuclear vibrational spectra that we mentioned in Section 5.3: the oscillations in shape of a nucleus that give rise to typical vibrational energy levels are not oscillations of a single particle about some centre, but are due to the collective motion of the nucleons. (The usual vibrational levels of a nucleus are associated with quadrupole oscillations of shape.) The idea of quantized lattice vibrations also explains departures from the law of Dulong and Petit at low temperatures [7]. The specific heat of a solid, c_v, remains close to Dulong and Petit's constant value $3N_A k_B$ (where N_A is Avogadro's number) provided the temperature T is sufficiently high, but decreases towards zero as T approaches zero kelvin, as shown in Figure 5.7. Dulong and Petit's value can be obtained by assigning an average energy $\langle E \rangle = k_B T$ to each normal mode of vibration of the lattice (which is just the same as the average energy of each oscillator in the classical calculation of the black body spectrum). But this is incorrect if the vibrational energy is quantized, and the average energy per mode should be given by Planck's expression (1.3):

$$\langle E \rangle = \frac{hv}{e^{hv/k_B T} - 1}$$

This is the essential ingredient for a correct description of specific heats, and the data can be fitted extremely well, provided a sophisticated

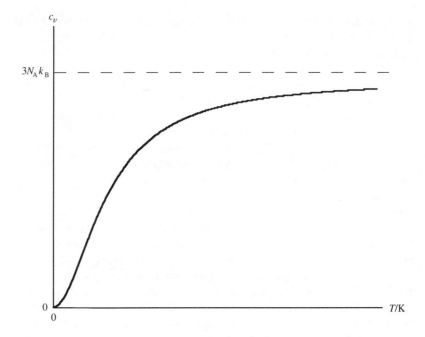

Figure 5.7 Typical dependence of the specific heat of a solid on temperature. The Dulong and Petit value $3N_A k_B$ is approached at sufficiently high T.

enough method is used to assess the number of normal modes and their distribution with frequency.

The vibrational energy of a lattice is identified in classical physics as elastic energy of the solid, and disturbances in the vibrational state propagate as elastic waves (for example sound waves). The quantization of the energy of the normal modes of the lattice implies that the elastic waves carry energy in quanta, just as Einstein suggested in the case of electromagnetic waves (Section 1.3). The quanta of elastic energy are called **phonons**, in analogy with the term 'photon' for a quantum of electromagnetic energy. According to Einstein's Assumption B (Section 1.3), each photon with frequency ν carries just one unit of energy $h\nu$, and similarly each phonon associated with a lattice vibration at frequency ν carries one unit of energy $h\nu$. So if the lattice is in the nth energy level associated with the normal mode of frequency ν, with energy determined by (5.15), the number of phonons present is n, and the total energy of that normal mode is the zero-point energy plus the energy of n phonons. In other words, the quantum number n denoting the energy level of the single oscillator is reinterpreted when we deal with many-body oscillations, and becomes the number of energy quanta of a particular frequency in the many-body system. The ladder operators that we used to derive the oscillator spectrum are also reinterpreted, and become operators that create or destroy one phonon.

If each photon carries one unit of energy hv, the energy spectrum of monochromatic electromagnetic radiation is also an oscillator spectrum of the form (5.15). (In classical electromagnetic theory the field in a plane wave is expanded in terms of Fourier components, each of which varies with position and time in the same way as the amplitude of a single normal mode in the vibration of a many-body system.) Just as in the case of lattice vibrations, the ladder operators associated with the electomagnetic oscillator-like spectrum can be reinterpreted as operators that create and destroy one photon.

The crucial characteristics of the operators $\hat{A}$ and $\hat{A}^{\dagger}$ of (5.4) that make them ladder operators are first their commutation relation (5.6) and secondly their relationship to the Hamiltonian (5.5). Let us define a corresponding pair of **annihilation** and **creation operators** $\hat{a}$ and $\hat{a}^{\dagger}$, which respectively destroy and create a quantum of energy (a phonon or a photon). We define the symbol $|n\rangle$ to represent a state containing n quanta of frequency v. The operators $\hat{a}$ and $\hat{a}^{\dagger}$ must act on the state $|n\rangle$ to produce new states containing respectively $n-1$ and $n+1$ quanta of the same frequency. To act in this way, $\hat{a}$ and $\hat{a}^{\dagger}$ must satisfy a commutation relation corresponding to (5.6):

$$[\hat{a}, \hat{a}^{\dagger}] = 1 \qquad (5.37)$$

and the total energy, of either the lattice oscillations or the monochromatic electromagnetic radiation, must be given by the Hamiltonian

$$\hat{H} = (\hat{a}^{\dagger}\hat{a} + \tfrac{1}{2})hv \qquad (5.38)$$

(In the electromagnetic case this refers to radiation in a particular state of polarization, as well as at a particular frequency.)

The energy eigenvalue equation for the Hamiltonian (5.38) can be written as

$$(\hat{a}^{\dagger}\hat{a} + \tfrac{1}{2})hv|n\rangle = E_n|n\rangle \qquad (5.39)$$

The commutation relation (5.37) guarantees that the operators $\hat{a}$ and $\hat{a}^{\dagger}$ act as ladder operators, and (5.39) can be solved in the same way as the energy eigenvalue equation for the oscillator in Section 5.1. As before, we assume that there is a ground state with non-negative energy, and obtain the same spectrum $E_n = (n+\tfrac{1}{2})hv$ ($n = 0, 1, 2, 3, \ldots$). Since in this case the integer n represents the number of quanta in the state $|n\rangle$, we can define a phonon or a photon **number operator** $\hat{N}$:

$$\hat{N} = (\hat{H} - \tfrac{1}{2})/hv = \hat{a}^{\dagger}\hat{a} \qquad (5.40)$$

Instead of solving the energy eigenvalue equation (5.39), we could solve the eigenvalue equation for the number operator $\hat{N}$ (5.40), with the requirement that the eigenvalues of $\hat{N}$ be non-negative. The spectrum of $\hat{N}$ is the set of integers $n = 0, 1, 2, 3, \ldots$, and the energy eigenstates $|n\rangle$ are also eigenstates of the number operator.

We have treated the normal modes of a lattice as quantum oscillations, and derived the energy spectrum using creation and annihilation operators corresponding to the single-oscillator ladder operators. But what is the significance of the oscillator wave function in this picture? What does the coordinate x refer to? It cannot be the position of any individual atom in the lattice, nor of the centre of mass of the solid (since the thermal motion is not an oscillation of the piece of material as a whole, but is due to internal vibrations occurring even when the solid itself is at rest). Similarly, in the case of electromagnetic radiation there is a problem if we try to find an interpretation of the oscillator wave function. This accounts for our use of $|n\rangle$ instead of the wave function $\psi_n(x)$ to represent the energy eigenstates – unlike $\hat{A}$ and $\hat{A}^{\dagger}$, the operators $\hat{a}$ and $\hat{a}^{\dagger}$ do not act on functions of x, nor can their definitions depend on the spatial variable. This difficulty signals a fundamental limitation on the analogy we have been developing between the energy eigenstates of a single non-relativistic quantum oscillator and those of lattice vibrations or electromagnetic radiation. Although we have been able to identify the correct energy spectrum and to define annihilation and creation operators, a full and rigorous discussion of energy quanta requires the use of **quantum field theory**.

References

[1] Synge J.L. and Griffith B.A. (1959). *Principles of Mechanics* 3rd edn, Chap. 17. New York: McGraw-Hill.

[2] Boas M.L. (1983). *Mathematical Methods in the Physical Sciences* 2nd edn, Chap. 12, Section 22. New York: Wiley.

[3] For further reading see Hollas J.M. (1992). *Modern Spectroscopy* 2nd edn, Chap 6. Chichester: Wiley.

[4] Herzberg G. (1950). *Spectra of Diatomic Molecules*. New York: Van Nostrand.

[5] Blin-Stoyle R.J. (1991). *Nuclear and Particle Physics*, Section 4.3. London: Chapman & Hall.

[6] Lederer C. and Shirley V.S., eds (1978). *Tables of Isotopes* 7th edn. New York: Wiley.

[7] Omar M.A. (1975). *Elementary Solid State Physics*, Section 3.4. Reading MA: Addison-Wesley.

Problems

*5.1 Use the program *Harmonic Oscillator* to investigate the energy eigenstates of the one-dimensional simple harmonic oscillator:

(a) Verify that there are no energy eigenvalues other than those given by (5.15). For the default parity (even) and energy ($\epsilon = 0.4$ in units of $\hbar\omega$) obtain a plot of the wave function (the solution of (5.3) for that value of ϵ) and see that it diverges outside the classically allowed region. Change the energy to 0.6, and see that the wave function diverges in the opposite way. Changing ϵ by 0.02 at each step, examine the wave function in the range between 0.4 and 0.6 to verify that the boundary conditions are only correctly satisfied at $\epsilon = 0.5$. Now increase the energy in steps of 0.1, and check that there are no values for which the boundary conditions are satisfied (for even parity) between 0.5 and 2.5. You can repeat the process for higher even-parity states, and for odd-parity states.

(b) Verify that the number of zeros inside the classically allowed region is n for the nth eigenfunction, and that the parity eigenvalue of the nth level is $(-1)^n$.

(c) Examine the probability distribution in a range of eigenstates (for example $n = 0, 1, 5, 10, 20$ and 30). Measure roughly the distance that the distribution penetrates into the classically forbidden region $|x| > L_n$ in each of these states. (Hold a ruler against the screen, and find the penetration distance as a fraction of L_n.) Measure also the mean separation between zeros of the distribution as a fraction of L_n. Prove analytically that this should decrease approximately as $2/\sqrt{n}$. Note how the shape of the envelope of the distribution changes as n is increased (compare with Figures 5.2 and 5.3).

5.2 Show, by direct substitution into (5.3), that each of the following functions is an energy eigenfunction for the one-dimensional simple harmonic oscillator, and deduce the corresponding eigenvalues:

$$\psi_0 = C_0 e^{-\xi^2/2}, \qquad \psi_1 = C_1 2\xi\, e^{-\xi^2/2}, \qquad \psi_2 = C_2(4\xi^2 - 2)\, e^{-\xi^2/2}$$

Normalize each of these functions and show that they are orthogonal to one another. (The relations

$$\int_{-\infty}^{\infty} d\xi\, \xi^n e^{-\xi^2} = 2(n-1)\int_{-\infty}^{\infty} d\xi\, \xi^{n-2}\, e^{-\xi^2} \quad (n \geqslant 2)$$

$$\int_{-\infty}^{\infty} d\xi\, e^{-\xi^2} = \sqrt{\pi}$$

may be used.)

5.3 (a) Following the method used in the text to demonstrate that $\hat{A}$ is a ladder operator for the simple harmonic oscillator, show that $\hat{A}^\dagger$ is also a ladder operator. (In this case start by using $\hat{A}^\dagger$ on (5.8).)

(b) Work through the derivation of (5.5) from (5.4), and deduce the commutator of the ladder operators, (5.6). Prove that

$$\hat{X}^2 = x^2 = \frac{\hbar}{2m\omega} [(\hat{A}^\dagger)^2 + \hat{A}^2 + 2\hat{A}^\dagger\hat{A} + 1]$$

and

$$\hat{P}_x^2 = -\tfrac{1}{2}\hbar m\omega [(\hat{A}^\dagger)^2 + \hat{A}^2 - 2\hat{A}^\dagger\hat{A} - 1]$$

(c) Use the commutator (5.6) to prove that $[\hat{A}^\dagger\hat{A}, \hat{A}^\dagger] = \hat{A}^\dagger$ and that $[\hat{A}^\dagger\hat{A}, \hat{A}] = -\hat{A}$.

5.4   (a) Use the definition (5.4) of the ladder operators to generate the un-normalized energy eigenfunctions for the $n = 2$ and $n = 4$ states from $\psi_3 = C_3(8\xi^3 - 12\xi)e^{-\xi^2/2}$. Assuming the expression given in (5.28) for the normalization constant, show that, in the particular case $n = 3$, (5.29) and (5.30) are satisfied.

(b) Check that the expressions for the Hermite polynomials H_3 and H_4 given in Table 5.1 satisfy the recurrence relation (5.24). Use (5.25) to generate H_5, and the the eigenfunction ψ_5.

(c) Using (5.29) and (5.30), evaluate the integrals

$$\int_{-\infty}^{\infty} dx\, \psi_n^* \hat{A}^\dagger \psi_k, \qquad \int_{-\infty}^{\infty} dx\, \psi_n^* \hat{A} \psi_k$$

for $k = n + 1$, $k = n - 1$ and $k \neq n \pm 1$.

5.5   Treat a simple pendulum of mass 0.001 kg and period 1 s quantum mechanically, and calculate its zero-point energy. If it is oscillating with amplitude 0.01 m, use classical mechanics to find its energy. If this energy is an eigenvalue E_n, estimate the value of n.

5.6   From the position of the broad vibrational spectral line of hydrogen chloride, centred on 2886 cm^{-1}, calculate the zero-point energy of the H ^{35}Cl molecule. (See Problem 4.10 for the mass of the molecule containing this isotope of chlorine.) Calculate the amplitude of a classical oscillator with this energy. (For comparison, the mean separation of the atoms in the molecule was calculated in Problem 4.10.)

5.7   A monoenergetic beam of electrons with kinetic energy E_0 is passed through carbon monoxide gas in a chamber, and the energies of the electrons that emerge is measured. The number of electrons emerging per unit time that have lost an amount of energy ΔE is $n(\Delta E)$. If $n(\Delta E)$ is plotted against the energy loss ΔE, a very large peak is observed at $\Delta E = 0$, followed by a set of smaller peaks with equal spacing 0.266 eV. Explain the significance of these peaks. (*Hint*: remember that the vibrational spectrum of CO is a broad line centred on 2146 cm^{-1}.)

5.8   Draw a sketch to illustrate the vibrational energy levels of a diatomic molecule, for $n = 0$ and $n = 1$, showing the fine structure due to rotational levels from $l = 0$ to 4. Indicate on your diagram the electric dipole transitions that produce

(a) the purely rotational millimetre and submillimetre spectrum;
(b) the vibrational–rotational lines in the fundamental band.

5.9   Raman scattering (see Problem 4.11) can occur when light changes the vibrational state of a molecule from which it is scattered. If such scattering is due only to transitions between adjacent vibrational levels, show that the spectrum of the scattered light should show a line on either side of the Rayleigh line (which corresponds to scattering with no change in the internal state of the molecule). Estimate the separation of these vibrational Raman lines from the Rayleigh line in the scattering of light by carbon monoxide, and compare it with the corresponding separation in the case of rotational Raman lines. (You will need to use Table 4.2.)

*5.10 Use the program *Anharmonic Oscillator* to find the first few energy eigenvalues of a particle in the potential

$$\hat{V} = \tfrac{1}{2}m\omega^2\left(x^2 + C\frac{m\omega}{\hbar}x^4\right) = \tfrac{1}{2}\hbar\omega(\xi^2 + C\xi^4)$$

for different values of the parameter C: try $C = 0.0001, 0.01, 0.1, 0.5$ and 1. Compare each of the eigenvalues, and the shape of the corresponding eigenfunction and probability distribution, with those for the harmonic oscillator $(C = 0)$.

5.11   The Hamiltonian for a two-dimensional simple harmonic oscillator is

$$\hat{H} = -\frac{\hbar^2}{2m}\left(\frac{\partial^2}{\partial x^2} + \frac{\partial^2}{\partial y^2}\right) + \tfrac{1}{2}m\omega^2(x^2 + y^2)$$

If $\psi_n(x)$ and $\psi_n(y)$ are energy eigenfunctions of a one-dimensional simple harmonic oscillator with the same mass and frequency, prove that $\Psi(x, y) = \psi_j(x)\psi_k(y)$ is an eigenfunction of $\hat{H}$ and find the corresponding eigenvalue. Show that the energy spectrum of the two-dimensional simple harmonic oscillator is $E_n = (n + 1)\hbar\omega$ $(n = 0, 1, 2, \ldots)$, and that each energy level except the lowest is degenerate. Find the pair of numbers (j, k) for each different product $\psi_j(x)\psi_k(y)$ that corresponds to the same energy eigenstate of the two-dimensional oscillator, in each of the cases $n = 1, 2, 3$ and 4.

5.12   In a three-dimensional lattice the density of vibrational modes corresponding to oscillations along one direction is $g(v) = (4\pi/c_s^3)v^2$ per unit volume and per unit frequency interval, where c_s is the velocity of sound (phonon velocity) in the lattice. (Compare (1.2) for the density of modes of electromagnetic radiation in a cavity.) Making the assumptions that the speed of sound is constant, that there is a maximum frequency v_D, and that vibrations can occur along all three axes, show that the average vibrational energy per unit volume of the lattice at temperature T is

$$\langle E \rangle_L = \frac{12h\pi}{c_s^3}\int_0^{v_D}dv\,\frac{v^3}{e^{hv/k_BT} - 1}$$

Show that the specific heat $c_v = \partial\langle E\rangle_L/\partial T$ is

$$\frac{12\pi k_B^4}{h^3 c_s^3}T^3\int_0^{x_D}dx\,\frac{x^4 e^x}{(e^x - 1)^2}$$

where $x_D = h\nu_D/k_B T$. This is the **Debye model**. Compare these results with those of the **Einstein model**, in which the molecules in the solid are assumed to vibrate independently (Problem 1.3).

5.13 (a) Find the spectrum of the number operator $\hat{N} = \hat{a}^\dagger \hat{a}$, assuming that its eigenvalues cannot be negative and that $[\hat{a}, \hat{a}^\dagger] = 1$. (*Hint*: write down the eigenvalue equation $\hat{N}|\alpha\rangle = \alpha|\alpha\rangle$, and show that $\hat{a}$ and $\hat{a}^\dagger$ act as ladder operators for this equation.)

(b) Creation and annihilation operators $\hat{a}_f$ and $\hat{a}_f^\dagger$ can be defined for fermions, such as electrons. $\hat{a}_f^\dagger$ creates a free fermion in a particular energy and momentum eigenstate, and $\hat{a}_f$ annihilates a fermion in that particular state. Instead of a commutation relation like that satisfied by phonon and photon creation and annihilation operators, the fermion operators satisfy $\hat{a}_f \hat{a}_f^\dagger + \hat{a}_f^\dagger \hat{a}_f = 1$. Show that the fermion number operator $\hat{N}_f = \hat{a}_f^\dagger \hat{a}_f$ satisfies

$$\begin{cases} \hat{a}_f^\dagger \hat{N}_f = (1 - \hat{N}_f)\hat{a}_f^\dagger \\ \hat{a}_f N_f = (1 - \hat{N}_f)\hat{a}_f \end{cases}$$

Hence prove that if one eigenvalue of $\hat{N}_f$ is $n = 0$, there is only one other eigenvalue, $n = 1$. (This means that there cannot be more than one fermion in the particular state associated with the operators $\hat{a}_f$ and $\hat{a}_f^\dagger$, which is the **Pauli exclusion principle**.)

Chapter 6

Ladder Operators: Angular Momentum

6.1 The ladder operator method for the angular momentum spectrum

Although we have already discussed orbital angular momentum in Chapter 4, angular momentum is such an important variable, in quantum as well as in classical mechanics, that it is worth examining it further. In Section 4.5 the eigenvalue equations for orbital angular momentum were expressed as the differential equations (4.33) and (4.38), whose solutions, subject to appropriate boundary conditions, gave the spectrum shown in (4.42) and (4.43). But it is also possible to find the angular momentum spectrum by a ladder operator method similar to that used to solve the energy eigenvalue problem for the simple harmonic oscillator in Section 5.1. This approach turns out to be more general than that of Chapter 4, and provides a way of describing spin. As we mentioned in Section 1.6, spin is a characteristic of the electron and other elementary particles that, unlike orbital angular momentum, has no classical counterpart.

The identification of ladder operators for the angular momentum spectrum depends on the commutation relations (4.10) satisfied by the Cartesian components of the angular momentum operator. (In Section 4.2 we derived these relations using the particular differential form for the angular momentum operators given in (4.8).) In this chapter we shall use $\hat{\boldsymbol{J}}$ rather than $\hat{\boldsymbol{L}}$ to represent the angular momentum operator (for reasons that will become clear in Section 6.2). Remember that the operator was defined to be dimensionless, and it is really $\hbar \hat{\boldsymbol{J}}$ that represents the angular momentum. We rewrite (4.10) as

$$[\hat{J}_x, \hat{J}_y] = i\hat{J}_z \qquad \text{and cyclic permutations} \qquad \textbf{(6.1)}$$

and put

$$\hat{J}^2 = \hat{J}_x^2 + \hat{J}_y^2 + \hat{J}_z^2 \qquad \textbf{(6.2)}$$

The components of $\hat{\boldsymbol{J}}$ are Hermitian operators, in accordance with Postulate 2 (Section 3.1), and since all the eigenvalues of a Hermitian operator are real, the eigenvalues of each of $\hat{J}_x^2$, $\hat{J}_y^2$ and $\hat{J}_z^2$, must be ≥ 0.

In Section 4.2 we deduced that $\hat{L}^2$ commutes with each component of $\hat{\boldsymbol{L}}$, (4.11). In exactly the same way, we can use (6.1) and (6.2) to show that

$$[\hat{J}^2, \hat{J}_i] = 0 \quad (i = x, y, z) \tag{6.3}$$

Following the argument of Section 4.2, the commutation relations (6.1) and (6.3) indicate that we should look for simultaneous eigenstates of $\hat{J}^2$ and of one component, $\hat{J}_z$ say. So we write eigenvalue equations corresponding to (4.13) and (4.14):

$$\hat{J}^2 \chi_{\alpha\beta} = \alpha^2 \chi_{\alpha\beta} \tag{6.4}$$

$$\hat{J}_z \chi_{\alpha\beta} = \beta \chi_{\alpha\beta} \tag{6.5}$$

where $\chi_{\alpha\beta}$ represents an eigenstate of angular momentum, corresponding to the eigenvalue α^2 of $\hat{J}^2$ and the eigenvalue β of $\hat{J}_z$. The operators that we shall identify as ladder operators are

$$\hat{J}_\pm = \hat{J}_x \pm i\hat{J}_y \tag{6.6}$$

Because of (6.1) and (6.3), $\hat{J}_+$ and $\hat{J}_-$ satisfy the commutation relations

$$[\hat{J}_z, \hat{J}_\pm] = \pm\hat{J}_\pm \tag{6.7}$$

$$[\hat{J}^2, \hat{J}_\pm] = 0 \tag{6.8}$$

First we multiply (6.4) on the left by $\hat{J}_+$ or $\hat{J}_-$:

$$\hat{J}_\pm \hat{J}^2 \chi_{\alpha\beta} = \alpha^2 \hat{J}_\pm \chi_{\alpha\beta}$$

Then (6.8) allows us to exchange the order of $\hat{J}^2$ and $\hat{J}_\pm$:

$$\hat{J}^2 \hat{J}_\pm \chi_{\alpha\beta} = \alpha^2 \hat{J}_\pm \chi_{\alpha\beta} \tag{6.9}$$

This is an eigenvalue equation of exactly the same form as (6.4), with the same eigenvalue α^2, but with the eigenstate represented by $\hat{J}_+\chi_{\alpha\beta}$ or $\hat{J}_-\chi_{\alpha\beta}$. All the different states that can be obtained by successive applications of $\hat{J}_+$ or $\hat{J}_-$ to $\chi_{\alpha\beta}$ are also eigenstates of $\hat{J}^2$, corresponding to the *same* eigenvalue α^2. The only other possibility is that the effect of $\hat{J}_+$ (or $\hat{J}_-$) on a state gives zero, and in that case no further eigenstates can be generated.

Now we multiply (6.5) by $\hat{J}_+$ or $\hat{J}_-$:

$$\hat{J}_\pm \hat{J}_z \chi_{\alpha\beta} = \beta \hat{J}_\pm \chi_{\alpha\beta}$$

Equation (6.7) can be used to reverse the order of the operators on the left-hand side, giving

$$(\hat{J}_z \hat{J}_\pm \mp \hat{J}_\pm)\, \chi_{\alpha\beta} = \beta \hat{J}_\pm \chi_{\alpha\beta}$$

or

$$\hat{J}_z \hat{J}_{\pm} \chi_{\alpha\beta} = (\beta \pm 1) \hat{J}_{\pm} \chi_{\alpha\beta} \tag{6.10}$$

This is an eigenvalue equation of the same form as (6.5), but with a new eigenstate corresponding to a new eigenvalue of $\hat{J}_z$. Equations (6.9) and (6.10) show that, provided the action of $\hat{J}_+$ does not make the eigenstate $\chi_{\alpha\beta}$ vanish, it produces a new eigenstate of $\hat{J}^2$ and $\hat{J}_z$, $\chi_{\alpha,\beta+1} = \hat{J}_+ \chi_{\alpha\beta}$, corresponding to the original eigenvalue α^2 of $\hat{J}^2$ and an eigenvalue $\beta + 1$ of $\hat{J}_z$. $\hat{J}_+$ is therefore a ladder operator that leaves the magnitude of the angular momentum unchanged, but steps the eigenvalue of $\hat{J}_z$ up by one unit. Similarly, $\hat{J}_-$ is a ladder operator that leaves the quantum number α unchanged, but steps the eigenvalue of $\hat{J}_z$ down by one unit. As in the case of the simple harmonic oscillator, successive applications of these ladder operators generate a set of states whose $\hat{J}_z$ eigenvalues form a discrete spectrum with unit spacing between each pair of levels. For a fixed eigenvalue α^2 of $\hat{J}^2$, if one eigenvalue of $\hat{J}_z$ is β_0 then the spectrum of $\hat{J}_z$ eigenvalues is of the form

$$\beta = \ldots, \beta_0 - 2, \beta_0 - 1, \beta_0, \beta_0 + 1, \beta_0 + 2, \ldots \tag{6.11}$$

In the simple harmonic oscillator case the eigenvalue spectrum was uniquely determined by imposing a lower limit on the energy. In the case of the spectrum of $\hat{J}_z$ there is an upper as well as a lower limit, which we can derive by examining the form of $\hat{J}^2$ given by (6.2). If $\hat{J}$ were an ordinary vector instead of a vector operator, this equation would imply that the length of a component cannot be greater than the length of the vector itself: $\hat{J}_z^2 \leqslant \hat{J}^2$. For a vector operator this means that the inequality applies to the eigenvalues:

$$\beta^2 \leqslant \alpha^2 \tag{6.12}$$

This implies that $-\alpha \leqslant \beta \leqslant +\alpha$, so that there is a maximum eigenvalue β_{max} and a minimum eigenvalue β_{min}. For the orbital angular momentum spectrum (4.42) and (4.43) the equality in (6.12) is never quite reached (except when $l = 0$), so we shall not prejudge the issue by assuming that β_{max} and β_{min} are equal to α and $-\alpha$ respectively. All we require is that the eigenvalues of $\hat{J}_z$ lie in the range determined by (6.12).

For a fixed value of α, let the eigenstates corresponding to β_{max} and β_{min} be $\chi_{\alpha,max}$ and $\chi_{\alpha,min}$ respectively. Using $\hat{J}_+$ on $\chi_{\alpha,max}$ or $\hat{J}_-$ on $\chi_{\alpha,min}$ would produce a state with an unacceptable eigenvalue of $\hat{J}_z$, violating the inequality (6.12). The only alternative is for the result in each case to be zero:

$$\hat{J}_+ \chi_{\alpha,max} = 0, \qquad \hat{J}_- \chi_{\alpha,min} = 0 \tag{6.13}$$

To derive β_{max} and β_{min} from these equations, we need the identities

$$\left. \begin{array}{l} \hat{J}_+ \hat{J}_- = \hat{J}_x^2 + \hat{J}_y^2 - i[\hat{J}_x, \hat{J}_y] = \hat{J}^2 - \hat{J}_z^2 + \hat{J}_z \\ \hat{J}_- \hat{J}_+ = \hat{J}_x^2 + \hat{J}_y^2 + i[\hat{J}_x, \hat{J}_y] = \hat{J}^2 - \hat{J}_z^2 - \hat{J}_z \end{array} \right\} \tag{6.14}$$

where we have used the definitions (6.2) and (6.6) and the commutation relation (6.1). Multiplying the first of (6.13) on the left by $\hat{J}_-$ and using the eigenvalue equations (6.4) and (6.5), we obtain

$$\hat{J}_-\hat{J}_+\chi_{\alpha,\text{max}} = (\hat{J}^2 - \hat{J}_z^2 - \hat{J}_z)\chi_{\alpha,\text{max}} = (\alpha^2 - \beta_{\text{max}}^2 - \beta_{\text{max}})\chi_{\alpha,\text{max}} = 0$$

This can only be satisfied for non-zero $\chi_{\alpha,\text{max}}$ by

$$\alpha^2 = \beta_{\text{max}}(\beta_{\text{max}} + 1) \qquad (6.15)$$

Similarly, using $\hat{J}_+$ on the second of (6.13), we obtain

$$\alpha^2 = \beta_{\text{min}}(\beta_{\text{min}} - 1) \qquad (6.16)$$

The last two equations are satisfied by

$$\beta_{\text{max}} = -\beta_{\text{min}} = j, \qquad j(j + 1) = \alpha^2 \qquad (6.17)$$

So, from (6.11), the spectrum of $\hat{J}_z$ for a given value of the quantum number j is

$$\beta = -j, -j + 1, \ldots, j - 1, j \qquad (6.18)$$

Starting from the lowest eigenstate of $\hat{J}_z$, all the higher eigenstates can be generated by using the ladder operator $\hat{J}_+$ an appropriate number of times. Suppose that n steps take us from $\chi_{\alpha,\text{min}}$ to $\chi_{\alpha,\text{max}}$. Then there are n eigenstates above the lowest one, making $n + 1$ states in all, and the eigenvalue of $\hat{J}_z$ in the highest state is n units greater than in the lowest state:

$$\beta_{\text{max}} = \beta_{\text{min}} + n \quad (n = 0, 1, 2, 3, \ldots)$$

or, from (6.17),

$$2j = n \quad (n = 0, 1, 2, 3, \ldots)$$

Therefore the spectrum (6.18) contains $2j + 1$ values of β, and then the allowed eigenvalues of $\hat{J}^2$ are

$$\alpha^2 = j(j + 1) \quad (j = 0, \tfrac{1}{2}, 1, \tfrac{3}{2}, 2, \ldots) \qquad (6.19)$$

Note that, according to (6.18), if j is an integer then β must also be an integer or zero, whereas if j is a half-integer then so is β.

When j is zero or an integer, (6.18) and (6.19) reproduce the orbital angular momentum spectrum (4.43) and (4.42). But the spectrum of (6.19) also includes half-integer values for the angular momentum quantum number, so our treatment of orbital angular momentum in Chapter 4 must have imposed an extra constraint that eliminated these. In fact, it was the condition (4.36), that the eigenfunction of $\hat{L}_z$ should be single-valued, that restricted β, and thus the quantum number l, to zero or integer values. Originally it was believed that the half-integer quantum numbers were not suitable for describing states of a physical particle, but, as we shall see in Section 6.2, the electron has certain intrinsic properties that can only be described in terms of an angular momentum character-ized by the quantum number $j = \frac{1}{2}$, called its **spin**.

6.2 Electron spin

Classical electromagnetic theory tells us that any rotating charged particle will produce a magnetic field like that due to a magnetic dipole, and that the corresponding magnetic dipole moment $\boldsymbol{\mu}$ is proportional to the angular momentum of the rotation. As we shall show in Section 8.1, the magnetic moment operator associated with the orbital angular momentum $\hbar\hat{\boldsymbol{L}}$ of an electron is $\hat{\boldsymbol{\mu}} = -\mu_B\hat{\boldsymbol{L}}$, where μ_B is the appropriate unit of magnetic moment, called the **Bohr magneton**. In Section 4.5 we found that one consequence of the quantization of angular momentum is that the angle made by the orbital angular momentum vector with a fixed axis in space is limited to certain discrete values, and for a state characterized by the quantum number l there are $2l + 1$ different allowed values of this angle. The relationship between $\hat{\boldsymbol{\mu}}$ and $\hat{\boldsymbol{L}}$ means that the orientation of the magnetic moment is restricted in the same way. Since the value of l is zero or an integer for orbital angular momentum, $2l + 1$ is odd, and so the number of possible orientations of the magnetic moment due to orbital motion relative to the fixed axis is also odd. However, experi-mental evidence, which we shall examine in Chapter 8, indicates that, *in addition* to the magnetic moment associated with its rotational motion in space, an electron has an intrinsic magnetic dipole moment with just two possible orientations. Even if the electron were at rest, it would produce a magnetic field through this intrinsic magnetic moment. We are used to thinking of the electron as a structureless point charge, but it also behaves as if it were a point magnetic dipole!

There is no way of avoiding the restriction to zero or integer values of the quantum numbers characterizing an angular momentum state associ-ated with motion in physical space. The condition (4.36) that the wave function should be single-valued (or equivalently, from (4.32), that it should be invariant under the square of the parity operator) *must* apply if the angular momentum operators are the differential operators (4.8) and the angular momentum eigenfunction is a function of the physical angles (θ, ϕ). On the other hand the existence of only two possible orientations for the intrinsic magnetic moment of the electron could be explained if this magnetic moment were proportional to an angular momentum with quantum number $\frac{1}{2}$, and this value was actually found to be a possibility when we solved the angular momentum eigenvalue problem by the ladder operator method. The way out of the difficulty is to assume that the electron has a fundamental property, called its **intrinsic spin** $\hbar\hat{\boldsymbol{S}}$, that

determines its intrinsic magnetic moment. The total magnetic dipole
moment of an electron is the (vector) sum of its intrinsic magnetic
moment and the magnetic moment due to its rotational motion. The spin
is an angular momentum in the sense that it determines the number of
allowed orientations of the intrinsic magnetic moment, but it is independent of the motion of the electron in space. So the spin angular momentum
operators cannot be differential operators of the form given for orbital
angular momentum by (4.8), and correspondingly the spin state of the
electron is *not* represented by a function of physical angles, so the
restriction imposed by (4.36) is not relevant.

The fundamental property of the angular momentum operator $\hat{\boldsymbol{J}}$ that
we used in Section 6.1 to determine its spectrum was the set of commutation relations (6.1). Although we originally derived commutation relations of this form for the particular differential operators (4.8), we shall
have to assume that the spin is a Hermitian operator of a different type
that satisfies the same commutation relations. We consequently *define* the
spin operator to be a linear Hermitian operator whose components satisfy

$$[\hat{S}_x, \hat{S}_y] = i\hat{S}_z \qquad \text{and cyclic permutations} \qquad (6.20)$$

$$\hat{S}^2 = \hat{S}_x^2 + \hat{S}_y^2 + \hat{S}_z^2 \qquad (6.21)$$

Mathematically, the spin operator $\hat{S}$ is just one particular example of the
general angular momentum operator $\hat{\boldsymbol{J}}$, and the orbital angular momentum operator $\hat{\boldsymbol{L}}$ is another. (We use $\hbar\hat{\boldsymbol{J}}$ to mean angular momentum in
general, and $\hbar\hat{S}$ or $\hbar\hat{\boldsymbol{L}}$ when we wish to refer specifically to spin or orbital
angular momentum.) The spin spectrum can be found by the ladder
operator method, just as we did for $\hat{\boldsymbol{J}}$. The possible eigenvalues of $\hat{S}^2$ and
$\hat{S}_z$ are therefore determined by (6.18) and (6.19), but because we want
the number of allowed orientations to be 2, we must choose $2s + 1 = 2$, so
that $s = \frac{1}{2}$ and

$$\alpha^2 = s(s + 1) = \tfrac{1}{2}(\tfrac{1}{2} + 1) = \tfrac{3}{4} \qquad (6.22)$$

$$\beta = m_s = \pm s = \pm\tfrac{1}{2} \qquad (6.23)$$

There are only two eigenstates of $\hat{S}_z$: $\chi_{+1/2}$, corresponding to the
eigenvalue $m_s = +\frac{1}{2}$, and $\chi_{-1/2}$, corresponding to $m_s = -\frac{1}{2}$. The eigenvalue
equations (6.4) and (6.5) become

$$\hat{S}^2\chi_{+1/2} = \tfrac{3}{4}\chi_{+1/2}, \qquad\qquad \hat{S}^2\chi_{-1/2} = \tfrac{3}{4}\chi_{-1/2} \qquad (6.24)$$

$$\hat{S}_z\chi_{+1/2} = +\tfrac{1}{2}\chi_{+1/2}, \qquad\qquad \hat{S}_z\chi_{-1/2} = -\tfrac{1}{2}\chi_{-1/2} \qquad (6.25)$$

It should be emphasized that the spin eigenstates are *not* represented by
functions of physical angles, since spin is not a physical rotation in space.
So $\chi_{+1/2}$ and $\chi_{-1/2}$ are not eigen *functions*. They are sometimes referred to
as **eigenspinors**.

All the sets of eigenfunctions we have encountered so far are normalizable and mutually orthogonal, and if two eigenfunctions are orthogonal,

it means that they are completely independent functions. Similarly, the eigenspinors $\chi_{+1/2}$ and $\chi_{-1/2}$ should be normalizable and independent of one another, but since they are not functions of spatial variables, orthonormality conditions like (2.2) and (2.5), which involve integrals over all space, are not applicable. The corresponding requirements are written

$$\chi_{m_s}^{\dagger}\,\chi_{m_s'} = \delta_{m_s m_s'} \tag{6.26}$$

Here $\chi_{m_s}^{\dagger}$ is the Hermitian conjugate of χ_{m_s}; we shall see exactly what is meant by this in Section 10.4.

In this section we have restricted our attention to the spin of the electron, characterized by the quantum number $s = \frac{1}{2}$. The proton and neutron also have intrinsic magnetic moments with two allowed orientations, so that they too must be assigned spins with $s = \frac{1}{2}$. (Note that the neutron has an intrinsic magnetic moment even though it has no net electric charge.) Some rather more exotic particles that appear in high-energy experiments using accelerators have larger half-integer spins. Yet other particles exist whose intrinsic magnetic moments have an odd number of allowed orientations, so that their spins are integers or zero. But, even though integer quantum numbers can describe orbital angular momentum states, the spin of such a particle is related to an intrinsic magnetic moment, and has nothing to do with the motion of the particle in space. The particles with half-integer spin are called **fermions**, while those with zero or integer spin are **bosons**. Fermions and bosons behave very differently, and in particular fermions obey the Pauli exclusion principle while bosons do not.

6.3 Addition of angular momenta

In Section 6.2 we stated that the total magnetic moment of an electron is the sum of its intrinsic and rotational magnetic moments. The number of allowed orientations of this total magnetic moment is determined by the vector sum of the spin $\hbar\hat{S}$ and the orbital angular momentum $\hbar\hat{L}$, and is called the **total angular momentum** $\hbar\hat{J}$:

$$\hat{J} = \hat{L} + \hat{S} \tag{6.27}$$

In discussing atomic structure, it is often necessary to find the resultant orbital angular momentum of several electrons, or their resultant spin, or to find the total angular momentum either of a single electron, or of the atom as a whole. Consequently we must learn how to add quantum mechanical angular momenta, which are not only vectors but also operators with quantized eigenvalues. The general problem is to find the eigenvalues of the resultant of two angular momentum operators of any

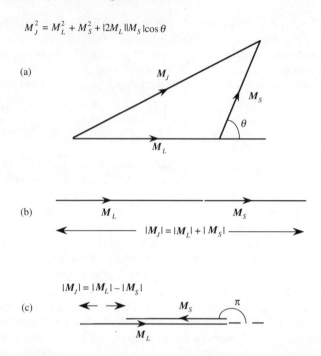

$$M_J^2 = M_L^2 + M_S^2 + |2M_L||M_S|\cos\theta$$

Figure 6.1 The classical addition of two angular momentum vectors, $M_L + M_S = M_J$, when the two vectors are (a) at an arbitrary angle θ; (b) parallel to one another; (c) in opposite directions.

type, $\hat{J} = \hat{J}_1 + \hat{J}_2$. But to simplify the notation and avoid introducing a lot of subscripts, let us consider the particular case of the resultant $\hat{J}$ of an orbital angular momentum $\hat{L}$ and a spin $\hat{S}$, as in (6.27). The results we obtain can easily be generalized to the case of two independent angular momenta of any type, such as the orbital angular momenta or the spins of two different particles.

If we add two classical angular momentum vectors, the resultant depends on the angle between them according to the cosine rule, as shown in Figure 6.1. The magnitude of the resultant is a maximum if the two vectors are parallel, and a minimum when the two vectors are in exactly opposite directions. From the diagram we can see that the maximum (minimum) magnitude of the resultant is the sum (difference) of the magnitudes of the vectors being added. Similarly, for quantum angular momenta we shall find that the maximum and minimum values of the quantum number j of the resultant are respectively the sum $l + s$ and the magnitude of the difference $|l - s|$ of the quantum numbers l and s of the two angular momenta being added. The magnitude of the classical resultant can take any value in the continuous range between the minimum and maximum values, depending on the angle between the vectors. But since the orientation of a quantum mechanical angular momentum is quantized, we shall find that the angle between two angular momenta is similarly restricted, so that the magnitude of the resultant can only take certain discrete values.

First of all we should check that $\hat{\boldsymbol{J}}$ in (6.27) satisfies the right commutation relations to be identified as an angular momentum operator. Equation (6.27) implies that each component of $\hat{\boldsymbol{J}}$ is of the form

$$\hat{J}_i = \hat{L}_i + \hat{S}_i \quad (i = x, y, z) \tag{6.28}$$

We assume that both $\hat{\boldsymbol{L}}$ and $\hat{\boldsymbol{S}}$ obey commutation relations of the form (6.1), so that

$$\begin{aligned}
[\hat{J}_x, \hat{J}_y] &= [\hat{L}_x + \hat{S}_x, \hat{L}_y + \hat{S}_y] \\
&= [\hat{L}_x, \hat{L}_y] + [\hat{S}_x, \hat{S}_y] + [\hat{L}_x, \hat{S}_y] + [\hat{S}_x, \hat{L}_y] \\
&= i\hat{L}_z + i\hat{S}_z + [\hat{L}_x, \hat{S}_y] + [\hat{S}_x, \hat{L}_y] \\
&= i\hat{J}_z + [\hat{L}_x, \hat{S}_y] + [\hat{S}_x, \hat{L}_y]
\end{aligned} \tag{6.29}$$

The last two terms are commutators of two completely different angular momenta: the operators representing $\hat{\boldsymbol{L}}$ are differential operators involving physical angles, while the spin operators do not depend on the spatial variables. Therefore any two components of $\hat{\boldsymbol{L}}$ and $\hat{\boldsymbol{S}}$ commute:

$$[\hat{L}_i, \hat{S}_j] = 0 \quad (i, j = x, y, z) \tag{6.30}$$

So the last two terms in (6.29) vanish, and the components of $\hat{\boldsymbol{J}}$ satisfy the angular momentum commutation relations (6.1). (Exactly the same would be true if we were adding the orbital angular momenta of two different particles, since in that case each operator would be defined in terms of angles associated with the position of one particular particle, and the two sets of angles (θ_1, ϕ_1) and (θ_2, ϕ_2) are totally independent. So the components of the orbital angular momentum operator of one particle, $\hat{\boldsymbol{L}}_1$, commute with all components of that of the other, $\hat{\boldsymbol{L}}_2$. Similarly, if $\hat{\boldsymbol{S}}_1$ and $\hat{\boldsymbol{S}}_2$ are the spin operators for two different particles, the components of one commute with those of the other.)

Since $\hat{\boldsymbol{J}}$ of (6.27) obeys the commutation relations (6.1), the spectrum of eigenvalues must be of the form given in (6.18) and (6.19). But we need to know the quantum numbers j and j_z in terms of the quantum numbers $\{l, m_l\}$ of $\hat{\boldsymbol{L}}$ and $\{s, m_s\}$ of $\hat{\boldsymbol{S}}$. Since $\hat{J}_z = \hat{L}_z + \hat{S}_z$, (6.28), an eigenstate of $\hat{L}_z$ and $\hat{S}_z$ is also an eigenstate of $\hat{J}_z$, with eigenvalue

$$j_z = m_l + m_s \tag{6.31}$$

We can now find all possible values of j_z, and since m_l runs from $-l$ to $+l$ in unit steps, and m_s from $-s$ to $+s$, the maximum and minimum values of j_z are given by

$$j_{z,\text{max}} = l + s = -j_{z,\text{min}} \tag{6.32}$$

Table 6.1

m_l	m_s	j_z
$+2$	$+\frac{3}{2}$	$+\frac{7}{2}$
	$+\frac{1}{2}$	$+\frac{5}{2}$
	$-\frac{1}{2}$	$+\frac{3}{2}$
	$-\frac{3}{2}$	$+\frac{1}{2}$
$+1$	$+\frac{3}{2}$	$+\frac{5}{2}$
	$+\frac{1}{2}$	$+\frac{3}{2}$
	$-\frac{1}{2}$	$+\frac{1}{2}$
	$-\frac{3}{2}$	$-\frac{1}{2}$
0	$+\frac{3}{2}$	$+\frac{3}{2}$
	$+\frac{1}{2}$	$+\frac{1}{2}$
	$-\frac{1}{2}$	$-\frac{1}{2}$
	$-\frac{3}{2}$	$-\frac{3}{2}$
-1	$+\frac{3}{2}$	$+\frac{1}{2}$
	$+\frac{1}{2}$	$-\frac{1}{2}$
	$-\frac{1}{2}$	$-\frac{3}{2}$
	$-\frac{3}{2}$	$-\frac{5}{2}$
-2	$+\frac{3}{2}$	$-\frac{1}{2}$
	$+\frac{1}{2}$	$-\frac{3}{2}$
	$-\frac{1}{2}$	$-\frac{5}{2}$
	$-\frac{3}{2}$	$-\frac{7}{2}$

Equation (6.17) indicates that the corresponding total angular momentum quantum number must be $j = l + s$ (similar to the classical case when the angular momenta to be added are parallel).

The easiest way to find what other values of j are possible is through a specific example: take $l = 2$ and $s = \frac{3}{2}$. (This spin might be the total spin of a three-electron atom, and l its total orbital angular momentum quantum number.) All possible values of j_z, calculated from (6.31), are given in Table 6.1. Note that the maximum and minimum values of j_z are $\pm\frac{7}{2}$, and again (6.17) indicates that a state with total angular momentum quantum number $j = \frac{7}{2} = 2 + \frac{3}{2}$ would have these maximum and minimum eigenvalues of $\hat{J}_z$. There are $2j + 1 = 8$ states with this total angular momentum quantum number, distinguished by their $\hat{J}_z$ eigenvalues, which are $\pm\frac{7}{2}$, $\pm\frac{5}{2}$, $\pm\frac{3}{2}$ and $\pm\frac{1}{2}$. But states other than the eight we have just specified occur in the table, and must correspond to other possible values of j. If we remove from the table the set of eight values of j_z correponding to $j = \frac{7}{2}$, the extreme values of j_z that remain are $\pm\frac{5}{2}$, each occurring once only. These are the maximum and minimum eigenvalues of $\hat{J}_z$ when $j = \frac{5}{2}$, and we remove one each of the six values $j_z = \pm\frac{5}{2}$, $\pm\frac{3}{2}$ and $\pm\frac{1}{2}$ from

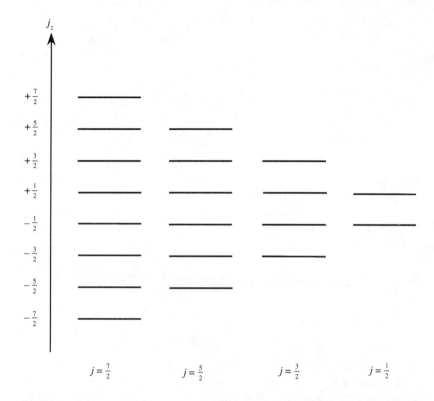

Figure 6.2 All the possible states resulting from the addition of two quantum mechanical angular momenta $\hat{L}$ and $\hat{S}$ with quantum numbers $l = 2$ and $s = \frac{3}{2}$ respectively. The j_z states naturally fall into groups, each of which contains the $2j + 1$ states corresponding to a particular total angular momentum quantum number j, with j ranging from $l + s$ to $l - s$ in integer steps.

the table, to complete the $\hat{J}_z$ spectrum corresponding to this value of j. Now we are left with no j_z value greater than $+\frac{3}{2}$, and the set of values $j_z = \pm\frac{3}{2}, \pm\frac{1}{2}$ corresponds to $j = \frac{3}{2}$. There are only two entries remaining, with $j_z = \pm\frac{1}{2}$, which are the $\hat{J}_z$ eigenvalues associated with $j = \frac{1}{2}$. So we have found eigenvalues of $\hat{J}_z$ corresponding to the following values of the quantum number j:

$$j = \tfrac{7}{2} = 2 + \tfrac{3}{2}$$
$$j = \tfrac{5}{2} = (2 + \tfrac{3}{2}) - 1$$
$$j = \tfrac{3}{2} = (2 + \tfrac{3}{2}) - 2$$
$$j = \tfrac{1}{2} = (2 + \tfrac{3}{2}) - 3 = 2 - \tfrac{3}{2}$$

All the states that can result from the addition that we have performed are shown schematically in Figure 6.2.

The values of j that we have found range from $l + s$ to $l - s$ in integer steps, so that, as we anticipated, the angle between $\hat{L}$ and $\hat{S}$ is restricted to a discrete set of values. (Note that this angle can never be exactly 0 or π, since the magnitudes of the angular momenta are $\sqrt{[l(l+1)]}$ and $\sqrt{[s(s+1)]}$, rather than l and s.) In our example we chose $s < l$. If we had chosen $s > l$ then we should have found the minimum value of j to be $s - l$. So in general the result of adding two angular momenta with quantum numbers l and s is

$$j = l + s, l + s - 1, l + s - 2, \ldots, |l - s| \qquad (6.33)$$

Although the table we constructed allows us to identify the possible resultant states correctly, each line of the table does *not* correspond to a unique value of j, with the exception of the first and last entries, which can only belong to $j = \frac{7}{2}$. For example, the state with $j = \frac{5}{2}$, $j_z = +\frac{5}{2}$, is actually a linear combination of two terms, one corresponding to $m_l = +2$, $m_s = +\frac{1}{2}$ and the other to $m_l = +1$, $m_s = +\frac{3}{2}$. A second, independent, combination of these two terms gives the final state with $j = \frac{7}{2}$, $j_z = +\frac{5}{2}$:

$$\left. \begin{aligned}
\chi_{5/2, +5/2} &= \sqrt{\tfrac{4}{7}}\, Y_{2,+2}\, \chi_{3/2,+1/2} - \sqrt{\tfrac{3}{7}}\, Y_{2,+1}\, \chi_{3/2,+3/2} \\
\chi_{7/2, +5/2} &= \sqrt{\tfrac{3}{7}}\, Y_{2,+2}\, \chi_{3/2,+1/2} + \sqrt{\tfrac{4}{7}}\, Y_{2,+1}\, \chi_{3/2,+3/2}
\end{aligned} \right\} \qquad (6.34)$$

Here χ_{jj_z} and χ_{sm_s} represent eigenstates of the resultant angular momentum and spin respectively, and Y_{lm_l} is the spherical harmonic that is the eigenfunction of orbital angular momentum. (Note that each term on the right-hand side involves the product of an orbital angular momentum eigenfunction and an eigenspinor.) These expressions show that although an eigenstate of $\hat{L}_z$ and $\hat{S}_z$ is an eigenstate of $\hat{J}_z$ (because of (6.31)), the converse is not necessarily true: the two terms in the expansion of $\chi_{5/2,+5/2}$, for example, contain different eigenstates of $\hat{L}_z$ and $\hat{S}_z$. However, both terms are eigenstates of $\hat{L}^2$ and $\hat{S}^2$, with $l = 2$ and $s = \frac{3}{2}$.

In general, an eigenstate of $\hat{J}^2$ and $\hat{J}_z$ is *not* an eigenstate of $\hat{L}_z$ and $\hat{S}_z$, although it may be an eigenstate of $\hat{L}^2$ and $\hat{S}^2$. We can prove this by examining the commutators of $\hat{J}^2$ and $\hat{J}_z$ with $\hat{L}^2$, $\hat{S}^2$, $\hat{L}_z$ and $\hat{S}_z$. Squaring (6.27), we find

$$\hat{J}^2 = \hat{L}^2 + \hat{S}^2 + 2\hat{L} \cdot \hat{S} \qquad (6.35)$$

(We have used the fact that $\hat{L} \cdot \hat{S} = \hat{S} \cdot \hat{L}$, which follows from (6.30).) Both $\hat{L}$ and $\hat{S}$ satisfy a commutation relation of the form (6.3), and so both $\hat{L}^2$ and $\hat{S}^2$ commute with each term in (6.35):

$$[\hat{L}^2, \hat{J}^2] = 0, \qquad [\hat{S}^2, \hat{J}^2] = 0 \qquad (6.36)$$

Similarly, using (6.28) for $\hat{J}_z$, we obtain

$$[\hat{L}^2, \hat{J}_z] = 0, \qquad [\hat{S}^2, \hat{J}_z] = 0 \qquad (6.37)$$

Equations (6.36) and (6.37) show that an eigenstate of $\hat{J}^2$ and $\hat{J}_z$ can simultaneously be an eigenstate of $\hat{L}^2$ and $\hat{S}^2$. However, because of the third term in (6.35), $\hat{L}_z$ and $\hat{S}_z$ do not commute with $\hat{J}^2$:

$$[\hat{L}_z, \hat{\boldsymbol{L}} \cdot \hat{\boldsymbol{S}}] = [\hat{L}_z, \hat{L}_x \hat{S}_x] + [\hat{L}_z, \hat{L}_y \hat{S}_y] + [\hat{L}_z, \hat{L}_z \hat{S}_z] \qquad (6.38)$$

Using (6.30) and commutation relations of the form (6.1) between different components of $\hat{\boldsymbol{L}}$, we find that the last term in (6.38) vanishes, but the first two give the result

$$[\hat{L}_z, \hat{\boldsymbol{L}} \cdot \hat{\boldsymbol{S}}] = i(\hat{L}_y \hat{S}_x - \hat{L}_x \hat{S}_y) \qquad (6.39)$$

Similarly

$$[\hat{S}_z, \hat{\boldsymbol{L}} \cdot \hat{\boldsymbol{S}}] = i(\hat{S}_y \hat{L}_x - \hat{S}_x \hat{L}_y) = -[\hat{L}_z, \hat{\boldsymbol{L}} \cdot \hat{\boldsymbol{S}}] \qquad (6.40)$$

These last two equations show that

$$[\hat{L}_z, \hat{J}^2] = -[\hat{S}_z, \hat{J}^2] \neq 0 \qquad (6.41)$$

This means that an eigenstate of $\hat{J}^2$ cannot simultaneously be an eigenstate of $\hat{L}_z$ or $\hat{S}_z$ separately, though it can be an eigenstate of $\hat{L}_z + \hat{S}_z = \hat{J}_z$.

An eigenstate of the resultant angular momentum can be written as a linear combination of products of eigenstates of the angular momenta that are being added, as we showed for two particular cases in (6.34). In general, we can write an expression of the following form for a simultaneous eigenstate of $\hat{L}^2$, $\hat{S}^2$, $\hat{J}^2$ and $\hat{J}_z$:

$$\chi_{lsjj_z} = \sum_{\substack{m_l \\ m_s = j_z - m_l}} C(j, j_z; l, m_l, s, m_s) \, Y_{lm_l} \chi_{sm_s} \qquad (6.42)$$

where Y_{lm_l} and χ_{sm_s} represent eigenstates of the angular momenta that are being added, and for each set of values of the quantum numbers in the parentheses $C(j, j_z; l, m_l, s, m_s)$ is a number called the **Clebsch–Gordan coefficient**. The values of the Clebsch–Gordan coefficients are uniquely defined by the mathematical properties of the angular momentum operators.

Reference

For a complete discussion of the mathematical properties of angular momentum see Edmonds A.R. (1974). *Angular Momentum in Quantum Mechanics* 2nd edn. Princeton University Press.

Problems

6.1   Prove the commutation relations (6.3), (6.7) and (6.8). Prove the expansions (6.14) for the products $\hat{J}_+\hat{J}_-$ and $\hat{J}_-\hat{J}_+$.

6.2   Use the expressions for the components of the orbital angular momentum operator given in (4.8) to construct the ladder operators for orbital angular momentum:

$$\hat{L}_+ = e^{i\phi}\left(\frac{\partial}{\partial\theta} + i\cot\theta\,\frac{\partial}{\partial\phi}\right)$$

$$\hat{L}_- = e^{-i\phi}\left(-\frac{\partial}{\partial\theta} + i\cot\theta\,\frac{\partial}{\partial\phi}\right)$$

Show, by explicit calculation of the relevant products, that these operators satisfy the commutation relations (6.7) and (6.8). Prove that $[\hat{L}_+, \hat{L}_-] = 2\hat{L}_z$.

6.3   (a)   Using the explicit form for the ladder operators $\hat{L}_\pm$ given in Problem 6.2, with the definition (4.45) of the spherical harmonics, and the associated Legendre functions in Table 4.1, prove that

$$\hat{L}_\pm Y_{10}(\theta, \phi) = \sqrt{2}\,Y_{1,\pm 1}(\theta, \phi), \quad \hat{L}_\pm Y_{1,\pm 1}(\theta, \phi) = 0$$

$$\hat{L}_\mp Y_{1,\pm 1}(\theta, \phi) = \sqrt{2}\,Y_{10}(\theta, \psi)$$

$$\hat{L}_+\hat{L}_- Y_{1,+1}(\theta, \phi) = 2Y_{1,+1}(\theta, \phi), \quad \hat{L}_-\hat{L}_+ Y_{1,-1}(\theta, \phi) = 2Y_{1,-1}(\theta, \phi)$$

$$\hat{L}_+\hat{L}_- Y_{10}(\theta, \phi) = 2Y_{10}(\theta, \phi)$$

(b)   Look for a solution of the form $\psi(\theta, \phi) = f(\theta)e^{2i\phi}$ to the equation $\hat{L}_+\psi(\theta, \phi) = 0$ (see (6.13)). Hence find (to within a normalization constant) the set of orbital angular momentum eigenfunctions corresponding to $l = 2$.

6.4 $\hat{S}_+$ and $\hat{S}_-$ are the ladder operators for spin, defined in a corresponding way to $\hat{J}_+$ and $\hat{J}_-$, (6.6). Prove that

$$\hat{S}_+\hat{S}_- = \hat{S}^2 - \hat{S}_z^2 + \hat{S}_z, \qquad \hat{S}_-\hat{S}_+ = \hat{S}^2 - \hat{S}_z^2 - \hat{S}_z$$

If χ_+ and χ_- are eigenspinors of $\hat{S}^2$ corresponding to the quantum number $s = \frac{1}{2}$, (6.24) and (6.25), prove that

$$\hat{S}_+\chi_+ = 0, \qquad \hat{S}_-\chi_+ = \chi_-, \qquad \hat{S}_+\chi_- = \chi_+, \qquad \hat{S}_-\chi_- = 0$$

(*Hint*: Since $\hat{S}_\pm$ are ladder operators, we know that $\hat{S}_-\chi_+ = c_1\chi_-$ and $\hat{S}_+\chi_- = c_2\chi_+$, where c_1 and c_2 are constants. To find their values, multiply the first expression by $\hat{S}_+$ and the second by $\hat{S}_-$.)

6.5   Using the method applied in the text to the case of angular momenta with quantum numbers 2 and $\frac{3}{2}$, find

(a)   all possible total spin states of a pair of particles each with spin half;

(b) all possible total orbital angular momentum states for a pair of particles in states with $l = 1$ and $l = 2$.

6.6   Show that the number of different quantum numbers j of the total angular momentum $\hat{\boldsymbol{J}} = \hat{\boldsymbol{L}} + \hat{\boldsymbol{S}}$, formed by combining an orbital angular momentum eigenstate with quantum number l and a spin eigenstate with quantum number s, is

$$2s + 1 \quad \text{if } s < l, \qquad 2l + 1 \quad \text{if } s > l$$

Show that the total number of different total angular momentum states is

$$(2l + 1)(2s + 1) = \sum_{j=|l-s|}^{l+s} (2j + 1)$$

6.7   Work through the derivation of (6.40).

6.8   The total angular momentum quantum numbers for a single-electron atom are $j = l + \frac{1}{2}$ and $j = l - \frac{1}{2}$ (unless $l = 0$). The values of the relevant Clebsch–Gordan coefficients $C(j, j_z; l, j_z \mp \frac{1}{2}, \frac{1}{2}, \pm \frac{1}{2})$ are, for $j = l + \frac{1}{2}$

$$C(j, j_z; l, j_z - \tfrac{1}{2}, \tfrac{1}{2}, +\tfrac{1}{2}) = \sqrt{\left(\frac{1}{2} + \frac{j_z}{2l + 1}\right)}$$

$$C(j, j_z; l, j_z + \tfrac{1}{2}, \tfrac{1}{2}, -\tfrac{1}{2}) = \sqrt{\left(\frac{1}{2} - \frac{j_z}{2l + 1}\right)}$$

and for $j = l - \frac{1}{2}$

$$C(j, j_z; l, j_z - \tfrac{1}{2}, \tfrac{1}{2}, +\tfrac{1}{2}) = -\sqrt{\left(\frac{1}{2} - \frac{j_z}{2l + 1}\right)}$$

$$C(j, j_z; l, j_z + \tfrac{1}{2}, \tfrac{1}{2}, -\tfrac{1}{2}) = \sqrt{\left(\frac{1}{2} + \frac{j_z}{2l + 1}\right)}$$

Write down expressions for all possible total angular momentum eigenstates of a one-electron atom with (a) $l = 1$ and (b) $l = 2$.

6.9   Prove that $\hat{\boldsymbol{L}} \cdot \hat{\boldsymbol{S}} = \hat{L}_z \hat{S}_z + \frac{1}{2}(\hat{L}_+ \hat{S}_- + \hat{L}_- \hat{S}_+)$. Hence, using the results of Problems 6.3(a) and 6.4, prove that the effects of $\hat{\boldsymbol{L}} \cdot \hat{\boldsymbol{S}}$ on the states with orbital angular momentum $l = 1$ and spin $s = \frac{1}{2}$, represented by products of the form $Y_{1m}(\theta, \phi)\chi_\pm$, are given by

$$\hat{\boldsymbol{L}} \cdot \hat{\boldsymbol{S}} \, Y_{1,+1}(\theta, \phi)\chi_+ = \tfrac{1}{2} Y_{1,+1}(\theta, \phi)\chi_+$$

$$\hat{\boldsymbol{L}} \cdot \hat{\boldsymbol{S}} \, Y_{1,+1}(\theta, \phi)\chi_- = -\tfrac{1}{2} Y_{1,+1}(\theta, \phi)\chi_- + \sqrt{\tfrac{1}{2}} Y_{10}(\theta, \phi)\chi_+$$

$$\hat{\boldsymbol{L}} \cdot \hat{\boldsymbol{S}} \, Y_{10}(\theta, \phi)\chi_+ = \sqrt{\tfrac{1}{2}} Y_{1,+1}(\theta, \phi)\chi_-$$

$$\hat{\boldsymbol{L}} \cdot \hat{\boldsymbol{S}} \, Y_{10}(\theta, \phi)\chi_- = \sqrt{\tfrac{1}{2}} Y_{1,-1}(\theta, \phi)\chi_+$$

$$\hat{\boldsymbol{L}} \cdot \hat{\boldsymbol{S}} \, Y_{1,-1}(\theta, \phi)\chi_+ = -\tfrac{1}{2} Y_{1,-1}(\theta, \phi)\chi_+ + \sqrt{\tfrac{1}{2}} Y_{10}(\theta, \phi)\chi_-$$

$$\hat{\boldsymbol{L}} \cdot \hat{\boldsymbol{S}} \, Y_{1,-1}(\theta, \phi)\chi_- = \tfrac{1}{2} Y_{1,-1}(\theta, \phi)\chi_-$$

Identify eigenstates of $\hat{J}^2$ and $\hat{J}_z$ (of the general form given in (6.42)), and check that your expressions agree with those you derived in Problem 6.8 in the case $l = 1$. (*Hint*: Use (6.35).)

CHAPTER 7

Symmetry and the Solution of the Schrödinger Equation

7.1 Three-dimensional systems with spherical symmetry

So far we have solved the Schrödinger equation only for particles constrained to move one-dimensionally, along a fixed axis. In general, it is much more difficult to solve the fully three-dimensional equation (3.20), and we must use any clues to the nature of the solutions that the characteristics of the particular physical system we are investigating can provide. The characteristics that are important in this respect are the symmetry properties of the Hamiltonian, determined by the symmetry of the potential in which the particle moves.

We have already mentioned symmetry under parity inversion: when we solved the Schrödinger equation for a particle in a one-dimensional square or simple harmonic potential well we noticed that the energy eigenfunctions are alternately symmetric and antisymmetric about the centre of the well, and in Section 4.4 we attributed this to the symmetry of the potential under parity inversion. Since the kinetic energy operator is also symmetric under parity, the complete Hamiltonian is symmetric. This means that the Hamiltonian commutes with the parity operator, and so the energy eigenfunctions are simultaneously eigenfunctions of parity.

In three dimensions a particle moving in a central potential is an example of a system that is not only symmetric under parity inversion, but also spherically symmetric. A **central potential** $V(r)$ depends only on the separation r of the particle from the centre of the potential, and not on the spherical polar angles θ and ϕ that determine particular directions in space. Such a potential is spherically symmetric, since $V(r)$ is constant over the surface of any sphere centred on the origin, and the magnitude of the constant depends only on the radius r of the sphere.

Many systems of interest in both classical and quantum physics are of this type, or may be treated as spherically symmetric to a first approximation. The force binding an electron in an atom is the Coulomb attraction to the nucleus, and the Coulomb interaction between two charged particles is represented by a central potential, varying as $1/r$. The three-dimensional simple harmonic oscillator potential is also central. In quantum physics it has applications in the study of the vibrational motion of nuclei and of molecules, and, according to one model describing the quark structure of the nucleon, each of the three quarks is bound to each of the other two through a three-dimensional simple harmonic oscillator potential. Another useful central potential is the three-dimensional version of the one-dimensional square well. In this so-called **spherical square well** the potential is constant within a sphere of radius r_0, and takes a different constant value outside. It has been used in simple models describing the binding of a nucleon within a nucleus by the strong nuclear force, and in particular it is the basis for elementary descriptions of the **deuteron** (the bound state of a proton and a neutron). The spherical square well has also been used to describe an electron confined three-dimensionally within a quantum dot (which we mentioned in Section 3.3), though in this case the potential outside the sphere is not constant but depends on the structure of the material within which the dot is embedded.

The Hamiltonian for a particle moving in a spherically symmetric potential is

$$\hat{H} = -\frac{\hbar^2}{2m}\nabla^2 + V(r) \tag{7.1}$$

where [1]

$$\nabla^2 = \frac{\partial^2}{\partial r^2} + \frac{2}{r}\frac{\partial}{\partial r} + \frac{1}{r^2}\left(\frac{\partial^2}{\partial\theta^2} + \cot\theta\frac{\partial}{\partial\theta} + \frac{1}{\sin^2\theta}\frac{\partial^2}{\partial\phi^2}\right) \tag{7.2}$$

The term in parentheses, which involves the angular variables only, is, from (4.9), equal to $-\hat{L}^2$, where $\hat{L}$ is the orbital angular momentum operator. So the Schrödinger equation that we wish to solve is of the form

$$\left[-\frac{\hbar^2}{2m}\left(\frac{\partial^2}{\partial r^2} + \frac{2}{r}\frac{\partial}{\partial r} - \frac{1}{r^2}\hat{L}^2\right) + V(r)\right]\psi(r,\theta,\phi) = E\psi(r,\theta,\phi)$$

$$\tag{7.3}$$

In classical mechanics spherical symmetry of the potential results in the conservation of orbital angular momentum [2]. In quantum mechanics, if the potential is spherically symmetric, the Hamiltonian commutes with the orbital angular momentum operators $\hat{L}^2$ and $\hat{L}_z$. To show this, note that since the expression for $\hat{L}^2$ does not contain r or derivatives with respect to r, $\hat{L}^2$ commutes with the potential $V(r)$, and with the complete Hamiltonian operator on the left-hand side of (7.3). (This would not be the case for a potential that varied with the angles θ and ϕ.) Similarly, $\hat{L}_z$

(defined in (4.8)) commutes with $\hat{L}^2$, (4.11), and with the complete Hamiltonian. Because the orbital angular momentum operators commute with the Hamiltonian, the particle may be in a state that is simultaneously an eigenstate of both energy and angular momentum. This is the quantum mechanical equivalent of angular momentum conservation in classical mechanics.

So, to solve the Schrödinger equation in three dimensions when the potential is central, we look for energy eigenfunctions that are simultaneously orbital angular momentum eigenfunctions. We have already solved the orbital angular momentum eigenvalue problem in Section 4.5, and found that the eigenfunctions are the spherical harmonics $Y_{lm_l}(\theta, \phi)$, (4.45). This determines the angular dependence of a simultaneous eigenfunction of energy and angular momentum, and so we assume that the energy eigenfunctions of (7.3) are of the form

$$\psi(r, \theta, \phi) = R(r)Y_{lm_l}(\theta, \phi) \tag{7.4}$$

Remember that, from (4.42) and (4.43), the eigenvalues of $\hat{L}^2$ and $\hat{L}_z$ are $l(l+1)$ and m_l respectively, where the allowed values of the quantum number l are

$$l = 0, 1, 2, \ldots \tag{7.5}$$

and, for a given l, there are $2l + 1$ possible values of m_l:

$$m_l = -l, -(l-1), \ldots, 0, \ldots, l-1, l \tag{7.6}$$

The angular part of the wave function, $Y_{lm_l}(\theta, \phi)$, does not depend on the shape of the central potential $V(r)$: the orbital angular momentum spectrum is *always* the spectrum found in Section 4.5, characterized by the quantum numbers l and m_l. (It is the same even if the potential is not central, but in that case the particle cannot be in an orbital angular momentum eigenstate when it is in an energy eigenstate, so the energy eigenfunctions cannot be written as simple products of the form (7.4).)

A wave function of the form (7.4) is an eigenfunction of $\hat{L}^2$ corresponding to the eigenvalue $l(l+1)$, so, substituting this form for $\psi(r, \theta, \phi)$ in (7.3) (and cancelling the identical spherical harmonics that appear on each side of the equation), we get the **radial equation**.

$$\left\{ -\frac{\hbar^2}{2m} \left[\frac{d^2}{dr^2} + \frac{2}{r} \frac{d}{dr} - \frac{l(l+1)}{r^2} \right] + V(r) \right\} R(r) = ER(r) \tag{7.7}$$

If we solve this for the **radial wave function** $R(r)$ that describes a bound state, we shall find that, as usual, solutions satisfying the boundary conditions that $R(r)$ be finite for all r exist only for a particular set of discrete energy eigenvalues. So the energy spectrum is determined by the radial equation, and will clearly depend on the particular shape of the potential $V(r)$, as we might expect.

Since l appears in (7.7) but m_l does not, the radial wave function $R(r)$

and the energy eigenvalue E may depend on the eigenvalue of $\hat{L}^2$ but not on that of $\hat{L}_z$. This is to be expected for a spherically symmetric system: if the energy of the particle were to depend on a particular component of the angular momentum, $\hat{L}_z$, then the z axis would be distinguished from other directions in space by physical effects – the energy would change with the orientation of the angular momentum vector. In a spherically symmetric system all directions in space are equivalent, so there can be no physical means of distinguishing one direction from another, and no m_l dependence of the energy eigenvalues.

7.2 The hydrogen atom

As an example of the solution of the three-dimensional Schrödinger equation with a spherically symmetric potential, we shall look at the Coulomb interaction between two electrically charged particles. This problem is very important because it forms the basis for all discussions of atomic structure, and provides a proper quantum mechanical derivation of Bohr's expression (1.6) for the energy levels of the hydrogen atom.

The hydrogen atom consists of an electron and a proton interacting through the potential

$$V(r) = - \frac{e^2}{4\pi\varepsilon_0 r} \tag{7.8}$$

Here r is the separation of the two particles. Just as in the classical case discussed in Appendix A, we can separate the total energy into two parts, one associated with the motion of the centre of mass of the system and the other with the relative motion of the two particles. Only the second part is of interest in investigating the internal structure of the two-particle system, and so we use the Hamiltonian for the relative motion of electron and proton. This effectively reduces the two-particle problem to a one-particle problem with the potential energy given by (7.8) and a kinetic energy operator of the usual form, $-(\hbar^2/2m)\nabla^2$. The operator ∇ has components $(\partial/\partial x, \partial/\partial y, \partial/\partial z)$, where x, y and z are the position coordinates of the electron relative to the proton, and the mass m is the reduced mass of the system, (A1.7). In terms of the masses of the free electron and proton, m_e and m_p, this is

$$m = \frac{m_e m_p}{m_e + m_p} \tag{7.9}$$

Since the mass of the proton is about 1836 times the mass of the electron, m is not very different from m_e. However, the slight difference between them has an experimentally detectable effect on the frequencies of the

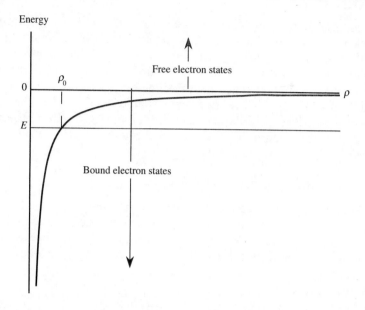

Figure 7.1 The Coulomb potential for an electron in the field of a proton. For total energy $E \geqslant 0$ the electron is free, while for bound states $E < 0$. The classical turning point for energy $E = -|E|$ is at $\rho = \rho_0$ ($\rho = r/r_B$).

spectral lines of hydrogen, which confirms that it is correct to use the reduced mass.

Since the potential (7.8) is central, we look for simultaneous eigenfunctions of energy and angular momentum of the form given in (7.4), and the radial equation (7.7) becomes

$$\left\{ -\frac{\hbar^2}{2m} \left[\frac{d^2}{dr^2} + \frac{2}{r} \frac{d}{dr} - \frac{l(l+1)}{r^2} \right] - \frac{e^2}{4\pi\varepsilon_0 r} \right\} R(r) = ER(r) \quad (7.10)$$

In order to have as few constants as possible appearing in this equation, we introduce a dimensionless variable and eigenvalue, as we did in the case of the simple harmonic oscillator. In this case we define

$$\rho = \frac{r}{r_B}, \qquad r_B = \frac{\hbar^2}{m} \frac{4\pi\varepsilon_0}{e^2} \qquad \rightarrow a_0 \text{ in most other books.}$$

$$\epsilon = \frac{E}{E_R}, \qquad E_R = \frac{m}{2\hbar^2} \left(\frac{e^2}{4\pi\varepsilon_0} \right)^2 \qquad (7.11)$$

where r_B is the **Bohr radius** and E_R the **Rydberg energy**.

In terms of ρ, the potential (7.8) becomes $V(\rho) = -2E_R/\rho$. The shape of this potential is shown in Figure 7.1, and we see that as $\rho \rightarrow \infty$ it approaches zero. A particle with total energy $E < 0$ is bound, since there is a position ρ_0 beyond which $V > E$, so that $\rho > \rho_0$ is a classically forbidden region into which the wave function for the quantum particle can only penetrate a little way. On the other hand, if the total energy is $E > 0$ then $E > V$ everywhere and the particle is free to move anywhere. We shall consider bound states, and find the energy levels, $E < 0$, of the neutral hydrogen atom. The free state solutions, $E > 0$, to the same Schrödinger equation describe the scattering of an electron by a proton.

With the substitutions of (7.11), the radial equation (7.10) becomes

$$\left[\frac{d^2}{d\rho^2} + \frac{2}{\rho} \frac{d}{d\rho} - \frac{l(l+1)}{\rho^2} + \frac{2}{\rho} + \epsilon \right] R(\rho) = 0 \qquad (7.12)$$

For a bound state ϵ is negative, and this equation must be solved subject to the boundary conditions that the wave function remain finite as $\rho \to \infty$ and at $\rho = 0$. This is done analytically in Appendix B3, and once again we see that solutions only exist for a discrete spectrum of eigenvalues, given in this case by

$$\epsilon = -\frac{1}{n^2} \quad (n = 1, 2, 3, \ldots)$$

The energy eigenvalues are determined by the value of the integer n, which is called the **principal quantum number**, and the energy spectrum is

$$E_n = -E_R \frac{1}{n^2} \quad (n = 1, 2, 3, \ldots) \qquad (7.13)$$

This is identical to Bohr's equation (1.6), since a comparison of the expression for E_R in (7.11) with (1.7) shows that $E_R = hcR_H$ (where R_H is the Rydberg constant). The spectrum of energy levels is shown in Figure 1.2.

The solution corresponding to the particular energy eigenvalue E_n and angular momentum quantum number l is the radial wave function

$$R_{nl}(\rho) = C_{nl} e^{-\rho/n} \rho^l L_{n-l-1}^{2l+1}(2\rho/n) \qquad (7.14)$$

Here C_{nl} is a constant and $L_{n-l-1}^{2l+1}(2\rho/n)$ is an **associated Laguerre polynomial** (see Appendix B3 and [3]), which can be written as a real finite series in powers of $2\rho/n$:

$$L_k^q\left(\frac{2\rho}{n}\right) = a_0 + a_1 \frac{2\rho}{n} + a_2 \left(\frac{2\rho}{n}\right)^2 + \ldots + a_k \left(\frac{2\rho}{n}\right)^k \qquad (7.15)$$

The coefficients a_i depend on the values of $q = 2l + 1$ and $k = n - l - 1$. The radial wave functions for the first three energy levels are given in Table 7.1, and shown graphically in Figure 7.2(a).

Substitution of the form (7.14) for the radial wave function into (7.4) gives the complete spatial dependence of the energy eigenfunctions:

$$\psi_{nlm_l}(r, \theta, \phi) = R_{nl}(r) Y_{lm_l}(\theta, \phi) \qquad (7.16)$$

Table 7.1 Radial wave functions for hydrogen.

$$R_{10}(r) = \left(\frac{1}{r_B}\right)^{3/2} 2e^{-r/r_B}$$

$$R_{20}(r) = \left(\frac{1}{2r_B}\right)^{3/2} \left(2 - \frac{r}{r_B}\right) e^{-r/2r_B}$$

$$R_{21}(r) = \left(\frac{1}{2r_B}\right)^{3/2} \frac{1}{\sqrt{3}} \frac{r}{r_B} e^{-r/2r_B}$$

$$R_{30}(r) = \left(\frac{1}{3r_B}\right)^{3/2} 2\left(1 - \frac{2}{3}\frac{r}{r_B} + \frac{2}{27}\left(\frac{r}{r_B}\right)^2\right) e^{-r/3r_B}$$

$$R_{31}(r) = \left(\frac{1}{3r_B}\right)^{3/2} \frac{8}{9\sqrt{2}} \frac{r}{r_B} \left(1 - \frac{r}{6r_B}\right) e^{-r/3r_B}$$

$$R_{32}(r) = \left(\frac{1}{3r_B}\right)^{3/2} \frac{4}{27\sqrt{10}} \left(\frac{r}{r_B}\right)^2 e^{-r/3r_B}$$

These functions are normalized according to (7.17)

Functions $\psi_{nlm_l}(r, \theta, \phi)$ corresponding to different sets of quantum numbers $\{n, l, m_l\}$ are orthogonal, and can be normalized so that

$$\int d^3r\, \psi^*_{nlm_l}(r, \theta, \phi)\psi_{n'l'm'_l}(r, \theta, \phi) = \delta_{nn'}\delta_{ll'}\delta_{m_lm'_l}$$

The volume element is $d^3r = r^2\,dr\,d\Omega$, where $d\Omega = d\phi\,d(\cos\theta)$ is the element of solid angle, and so the integral becomes

$$\int_0^\infty dr\, r^2 R_{nl}(r)R_{n'l'}(r)\int_0^{2\pi} d\phi \int_{-1}^1 d(\cos\theta)\, Y^*_{lm_l}(\theta, \phi)Y_{l'm'_l}(\theta, \phi)$$

(We have chosen the normalization constants C_{nl} in (7.14) to be real, so the radial wave function is real.) Since the spherical harmonics satisfy the orthonormality condition (4.46), this integral is zero unless $l = l'$ and $m_l = m'_l$, and the radial wave function must satisfy

$$\int_0^\infty dr\, r^2 R_{nl}(r)R_{n'l}(r) = \delta_{nn'} \tag{7.17}$$

The form of the radial probability distribution depends on both the principal quantum number n and the orbital angular momentum quantum number l, but not on m_l. The energy eigenvalues (7.13) depend only on n and not on l or m_l, so that, except for the ground state ($n = 1$, $l = 0$, $m_l = 0$), there are several different angular momentum states with the same energy – the energy levels are degenerate. The different possible values of l corresponding to the same n can be found by inspecting (7.14) and (7.15): the Laguerre polynomials differ in the highest power of $2\rho/n$ that they contain, and this power is

$$k = n - l - 1 = 0, 1, 2, 3, \ldots \tag{7.18}$$

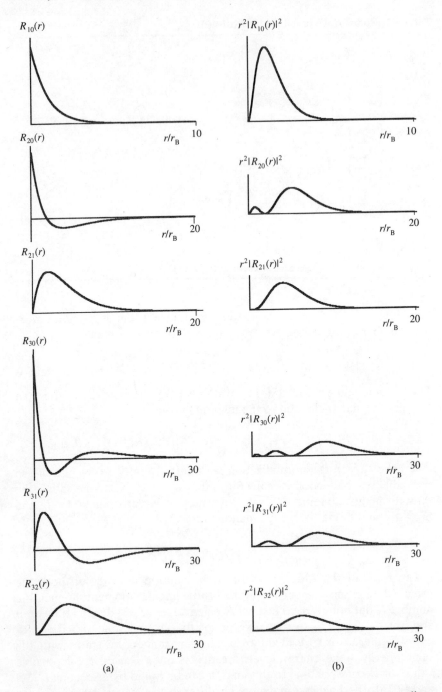

Figure 7.2 The radial wave functions for hydrogen $R_{nl}(r)$ and the corresponding probability distributions $r^2|R_{nl}|^2$, for the states with $(n, l) = (1, 0)$, $(2, 0)$, $(2, 1)$, $(3, 0)$, $(3, 1)$ and $(3, 2)$.

For $n = 1$ we can only have $l = 0$ (and $k = 0$), but for higher values of n, l can range from 0 (with $k = n - 1$) to $n - 1$ (with $k = 0$). So for each n, there are n different possible values of l:

$$l = 0, 1, \ldots, n - 1 \tag{7.19}$$

Each state with a particular value of l is $(2l + 1)$-fold degenerate, (7.6), so altogether the number of different orbital angular momentum states with the same energy E_n is

$$\sum_{l=0}^{n-1} (2l + 1) = n^2 \tag{7.20}$$

If we include the spin of the electron, the total degeneracy of the nth energy level is $2n^2$. This is because there are two possible spin states for each of the n^2 states with different orbital angular momentum quantum numbers, and these states are distinguished by the eigenvalue of $\hat{S}_z$, $m_s = \pm\frac{1}{2}$, introduced in Section 6.2:

$$\psi_{nlm_l m_s}(r, \theta, \phi) = R_{nl}(r)Y_{lm_l}(\theta, \phi)\chi_{m_s} \tag{7.21}$$

Alternatively (Problem 7.10), we could label the state of the electron by the set of quantum numbers $\{n, l, j, j_z\}$, where j and j_z are the quantum numbers of the total angular momentum of the electron, which is the resultant of its orbital angular momentum and spin. Such a state is described by a radial wave function $R_{nl}(r)$ of the form given in (7.14), but its angular dependence is of the form χ_{lsjj_z} given in (6.42), with $s = \frac{1}{2}$.

It can be shown (see Problem 7.5) that none of the components of the position operator $\hat{R} = (r, \theta, \phi)$ commute with the hydrogen atom Hamiltonian, nor does the radial momentum operator $\hat{P}_r$ of (4.22). So the position and radial momentum of the electron relative to the proton are both indeterminate in an energy eigenstate. The probability density $\mathbb{P}(r)$ for finding an electron at a distance r from the proton is the integral of $|R_{nl}(r)Y_{lm_l}(\theta, \phi)|^2$ over the surface of a sphere of radius r centred on the origin. The element of surface area is $r^2\,d\Omega$, and the spin state of the electron does not affect this probability distribution. So, for an electron in the state with quantum numbers (n, l, m_l, m_s),

$$\mathbb{P}(r) = \int d\Omega\, r^2 |R_{nl}(r)Y_{lm_l}(\theta, \phi)|^2 = r^2(R_{nl})^2$$

(where we have used (4.46) again). This is independent of m_l and m_s, and it is plotted for a few states in Figure 7.2(b). There are several important points to notice about these distributions:

(1) the factor r^2 makes the radial probability density vanish at the origin even for $l = 0$ states, for which the radial wave function is non-zero (though finite) at $r = 0$;

(2) for each state, characterized by n and l, there are $n - l - 1$ nodes in the distribution, in addition to the zero at the origin;

(3) the distributions for states with $l = 0$ have n maxima, with height increasing with the distance of the peak from the origin; similarly, for $l \neq 0$ there are $n - l$ maxima, increasing in height with distance from the origin.

This last characteristic provides a link with the Bohr model, since although an electron in an energy eigenstate does not follow a well-defined orbit about the proton, its most probable radial distance in the nth energy eigenstate when $l = n - 1$ is equal to the radius of the nth Bohr orbit. In particular, the maximum probability density in the ground state occurs at $r = r_B$, where r_B is the radius of the first Bohr orbit (Problem 7.8b).

7.3 Atomic structure [4]

According to Bohr's model, the electrons in an atom can only exist in certain discrete orbits about the nucleus, and the pattern of these allowed orbits is the same for all elements. So the energy level spectrum for hydrogen, (1.6), sets the pattern for every other element. Similarly, quantum mechanical models of the structure of atoms containing more than one electron are based on the solution to the energy eigenvalue problem for hydrogen. The atomic number Z is the number of units of positive charge on the atomic nucleus, which is equal to the number of orbiting electrons in the neutral atom, and atoms of different elements are distinguished by this number. The energy levels of an atom containing Z electrons are determined by a set of Z coupled Schrödinger equations, since the potential in which a given electron moves includes the effects of the repulsive interaction with each of the other electrons as well as the attraction to the nucleus. This many-body problem can only be solved by making some suitable approximations.

The simplest approximation of all is to treat each electron as if it moved independently in the Coulomb potential $V(r) = - Ze^2/4\pi\varepsilon_0 r$ of the nucleus, and to neglect the interaction with the other electrons completely. The Schrödinger equation for each electron is then the same as the hydrogen atom equation, apart from the magnitude of the potential energy. The energy eigenfunctions are of the same form as those for hydrogen, (7.16), but with r_B replaced everywhere by r_B/Z, and the energy spectrum for each electron is the same as the hydrogen spectrum, (7.13), with the Rydberg energy E_R replaced by $Z^2 E_R$. But this implies that in a multi-electron atom all electrons in states with the same principal quantum number n should have the same energy, whereas in practice it is found that the energy also depends on the orbital angular momentum quantum number l. Electrons in states with the same value, n, for the principal quantum number are said to occupy the same, nth, **shell** of the atom, and each shell contains n **subshells**, in each of which the quantum number l has a fixed value.

An improved approximation, taking account of the mutual repulsion of the electrons, is needed to calculate the way the energy of an electron in a multi-electron atom varies with l. In an **independent electron model** we solve a Schrödinger equation for each of the electrons individually, using an effective potential that takes at least some account of the repulsion due to the other electrons as well as the attraction of the nucleus. As long as the effective potential is spherically symmetric, the energy eigenfunctions will be simultaneously eigenfunctions of orbital angular momentum, as explained in Section 7.1, and the states of individual electrons can therefore be labelled by the set of quantum numbers $\{n, l, m_l, m_s\}$. (Such an approximation is referred to as a **central field approximation**.) Any distortion of the $1/r$ behaviour of the (central) potential is sufficient to make the energy eigenvalues depend on l as well as n, but states differing only in their values of m_l and m_s remain degenerate. In a multi-electron atom the energy of an electron in a particular shell increases with increasing l. States with $l = 0$ are referred to as **s states**, and those with $l = 1$, 2 and 3 are called respectively **p, d and f states**. The electronic levels in an atom are labelled by the principal quantum number n and a letter indicating the value of l, and their energies increase with n and l: the 1s state is the most tightly bound, followed in order by the 2s, 2p, 3s and 3p. But the next higher energy level is the 4s instead of the 3d as we might have expected. This is because the probability density of the 4s electrons is larger close to the nucleus than that of the 3d electrons, and so the effective Coulomb attraction to the nucleus is greater. The order of the levels after 3p is 4s, 3d, 4p, 5s, 4d, ..., though there are certain departures from this among the transition elements, as we shall see below.

The Pauli exclusion principle must be invoked, as in the Bohr model, to explain why, in the ground state of a multi-electron atom, not all the electrons are in the same $n = 1$, $l = 0$ state. This principle states that only one electron may occupy each state defined by a unique set of quantum numbers $\{n, l, m_l, m_s\}$. Although it cannot be derived using non-relativistic quantum mechanics, the exclusion principle arises automatically in quantum field theory, where it is a necessary requirement on any system containing more than one electron. Since there are $2l + 1$ different values of m_l for a given l, and for each of these there are two different values of m_s, the number of electrons that can be accomodated in a particular subshell is $2(2l + 1)$. So an s state can hold at most two electrons, a p state at most 6, a d state at most 10, and so on. Thus, according to the exclusion principle, the two electrons in a helium atom can both occupy the lowest, 1s, energy level, but in the ground state of lithium (with $Z = 3$) two electrons fill the 1s level, and the third electron must occupy the next higher level, 2s, as indicated in Table 7.2. After beryllium ($Z = 4$), both the 1s and 2s levels are full, and in the elements from boron ($Z = 5$) to neon ($Z = 10$) electrons successively fill the six states in the 2p level. This accounts for the first two rows of the periodic table of elements (Table 7.2), and the third row is built up similarly, with the 3s and 3p levels full by the time we reach argon ($Z = 18$). The energies of the ten 3d states lie very close to, but mainly above, the 4s level (as mentioned in the last paragraph) but below the 4p level, so potassium and calcium ($Z = 19$ and 20) have electrons in the 4s level. Then in the 10 **transition elements** from scandium to zinc electrons occupy the 3d states. The next six electrons fill 4p states, so krypton ($Z = 36$) has the typical completely filled subshell structure of the noble gases. In a similar way the rest of the periodic table is built up, with 10 transition elements occurring between the second and third main columns, corresponding to the filling of d states. In addition, there are 14 elements between lanthanum and hafnium, the first and second transition elements of the 6th row, and in these so-called **lanthanides** the 4f states are progressively filled. The **actinides** occur in a corresponding position in the 7th row, and uranium, the last naturally occurring element in the table, belongs to this group. This accounts for the general structure of the periodic table, though there are slight variations from one element to the next in the precise ordering of levels. For example, in palladium ($Z = 46$) there are two 5s and eight 4d electrons in the ground state. However, in silver ($Z = 47$) the ten 4d states are completely filled, and there is only one 5s electron.

The probability distributions for the hydrogen atom in Figure 7.2(b) show that as n increases, the probability distribution peaks around larger and larger radial distances r. This is a general feature of atomic energy eigenfunctions, so that the probability density for an electron in a state with small n is negligible in the regions where an electron in a state with larger n has its maximum probability density. In this average sense, the electrons in the lower shells are mainly inside the region containing the electrons in higher shells, and the ground states of all atoms (except hydrogen) consist of a **core** of completely filled subshells, and a number (which is zero for the noble gases) of outer or **valence** electrons. The probability of a valence electron being found outside the region of space occupied by the core is large, but there are also subsidiary maxima of its probability distribution within the core. The degree to which the probability distribution for a valence electron penetrates the core tends to be larger for smaller l, and the effective potential in which an electron moves depends on this degree of penetration, and the extent to which the core electrons screen the full effect of the electric charge of the nucleus. This explains the filling of the 4s levels before the 3d, as mentioned

Table 7.2 The periodic table of elements.

Core / subshell																		
1s	H (1)																	He (2)
He core + 2s 2p	Li (1)	Be (2)											(2) B (1)	(2) C (2)	(2) N (3)	(2) O (4)	(2) F (5)	(2) Ne (6)
Ne core + 3s 3p	Na (1)	Mg (2)											(2) Al (1)	(2) Si (2)	(2) P (3)	(2) S (4)	(2) Cl (5)	(2) Ar (6)
Ar core + 4s 3d 4p	K (1)	Ca (2)	(2) Sc (1)	(2) Ti (2)	(2) V (3)	(2) Cr (4)	(2) Mn (5)	(2) Fe (6)	(2) Co (7)	(2) Ni (8)	(1) Cu (10)	(2) Zn (10)	(2)(10) Ga (1)	(2)(10) Ge (2)	(2)(10) As (3)	(2)(10) Se (4)	(2)(10) Br (5)	(2)(10) Kr (6)
Kr core + 5s 4d 5p	Rb (1)	Sr (2)	(2) Y (1)	(2) Zr (2)	(2) Nb (3)	(2) Mo (4)	(2) Tc (5)	(2) Ru (6)	(2) Rh (7)	(2) Pd (8)	(1) Ag (10)	(2) Cd (10)	(2)(10) In (1)	(2)(10) Sn (2)	(2)(10) Sb (3)	(2)(10) Te (4)	(2)(10) I (5)	(2)(10) Xe (6)
Xe core + 6s 4f 5d 6p	Cs (1)	Ba (2)	(2)(0) La (1)	(2)(14) Hf (2)	(2)(14) Ta (3)	(2)(14) W (4)	(2)(14) Re (5)	(2)(14) Os (6)	(2)(14) Ir (7)	(2)(14) Pt (8)	(1)(14) Au (10)	(2)(14) Hg (10)	(2)(14)(10) Tl (1)	(2)(14)(10) Pb (2)	(2)(14)(10) Bi (3)	(2)(14)(10) Po (4)	(2)(14)(10) At (5)	(2)(14)(10) Rn (6)
Rn core + 7s 6d	Fr (1)	Ra (2)	(2) Ac (1)															

La core + 4f electrons: 14 lanthanide elements

Ac core + 5f electrons: Th (1) Pa (2) U (3) + 11 elements which do not occur naturally

Numbers in parentheses next to the symbol for the element give the number of outermost, valence, electrons in the subshell indicated in the left hand column. The integer specifies n, and s, p d and f refer to subshells with $l = 0$, 1, 2 and 3 respectively. Numbers in parentheses above the symbol for the element give the number of electrons in sublevels outside the noble gas core, but lower sublevels than the valence electrons.

earlier, and accounts for the structure of the transition elements, the lanthanides and the actinides. The outer electrons are more affected than those in the core by external electromagnetic fields, such as those due to other atoms, and so it is the valence electrons that determine the chemical properties of an element.

In an independent electron approximation the use of an effective potential to represent the average effect of the other electrons means that some energy due to electron–electron repulsion is included in the energy eigenvalue of each electron. In this approximation the remaining effects of electron–electron repulsion are neglected, so that the electrons behave as though each moved independently, and the total energy of the atom is the sum of the energies of the individual electrons. Similarly, the total angular momentum of the atom as a whole is found by adding the angular momenta of the individual electrons according to the rules for addition of angular momenta discussed in Section 6.3. It is usual to use capital letters to denote angular momentum quantum numbers that refer to the state of the atom, reserving the lower case letters for the quantum numbers of individual electrons. Thus $\{L, M_L\}$ refers to the orbital angular momentum state of the atom found by adding the orbital angular momenta of all the electrons, and $\{S, M_S\}$ refers to the spin state found by adding the spins of all the atomic electrons. In general, these resultant orbital angular momentum and spin states give a good description of the angular momentum state for light atoms. Such atoms are said to exhibit L–S, or **Russell–Saunders, coupling**. However, in heavy atoms a magnetic component of the interaction between the nucleus and an electron becomes increasingly important (this is spin–orbit coupling, which we shall discuss in Section 8.4). This has the effect of making each electron behave as if it were in an eigenstate of the total angular momentum, represented by the quantum numbers $\{j, j_z\}$, which is the sum of its orbital angular momentum and spin, defined in Section 6.3. The angular momentum state of the atom as a whole, labelled by the quantum numbers $\{J, M_J\}$, is found by adding the total angular momenta of all the electrons, and this scheme is called J–J **coupling**. Whether the coupling is L–S, J–J, or some intermediate form, the total angular momentum of an atomic core (consisting of completely filled subshells) vanishes, so that the angular momentum state of the atom as a whole is determined by the valence electrons.

7.4 Periodic potentials and translational symmetry

In Section 7.3 we discussed atomic energy levels in terms of an independent electron model, where each electron moves in an effective central potential describing the average effect of the other electrons as well as the attraction to the nucleus. This system has spherical symmetry. The electronic energy levels of molecules are more difficult to calculate, since the outermost atomic electrons are no longer bound to one particular atom. If we ignore the interactions between these valence electrons, each moves in a potential due to the positive ionic cores of all the atoms in the molecule, which cannot, in general, be spherically symmetric. So the simple symmetry of a single atom is lost. Molecules *do* have symmetry properties, which depend on the shape of the molecule, and these are an important aid in determining molecular energy eigenstates, but in general some understanding of group theory is necessary to make use of them. However, in the extreme case of an infinite crystalline solid we again discover a simple symmetry that imposes restrictions on the energy eigenfunctions, and which can be implemented without a knowledge of group theory.

In a crystalline solid the atoms are arranged regularly in a lattice, such as the cubic lattice shown in Figure 7.3(a). The potential experienced by an electron due to the atomic nuclei is therefore periodic in space (Figure 7.3b). Again we make use of an independent electron model, and solve the Schrödinger equation for one electron moving in an effective potential that includes some of the effects due to other electrons by treating them as a static distribution of negative charge also with the periodicity of the lattice. The effective potential along a symmetry axis of the lattice will be of the general form shown in Figure 7.3(c). The full three-dimensional analysis is very complicated, but the essential characteristics of the energy spectrum are evident from the one-dimensional case. The Kronig–Penney model [5] was one of the first theoretical treatments of a one-dimensional crystal lattice, and used a periodic square well potential, like that illustrated in Figure 7.3(d), in the limit where the width of the potential barriers is zero and their height becomes infinite (Problems 7.13 and 7.15). The quantum well structures mentioned in Section 3.3 can provide real examples of one-dimensional periodic systems: in a specimen constructed from many alternate regularly spaced layers of two appropriate semiconductors, with barrier layers sufficiently thin that an electron can tunnel from one well to the next, an electron experiences a periodic square well potential.

The ideal one-dimensional lattice is composed of a set of identical unit cells of length equal to the **lattice constant** L, and the potential is the same at equivalent points in each cell. This means that the potential does not change as we move from the point x to a point $x + nL$ separated from it by an integer multiple n of the lattice constant:

$$V(x + nL) = V(x) \quad (n = 0, 1, 2, 3, \ldots) \tag{7.22}$$

The kinetic energy operator is unchanged by *any* translation, since for any constant x_0

$$\frac{d}{d(x + x_0)} = \frac{d}{dx} \tag{7.23}$$

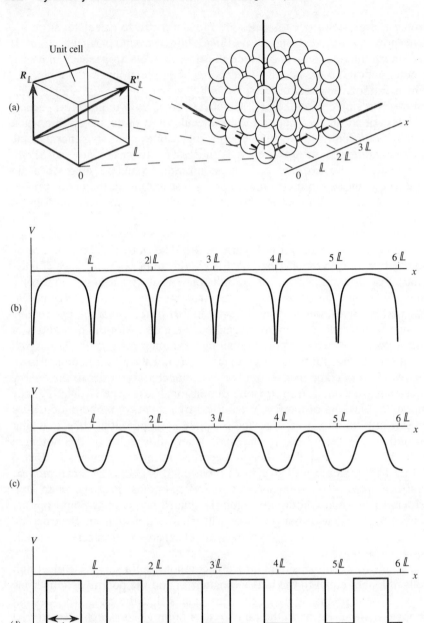

Figure 7.3(a) A three-dimensional cubic lattice in physical space, with lattice spacing L along the symmetry axis x, with an enlargement of a unit cell showing two different lattice vectors, R_L and R'_L. (b) The form of the potential due to the nuclei of the atoms of the lattice. (c) The effective potential when electron-electron repulsion is included. (d) A periodic square well potential.

Thus the Hamiltonian for an electron in the periodic potential of (7.22) is invariant under the discrete set of translations $x \rightarrow x + n\mathbb{L}$ ($n = 0$, $1, 2, \ldots$):

$$\hat{H} = -\frac{\hbar^2}{2m}\frac{d^2}{dx^2} + V(x) = -\frac{\hbar^2}{2m}\frac{d^2}{d(x + n\mathbb{L})^2} + V(x + n\mathbb{L}) \quad \textbf{(7.24)}$$

Suppose that $\hat{T}_{\mathbb{L}}$ is a linear operator that, when acting on a function of x, produces a translation through a distance $\mathbb{L}$. A translation through $n\mathbb{L}$ is achieved by using the operator n times:

$$\psi(x + \mathbb{L}) = \hat{T}_{\mathbb{L}}\psi(x), \qquad \psi(x + n\mathbb{L}) = (\hat{T}_{\mathbb{L}})^n\psi(x) \qquad \textbf{(7.25)}$$

This translation operator commutes with the periodic Hamiltonian, since

$$\hat{T}_{\mathbb{L}}\left[-\frac{\hbar^2}{2m}\frac{d^2}{dx^2} + V(x)\right]\psi(x) = \left[-\frac{\hbar^2}{2m}\frac{d^2}{d(x + \mathbb{L})^2} + V(x + \mathbb{L})\right]\hat{T}_{\mathbb{L}}\psi(x)$$

That is, because of (7.24),

$$\hat{T}_{\mathbb{L}}\hat{H}\psi(x) = \hat{H}\hat{T}_{\mathbb{L}}\psi(x) \qquad \textbf{(7.26)}$$

Therefore the energy eigenfunctions can be simultaneously eigenfunctions of $\hat{T}_{\mathbb{L}}$. Just as in the spherically symmetric case we looked for simultaneous eigenfunctions of energy and angular momentum, in the case of the periodic potential we shall look for simultaneous eigenfunctions of the Hamiltonian and the discrete translation operator $\hat{T}_{\mathbb{L}}$.

Since the Hamiltonian has the same translational symmetry as the lattice, it seems reasonable to assume that in an energy eigenstate the probability distribution for the electron should also have this symmetry:

$$|\psi(x + n\mathbb{L})|^2 = |\psi(x)|^2 \quad (n = 0, 1, 2, 3, \ldots) \qquad \textbf{(7.27)}$$

One solution of this equation is that the eigenfunction itself should be periodic. However, the general solution is not so restrictive as this. **Bloch's theorem** states that a simultaneous eigenfunction of the energy and the translation operator must be of the form

$$\psi_k(x) = e^{ikx}u_k(x), \qquad u_k(x + \mathbb{L}) = u_k(x) \qquad \textbf{(7.28)}$$

where k can take any real value, and, as indicated, $u_k(x)$ has the same periodicity as the lattice. It is easy to see that this form satisfies (7.27).

Under the translation $x \rightarrow x + \mathbb{L}$ a wave function of the Bloch form (7.28) transforms as

$$\psi_k(x) \rightarrow \hat{T}_{\mathbb{L}}\psi_k(x) = \psi_k(x + \mathbb{L}) = e^{ik\mathbb{L}}\psi_k(x) \qquad \textbf{(7.29)}$$

This means that $\psi_k(x)$ is an eigenfunction of the translation operator $\hat{T}_{\mathbb{L}}$ corresponding to the eigenvalue $e^{ik\mathbb{L}}$. (Note that this eigenvalue is not real, so that $\hat{T}_{\mathbb{L}}$ is not a Hermitian operator, and thus, according to Section 3.1, it does not represent an observable variable. However, each eigenstate of $\hat{T}_{\mathbb{L}}$ is also an eigenstate of a Hermitian operator whose eigenvalues k are real.) Successive discrete translations along the same axis, through $n_1 \mathbb{L} = \mathbb{L}_1$ followed by $n_2 \mathbb{L} = \mathbb{L}_2$, should be equivalent to a single translation through $\mathbb{L}_1 + \mathbb{L}_2$. It is easy to show that this requirement is satisfied by the Bloch function (7.28): from (7.29) we see that

$$\hat{T}_{\mathbb{L}_2} \hat{T}_{\mathbb{L}_1} \psi_k(x) = \hat{T}_{\mathbb{L}_2} e^{ik\mathbb{L}_1} \psi_k(x)$$

Since $e^{ik\mathbb{L}_1}$ is a constant phase factor, independent of x, the translation operator has no effect on it, and so

$$\hat{T}_{\mathbb{L}_2} e^{ik\mathbb{L}_1} \psi_k(x) = e^{ik\mathbb{L}_1} \hat{T}_{\mathbb{L}_2} \psi_k(x) = e^{ik\mathbb{L}_1} e^{ik\mathbb{L}_2} \psi_k(x) = e^{ik(\mathbb{L}_1 + \mathbb{L}_2)} \psi_k(x)$$

According to (7.29), this last term is equal to $\psi_k(x + \mathbb{L}_1 + \mathbb{L}_2)$ and so we have shown that

$$\hat{T}_{\mathbb{L}_2} \hat{T}_{\mathbb{L}_1} \psi_k(x) = \hat{T}_{\mathbb{L}_2 + \mathbb{L}_1} \psi_k(x), \qquad \mathbb{L}_i = n_i \mathbb{L} \quad (i = 1, 2) \qquad \textbf{(7.30)}$$

That is, successive translations through integer multiples of the lattice constant have the same effect on the Bloch function as a single translation, as required. This would not be satisfied if the factor e^{ikx} in the wave function (7.28) were replaced by $e^{if(x)}$, where $f(x)$ is an arbitrary real function of x, although the modified wave function would satisfy (7.27). Indeed, the form of the wave function prescribed by Bloch's theorem is the only one that will satisfy both (7.27) and (7.30).

The variable whose eigenvalues are $\hbar k$ is called the **crystal momentum**. Of course it is not the true momentum of the electron, which cannot take a precise value in an energy eigenstate when the particle is not free. It is easy to check that, because of the x dependence of $u_k(x)$, the Bloch function is not an eigenfunction of the momentum operator $\hat{P}_x = -i\hbar \, d/dx$. However, in the limit as the lattice constant goes to zero, $\mathbb{L} \to 0$, the solid becomes a continuous distribution of matter, and (7.22) implies that the periodic potential is unchanged by a translation through any distance, no matter how small. This means that the potential is invariant under the continuous set of all possible translations, and must therefore be a constant, which means that the force on the electron is zero. So in this limit the electron behaves as if it were free. Similarly, the periodicity of $u_k(x)$ implies that it becomes constant as $\mathbb{L} \to 0$, and so $\psi_k(x) \to C e^{ikx}$ as $\mathbb{L} \to 0$, where C is constant. Thus as $\mathbb{L} \to 0$ the Bloch function (7.28) reduces to the momentum eigenfunction for a free particle, (3.10), and the crystal momentum $\hbar k$ reduces to the physical momentum.

Wave functions of the form (7.28) with values of k differing by integer multiples of $2\pi/\mathbb{L}$ correspond to the same eigenvalue of $\hat{T}_{\mathbb{L}}$, since

$$e^{i(k + 2n\pi/\mathbb{L})\mathbb{L}} = e^{ik\mathbb{L}} e^{i2n\pi} = e^{ik\mathbb{L}} \qquad \textbf{(7.31)}$$

In fact, eigenfunctions corresponding to the same eigenvalue $e^{ik\mathbb{L}}$ but different k describe the same state. To see this, we multiply the wave function (7.28) by $e^{i2n\pi x/\mathbb{L}}$:

$$e^{i2n\pi x/\mathbb{L}}\psi_k(x) = e^{i(k+2n\pi/\mathbb{L})x}u_k(x) = e^{ikx}[e^{i2n\pi x/\mathbb{L}}u_k(x)] \qquad (7.32)$$

The term in square brackets has the same periodicity as $u_k(x)$ itself, so we could define a new periodic function

$$u_{k'}(x) = e^{i2n\pi x/\mathbb{L}}u_k(x) = u_{k'}(x + \mathbb{L}), \qquad k' = k + \frac{2n\pi}{\mathbb{L}} \qquad (7.33)$$

Then

$$\psi_{k'}(x) = e^{i2n\pi x/\mathbb{L}}\psi_k(x) = e^{ikx}u_{k'}(x)$$

is of exactly the same form as $\psi_k(x)$, and corresponds to the same eigenvalue of $\hat{T}_{\mathbb{L}}$, and the same probability density $|\psi_{k'}(x)|^2 = |\psi_k(x)|^2 = |u_k(x)|^2$. So the functions $\psi_{k'}(x)$ and $\psi_k(x)$, with values of k differing by an integer multiple of $2\pi/\mathbb{L}$, represent the same physical state. This means that we need only consider k-values within a single range of width $2\pi/\mathbb{L}$. Different possible choices for this range are called **Brillouin zones**, and the first Brillouin zone is defined to be the range $-\pi/\mathbb{L} \leqslant k \leqslant \pi/\mathbb{L}$. The second Brillouin zone is $\pi/\mathbb{L} \leqslant |k| \leqslant 2\pi/\mathbb{L}$, and in general the nth Brillouin zone is determined by

$$\frac{(n-1)\pi}{\mathbb{L}} \leqslant |k| \leqslant \frac{n\pi}{\mathbb{L}} \quad (n = 1, 2, \ldots) \qquad (7.34)$$

7.5 Energy bands

Having decided that the simultaneous eigenfunctions of energy and the discrete translation operator $\hat{T}_{\mathbb{L}}$ must be Bloch functions of the form (7.28), we must now solve the Schrödinger equation to find the energy eigenvalues. Using the Hamiltonian (7.24), and a Bloch function corresponding to a particular value k, we obtain

$$\left[-\frac{\hbar^2}{2m}\frac{d^2}{dx^2} + V(x)\right]e^{ikx}u_k(x) = Ee^{ikx}u_k(x)$$

which gives

$$\left[-\frac{\hbar^2}{2m}\left(\frac{d^2}{dx^2} + 2ik\frac{d}{dx} - k^2\right) + V(x)\right]u_k(x) = Eu_k(x) \qquad (7.35)$$

Because $V(x)$ and $u_k(x)$ are both periodic, it is sufficient to solve the equation in one unit cell. The appropriate boundary conditions are that the wave function and its first derivative be continuous across the cell boundaries (as in the case of the single square well, Section 3.3). Consider the boundary at $x = 0$. According to (7.28), the wave function just to the left of this boundary is related to that at a corresponding point just to the left of $x = L$ by $\psi_k(0) = e^{-ikL}\psi_k(L)$. This must be equal to the wave function just to the right of $x = 0$. Similarly, the derivative of the wave function just to the left of $x = 0$ is determined using (7.28), and the complete set of boundary conditions can be written as

$$\psi_k(0) = e^{-ikL}\psi_k(L), \qquad \frac{d\psi_k}{dx}\bigg|_{x=0} = e^{-ikL}\frac{d\psi_k}{dx}\bigg|_{x=L} \qquad \text{(7.36)}$$

The corresponding conditions on the periodic function $u_k(x)$ are

$$u_k(0) = u_k(L), \qquad \frac{du_k}{dx}\bigg|_{x=0} = \frac{du_k}{dx}\bigg|_{x=L} \qquad \text{(7.37)}$$

In the case of a potential that has discontinuities within the unit cell, such as that illustrated in Figure 7.3(d), ψ_k and its derivative must also be made continuous across each point where there is a discontinuity in the potential.

The time dependence of any energy eigenfunction is given by (3.5), and so the complete form for a simultaneous eigenfunction of energy and the translation operator is

$$\psi_{k,E}(x, t) = e^{i(kx - Et/\hbar)}u_k(x) \qquad \text{(7.38)}$$

The first factor is of a similar form to the energy and momentum eigenfunction for a free particle, (3.1), but with the crystal momentum replacing the free particle momentum eigenvalue. It determines the way in which the electron propagates through the crystal as a whole, while the function $u_k(x)$ describes the details of electron behaviour that depend on the particular form of the potential in each unit cell.

All possible eigenvalues of the energy are found by solving (7.35) for each value of k within one Brillouin zone. The boundary conditions (7.37) restrict the energy spectrum for a particular value of k to a discrete set of levels $E_n(k)$ ($n = 1, 2, \ldots$), which must depend on k since k appears explicitly in the differential equation. For an infinite solid there is a continuum of possible k-values within each Brillouin zone (defined by (7.34) for a particular integer n). However, a real specimen of a crystalline solid will contain some very large but finite number N of unit cells. (The lattice spacing is usually of the order of 1 nm, so that even for a specimen as short as 10^{-5} m we have $N \approx 10^4$.) The effect of the finite length is to make the spectrum of k-values discrete, and, as we shall see, there are N distinct values within each Brillouin zone. Although it is important to be able to count the number of distinct k states, the energy spacing between one such state and the next is so small in a macroscopic specimen that the variation of energy with k is effectively continuous.

To calculate the discrete spectrum of k-values, we must apply boundary conditions to the wave function at either end of the solid specimen. The wave function describing an electron deep inside a macroscopic sample will not depend critically on the precise form of the conditions imposed at the boundaries, and the most convenient to use are **periodic boundary conditions**:

$$\psi_k(N\mathbb{L}) = \psi_k(0) \tag{7.39}$$

That is, the eigenfunction $\psi_k(x)$ is the same at the two ends of the sample, which we have taken to be $x = 0$ and $x = N\mathbb{L}$. For a Bloch function of the form (7.28) this requires $e^{ikN\mathbb{L}} = 1$, so that within one Brillouin zone of width $2\pi/\mathbb{L}$ there are N different possible values for k. For example, in the first Brillouin zone

$$k = j \frac{2\pi}{N\mathbb{L}} \quad \begin{cases} j = 0, \pm 1, \pm 2, \ldots, \pm \frac{1}{2}(N - 1) & (N \text{ odd}) \\[2mm] j = 0, \pm 1, \pm 2, \ldots, \pm (\frac{1}{2}N - 1), \frac{1}{2}N & (N \text{ even}) \end{cases} \tag{7.40}$$

Note that if N is even, the value $k = \pi/\mathbb{L}$ is included in the zone, but $k = -\pi/\mathbb{L}$ is not. Whether N is odd or even, there is one k state for each unit cell of the finite specimen. In a macroscopic sample the separation between different allowed values of k becomes so small that the spectrum (7.40) effectively becomes a continuous distribution over the range $-\pi/\mathbb{L}$ to $\pi/\mathbb{L}$. In this case the number of states within a range of k-values of width dk can be written (from (7.40)) as

$$dn(k) = \frac{N\mathbb{L}}{2\pi} dk \tag{7.41}$$

If, instead of (7.39), we had chosen to make the wave function vanish at the boundaries, we should have to deal with standing waves. But important properties of the solid, such as its electrical conductivity, depend on the transport of electrons through the crystal lattice, and so it is better to deal with individual Bloch functions that are of the required travelling wave form as shown by (7.38), instead of standing wave combinations of them.

In general, the solution of the eigenvalue equation (7.35) will give a different set of energy levels for each of the N values of k determined by (7.40). For each value of the quantum number n there is a set of N different energy values $E_n(k)$, each corresponding to a different value of k. The energy levels for a fixed n but different k form an **energy band**, labelled by the appropriate value of n. In quantum well structures the number N of unit cells can be made small, and the corresponding number of distinct energy levels in a band has been observed experimentally. The spectra in Figure 7.4 show that the $n = 1$ level in a simple quantum well (made from an InGaAs layer in GaAs) is split in two in a sample consisting of two InGaAs wells separated by a narrow barrier of GaAs, and into five in a five-well sample. The spectrum is produced by light emitted when an electron excited into an $n = 1$ state in the well in the conduction band recombines with a hole in an $n = 1$ state of the inverted

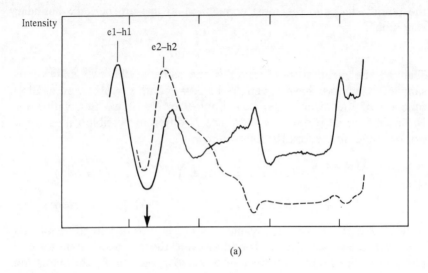

(a)

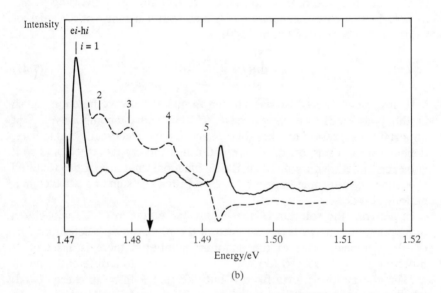

(b)

Figure 7.4 The splitting of energy levels in multiple InGaAs/GaAs quantum wells: (a) into two in a two-well sample; (b) into five in a five-well sample. The arrow marks the position at which a peak would occur in a similar single-well sample. (Adapted from Figure 1 of [6].) The peaks in the spectrum result from the recombination of an electron, in a state in the wells in the conduction band with $n = 1$ and $k = i$, with a hole, in a state in the inverted wells in the valence band, with the same quantum numbers $n = 1$ and $k = i$. The solid and dashed curves were obtained using slightly different experimental techniques.

well in the valence band. The transitions are labelled ei–hi, with the value of i distinguishing levels corresponding to the same n but different k.

As N increases, the spacing of the energy levels within each band decreases, and for electrons in a macroscopic crystalline solid, where N is very large, the energies within any band effectively span a continuous range of values. On the other hand, the different energy bands are separated by gaps, and in a perfect crystal lattice an electron cannot have an energy within one of these forbidden regions. A typical result for an electron in an effective potential with relatively shallow wells is shown in Figure 7.5(a). In the one-dimensional case there are no values of k for which different energy bands overlap, but this is not always the case for two- and three-dimensional lattices.

An alternative way of plotting the energy as a function of k is shown in Figure 7.5(b). To separate the different energy bands on the diagram, each is plotted as a function of values of k in a different Brillouin zone. Thus the $n = 1$ band is shown as a function of k in the first Brillouin zone, $|k| \leqslant \pi/\mathbb{L}$, the $n = 2$ band as a function of k in the second Brillouin zone, $\pi/\mathbb{L} \leqslant |k| \leqslant 2\pi/\mathbb{L}$, and so on. This is called the **extended zone scheme**, while the plot in which only the first Brillouin zone is used is called the **reduced zone scheme**. There is no difference in the physical information contained in Figures 7.5(a) and (b), since there is no physical significance in the particular choice of Brillouin zone.

The E/k plots for a free particle in both the reduced and extended zone schemes are shown in Figures 7.5(c) and (d). For the free particle the energy varies parabolically with k, since in this limit $\hbar k$ becomes the particle momentum, which is related to its energy by (3.22):

$$E(k) = \frac{\hbar^2 k^2}{2m} \tag{7.42}$$

If the periodic potential is weak (with shallow wells) the variation of energy with k is similar to this, as illustrated in Figure 7.5. The important difference is that in the free particle case the energy varies continuously over the range $(0, \infty)$, whereas in the periodic potential there are forbidden energy gaps at values of k marking the edge of a Brillouin zone. Provided N is not too small, these gaps are considerably larger than the spacing between levels corresponding to different k within the same band. If we think of k as the wave vector associated with the propagation of the electron through the whole lattice, the corresponding wavelength is $\lambda = 2\pi/k$. The values of k marking the edge of a Brillouin zone are such that an integer number of half-wavelengths fit into a unit cell of the lattice, so that states with equal but opposite values of k can interfere to produce standing waves in the cell. This is not possible for any other values of k. For these particular values the energy eigenfunction in each unit cell is similar to that for a bound state in a single potential well of

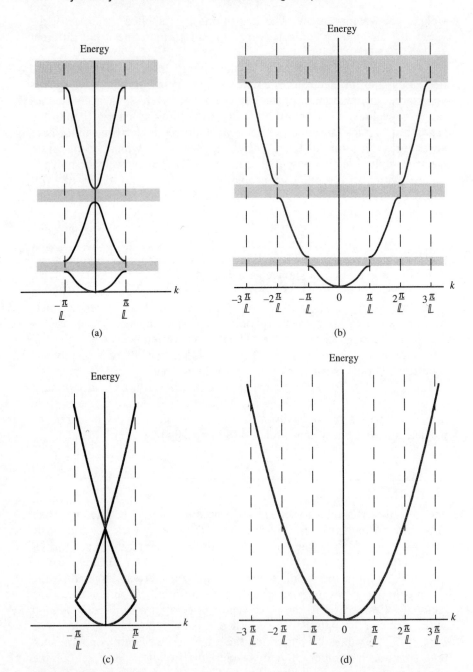

Figure 7.5 Plots of energy against crystal momentum for an electron in a periodic potential: (a) in the reduced zone scheme; (b) in the extended zone scheme. The shaded regions are the forbidden energy gaps. Corresponding plots of energy against the true momentum for a free electron are shown in (c) and (d).

infinite depth (illustrated in the case of an infinite square well in Figure 3.3).

7.6 Crystalline solids [7]

When we discussed the structure of many-electron atoms we started with the energy spectrum of a single electron in a spherically symmetric potential, and made use of the Pauli exclusion principle to allot the atomic electrons to different states. We proceed along similar lines in the case of the crystalline solid: so far we have discussed the energy spectrum of a single electron in a one-dimensional periodic potential. For a real solid we need to solve the corresponding three-dimensional problem, which is considerably more difficult. However, the general structure of the single-electron energy spectrum is the same in a three-dimensional periodic potential: it consists of a set of bands of very closely spaced levels, with relatively large ranges of forbidden energies, or band gaps, between them (though, as mentioned in Section 7.5, in the three-dimensional case the energy ranges of different bands may overlap). The closely spaced levels within a band are distinguished by the eigenvalues of the translation operator in three dimensions, which are of the form $\exp(i\mathbf{k} \cdot \mathbf{R}_l)$, where $\mathbf{k}$ is the crystal momentum vector within a Brillouin zone and $\mathbf{R}_l$ is a lattice vector connecting a point in one unit cell with the corresponding point in another cell, as shown in Figure 7.3(a). The Brillouin zone is now a volume of three-dimensional $\mathbf{k}$-space rather than a linear range of values of the one-dimensional vector k. The difference between two vectors $\mathbf{k}$ and $\mathbf{k}'$ that are in different Brillouin zones but correspond to the same eigenvalue of the translation operator is called a **reciprocal lattice vector** $\mathbf{g}$:

$$\mathbf{g} = \mathbf{k} - \mathbf{k}', \qquad \exp(i\mathbf{g} \cdot \mathbf{R}_l) = 1 \qquad (7.43)$$

Next we must invoke the Pauli exclusion principle to allot each of the many electrons in the solid to a different state. Each energy level in a particular band is labelled by $E(\mathbf{k})$, where $\mathbf{k}$ is a unique vector (within one Brillouin zone), and the band label n has been dropped. According to the Pauli exclusion principle, two electrons can occupy each of these states. The number of states within a band is therefore equal to twice the number of different $\mathbf{k}$ vectors within a Brillouin zone.

As in the one-dimensional case, the spectrum of values of $\mathbf{k}$ within a Brillouin zone can be found by taking account of the finite size of the solid and applying periodic boundary conditions, (7.39) (but this time in three dimensions). A set of values of the form given in (7.40) results for each Cartesian component of $\mathbf{k}$. In a cubic solid containing N^3 atoms, with the same lattice constant $\mathbb{L}$ along each Cartesian direction, the number of different crystal momentum states in the range $d^3k = dk_x \, dk_y \, dk_z$ is the product $dn = dn_x \, dn_y \, dn_z$, where each factor is of the form given by (7.41):

$$dn(\mathbf{k}) = \left(\frac{N\mathbb{L}}{2\pi}\right)^3 d^3k = \left(\frac{N\mathbb{L}}{2\pi}\right)^3 k^2 \, dk \, d\Omega = \frac{V}{(2\pi)^3} k^2 \, dk \, d\Omega \quad \textbf{(7.44)}$$

where $V = (N\mathbb{L})^3$ is the volume of the solid, and $d\Omega$ is the element of solid angle within which all $\mathbf{k}$ vectors in the range d^3k lie. The number of electrons that can be accommodated in these states is $2 \, dn(\mathbf{k})$. It is often useful to have a measure of the number of states in an energy range dE, and the **density of states** $g(E)$ is defined so that the number of single electron states per unit volume in the energy range between E and $E + dE$ is $g(E) \, dE$. This is related to $2 \, dn(\mathbf{k})$ through the $\mathbf{k}$ dependence of E, and if E depends on the magnitude of $\mathbf{k}$ but not on its direction (as in the free electron case, where (7.42) applies in three dimensions as well as one) then we obtain

$$g(E) = \frac{2}{V} \frac{dn(\mathbf{k})}{dE} = \frac{2}{(2\pi)^3} k^2 \, d\Omega \frac{dk}{dE} \quad \textbf{(7.45)}$$

The density of states for electrons in a crystalline solid is zero within an energy gap.

Equation (7.45) also gives the density of states for a free electron with momentum $\hbar\mathbf{k}$. If we think of the particle moving within some arbitrary but finite volume V, the number of states in a particular momentum range is given by (7.44). But the density of states is independent of V, and so we can use (7.45) even in the limit as V becomes infinite. (Remember that the factor of two takes account of the two possible spin states of an electron.)

The effect on an electron of the nuclei of the atoms forming the lattice can be represented by a periodic potential with deep, narrow wells as indicated in Figure 7.3(b). The lowest-energy electrons have large probability distributions close to the potential minima, and in each unit cell the periodic part of the Bloch function, $u_k(x, y, z)$, is very similar to the energy eigenfunction for a core electron in the isolated atom. These core electrons screen the higher-energy electrons from the full effect of the nuclear electric charge. The outer, valence, electrons have an appreciable probability density within the core, and so the effective potential due to the nuclei and core electrons should vary rapidly with penetration depth, becoming very deep close to each nucleus. However, an effective **pseudo-potential** can be defined [8], which averages over these rapid variations within the core, and allows the valence electrons to be treated as almost free.

The band structure of a solid determines whether it is an insulator, conductor or semiconductor. In an insulator the highest occupied band at zero kelvin (the **valence band**) is completely full, and the next-higher band, which is separated from it by a relatively large band gap, is completely empty, as shown in Figure 7.6(a). The band gap in a typical insulator such as sapphire or quartz is about 6 eV. Because of the Pauli exclusion principle, an electron can only be excited into an unoccupied level, and so the energy needed to excite a valence band electron is at least as great as the band gap energy. A semiconductor also has a completely full valence band at zero kelvin, but in this case the band gap is small (about 1.2 eV in silicon) so that at normal temperatures some electrons will be excited into the higher (**conduction**) band, as shown in Figure 7.6(b). (According to Boltzmann, the probability that at temperature T a particle will have energy E above its ground state is given by $e^{-E/k_B T}$, and $k_B T \approx 0.026$ eV at $T = 300$ K). Within a partly filled band, some filled levels are adjacent to unoccupied levels. Only a negligible amount of energy is needed to excite electrons in such levels, since the separation of levels within a band is so small (approaching zero as the size of the solid becomes infinite). Such excitation is the basis of electrical conduction, so electrons in the conduction band of a semiconductor at non-zero temperatures can carry electric current. Since the probability of thermal excitation of an electron into the conduction band increases with temperature, so does the conductivity of the semiconductor, and clearly any insulator becomes semiconducting at high enough temperatures (provided it remains solid). A true conductor, on the other hand, has its highest occupied band at 0 K only partly full, as shown in Figure 7.6(c). This band is therefore a conduction band, and since the population of electrons in it is not changed significantly by increasing the temperature, the conductivity is roughly temperature-independent. In a conductor the energy of the highest level that is occupied at zero kelvin is called the **Fermi energy** E_F. In Figure 7.6(c) the number of electrons in each state within the conduction band is plotted against the energy of the state, both at zero kelvin and at a finite temperature. At zero kelvin the curve is discontinuous at $E = E_F$, since all the states below this are filled, while all those above it are empty. At non-zero temperatures some electrons from just below E_F can be thermally excited to just above it, so that the curve decreases more gently to zero. (The shape of this curve cannot be predicted by classical Boltzmann statistics, but requires the use of quantum, Fermi–Dirac, statistics [9]. The Boltzmann probability distribution, proportional to $e^{-E/k_B T}$, approaches the Fermi–Dirac distribution at high values of $E/k_B T$.)

The independent electron model is useful for conduction electrons in semiconductors, where only a small proportion of the levels in the conduction band is populated, and electrons can be treated as nearly free. In the conduction band of a conductor the electrons can also be treated as free to a first approximation, but the large population of the band means that it is more appropriate to treat them as a gas of charged particles than

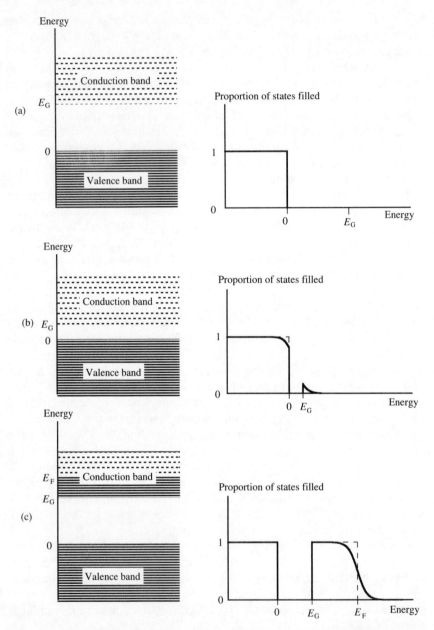

Figure 7.6 Band gap diagrams for (a) an insulator, (b) a semiconductor and (c) a conductor at a temperature of 0 K. The full stripes indicate filled levels, and the dashed stripes empty ones. The zero of the energy scale is chosen at the top of the valence band in each case. On the right of each band gap diagram the fraction of energy states that are filled is shown as a function of energy. In (b) and (c) the dashed line refers to 0 K and the solid line to room temperature. E_G is the band gap energy in each case, and in the case of the conductor the Fermi energy E_F is also indicated.

as individual electrons. Important many-particle effects occur, such as plasma oscillations of the negative charge density in the electron gas, and independent electron models cannot deal with such phenomena. The independent electron model also breaks down for insulators, where the occupied bands are completely full and many-body effects again become very important.

The interaction of an electron with other electrons and with the crystal lattice affects the way it responds to an applied force. This means that in a free electron model the inertial mass of an electron is not the same as for a free particle, and we introduce an **effective mass** m^*. A useful approximate expression for the effective mass of an electron within a band of energies $E(k)$ is

$$m^* = \hbar^2 \left(\frac{d^2 E}{dk^2} \right)^{-1} \tag{7.46}$$

(Note that this reduces to the free mass if the energy takes the form (7.42), which applies to a free particle.) The E/k plots in Figure 7.5 show that there are values of k, in particular near a Brillouin zone boundary, for which m^* is negative, so that the acceleration of the electron is in the opposite direction to the applied force! Since the way in which the energy varies with crystal momentum may be different along different directions through the crystal, the effective mass is really a tensor quantity rather than a scalar.

When an electron is excited from a full band to a higher empty or partially filled one it leaves an empty state, or **hole**, behind. A hole can move through the lattice, since an electron can move into the empty state, leaving a new empty state behind. It behaves like a particle with the opposite charge and effective mass to those of the electron that would occupy the state, and so in regions of k space where the effective mass is negative it is often more convenient to describe the behaviour of the solid in terms of positively charged holes with positive effective mass. As the Hall effect (see Section 8.5) first demonstrated, the electric current in some semiconductors is carried by positively charged holes rather than negatively charged electrons.

References

[1] Kreyszig E. (1988) *Advanced Engineering Mathematics* 6th edn, Section 8.11, Example 5. New York: Wiley.
[2] Berkeley Physics Course, Vol. 1 (1973) *Mechanics* 2nd edn, Chap. 6, pp. 185–92. New York: McGraw-Hill.
[3] Boas M.L. (1983) *Mathematical Methods in the Physical Sciences* 2nd edn, Chap. 12, Section 22. New York: Wiley.

[4] For further reading see Woodgate G.K. (1980). *Elementary Atomic Structure* 2nd edn. Oxford University Press.

[5] Kittel C. (1986) *Introduction to Solid State Physics* 6th edn, Chap. 7, pp. 164–6. New York: Wiley.

[6] Moore J.K. *et al.* (1990) *Physical Review* **B42**, 3024.

[7] The books mentioned in Refs [5] and [8] are good introductions to solid state physics.

[8] Omar M.A. (1975) *Elementary Solid State Physics*, Section 5.9. Reading MA: Addison-Wesley.

[9] Kittel C. and Kroemer H. (1980). *Thermal Physics*, Chaps 6 and 7. San Francisco: W.H. Freeman.

Problems

7.1   Verify the form of the radial equation (7.7) by substituting a wave function of the form (7.4) into the Schrödinger equation (7.3).

7.2   The potential energy of a charge q in the field of an electric dipole with moment $\mathcal{D}$ is

$$\hat{V} = -q\mathcal{D} \cdot \mathbf{r} = -q\mathcal{D} \, \frac{\cos \theta}{r^2}.$$

where θ is the angle between $\mathcal{D}$ and $\mathbf{r}$, the vector separation of the charge from the centre of the dipole. Find the commutators of this potential with $\hat{L}_z$ and $\hat{L}^2$. What can you deduce about the energy eigenstates of a charged particle in the field of an electric dipole?

7.3   Write down the Hamiltonian for the two-dimensional simple harmonic oscillator (see Problem 5.11) in polar coordinates (r, ϕ), and prove that it commutes with $\hat{L}_z$. Explain how this leads to a product form for the energy eigenfunctions, and find the appropriate radial equation. Show that each of the following functions are energy eigenfunctions for the oscillator, giving the energy and angular momentum eigenvalues in each case:

$$\psi_a(r, \phi) = C_a \exp\left(-\frac{m\omega r^2}{2\hbar}\right)$$

$$\psi_b(r, \phi) = C_b r e^{i\phi} \exp\left(-\frac{m\omega r^2}{2\hbar}\right)$$

$$\psi_c(r, \phi) = C_c r e^{-i\phi} \exp\left(-\frac{m\omega r^2}{2\hbar}\right)$$

How are these functions related to energy eigenfunctions of the form $\Psi(x, y) = \psi_j(x)\psi_k(y)$, used in Problem 5.11? Note that

$$\frac{\partial^2}{\partial x^2} + \frac{\partial^2}{\partial y^2} = \frac{\partial^2}{\partial r^2} + \frac{1}{r}\frac{\partial}{\partial r} + \frac{1}{r^2}\frac{\partial^2}{\partial \phi^2}$$

7.4   Write down the Schrödinger equation for a particle in a spherical square
well,

$$\hat{V} = \begin{cases} 0 & (r \leqslant r_0) \\ V_0 & (r > r_0) \end{cases}$$

Show that the ground state eigenfunction is of the form $(C_1 \sin kr)/r$ for
$r \leqslant r_0$, and $C_2 e^{-\alpha r}/r$ for $r > r_0$. By matching the wave function and its
derivative at the edge of the well, prove that the ground state energy is
determined by $k \cot kr_0 = -\alpha$, and that for the ground state kr_0 lies in the
range $(\frac{1}{2}\pi, \pi)$. Deduce that there can be no bound state in the well unless
$V_0 r_0^2 > \pi^2 \hbar^2/8m$.

7.5   Write down the full Hamiltonian, of the form given in (7.1), for the
hydrogen atom. Show that it does not commute with any component of the
position operator $\hat{R} = (r, \theta, \phi)$, or with the radial momentum operator
$\hat{P}_r = -i\hbar(\partial/\partial r + 1/r)$. Explain what this means for the Bohr picture of an
electron circling the proton in a well-defined orbit.

7.6   Substitute the form $R(\rho) = e^{-\alpha\rho}\rho^l f(\rho)$ in the radial equation (7.12) for
hydrogen and show that $f(\rho)$ satisfies (B3.3). (This becomes the Laguerre
equation for $\alpha^2 = -\epsilon = |\epsilon|$.)

7.7   Prove that the following wave functions are solutions of the radial equation
(7.10) for hydrogen, and find the corresponding values of n and l:

$$R(r) = C e^{-r/r_B}, \qquad R'(r) = C'\frac{r}{r_B}\left(1 - \frac{1}{6}\frac{r}{r_B}\right)e^{-r/3r_B}$$

Calculate the normalization constant in each case, and show explicitly that
the functions are orthogonal to one another.

7.8   (a) For a hydrogen atom in its ground state, calculate the width of the
classically allowed region, and find the probability for the electron to
be found outside this region.

(b) Show that the most probable radial distance of the electron from the
proton in the ground state of hydrogen is r_B, and find the total probability of finding the separation to be greater than this.

(c) Using the general form for the radial wave function for hydrogen given
in (7.14), find the most probable radial distance of the electron from
the proton in the state with quantum numbers n, $l = n - 1$. (For $k = 0$
the Laguerre polynomials L_k^q take the values $L_0^q = q!$.)

*7.9   (a) Use the computer program *Schrödinger equation: central potentials
Coulomb potential* to verify the properties (i), (ii) and (iii) of the
hydrogen atom probability distribution given at the end of Section 7.2.

(b) The radial equation for a three-dimensional harmonic oscillator is given
by (7.7) with $V = \frac{1}{2}m\omega^2 r^2$, and the energy eigenvalues are characterized by a quantum number $n \geqslant l$. Use the computer program *Schrödinger equation: central potentials harmonic oscillator* to prove that

(i) the energy spectrum is $E_n = (n + \frac{3}{2})\hbar\omega$ $(n = 0, 1, 2, \ldots)$;

(ii) there are no energy eigenfunctions with even l if n is odd, and none with odd l if n is even;

(iii) the number of zeros of the radial wave function R_{nl} and the probability distribution $r^2|R_{nl}|^2$ is $\frac{1}{2}(n - l)$ (not counting zeros at $r = 0$ and as $r \to \infty$);

(iv) for fixed n the region where the probability density is appreciable moves further from the centre $(r = 0)$ as l increases.

7.10 Show that the eigenstates of the hydrogen atom Hamiltonian

$$\hat{H} = -\frac{\hbar^2}{2m}\left(\frac{d^2}{dr^2} + \frac{2}{r}\frac{d}{dr} - \frac{\hat{L}^2}{r^2}\right) - \frac{e^2}{4\pi\varepsilon_0 r}$$

may be simultaneously eigenstates of $\hat{J}^2$ and $\hat{J}_z$, where $\hat{J} = \hat{L} + \hat{S}$ is the total angular momentum. Using the results of Problem 6.8, write down the complete set of degenerate eigenfunctions for the level $n = 2$ in (a) the $\{n, l, j, j_z\}$ classification scheme and (b) the $\{n, l, m_l, m_s\}$ scheme. Do the same for $n = 1$ and $n = 3$. Check that the degeneracy of the levels is the same in both schemes.

7.11 When an electron in a semiconductor is excited into the conduction band, it interacts with the hole left in the valence band through a Coulomb potential. The electron and hole may form a bound state (called an **exciton**), and its energy spectrum is of the same form as that of the hydrogen atom. However, in a quantum well material the electron and hole are confined in the z direction, and so the hydrogen-like Schrödinger equation is two-dimensional, of the form

$$\left[-\frac{\hbar^2}{2m}\left(\frac{d^2}{d\rho^2} + \frac{1}{\rho}\frac{d}{d\rho} - \frac{\hat{L}_z^2}{\rho^2}\right) - \frac{e^2}{4\pi\varepsilon_0\rho}\right]R(\rho) = ER(\rho)$$

$$\rho = \sqrt{(x^2 + y^2)}$$

where m is the reduced mass of the electron and hole, (A1.7). Show that the ground state energy is $-4E_R$, and that the most probable radial separation in the ground state is $2r_B$, where E_R and r_B are of the forms given in (7.11). If the effective mass of the electron is $0.07m_e$ and that of the hole is $0.08m_e$, calculate the ground state energy and the most probable radius of the exciton.

7.12 Find the effect of the operator $\hat{T}_L$ that produces a translation through L in the x direction on each of the following functions:

(a) $(x - L)(x + L)$ 　　(b) $\tan\dfrac{\pi x}{L}$

(c) $\sin\dfrac{\pi x}{L}$ 　　(d) $\sin\dfrac{\pi x}{2L}$

(e) $e^{ikx^2}u(x)$, where $u(x + L) = u(x)$ and k is constant

(f) $e^{i\sin(2\pi x/L)}u(x)$, where $u(x + L) = u(x)$

Identify any that are eigenfunctions of $\hat{T}_L$, and give the corresponding eigenvalue. Which of them satisfy (7.27)? and which satisfy (7.30)?

7.13 Write down the form of the solutions to the Schrödinger equation (7.35) when the potential within a unit cell consists of a well of width a and a barrier of width $b = L - a$ and height V_0, as shown in Figure 7.3(d). For convenience, measure all energies from the bottom of the wells, and put $x = 0$ at the edge of a well. Write down the set of four boundary conditions that the solution $u_k(x)$ must satisfy. These can be solved to give an equation determining the eigenvalue E:

$$\cos kL = \cos \alpha a \cosh \beta b + \frac{\beta^2 - \alpha^2}{2\alpha\beta} \sin \alpha a \sinh \beta b$$

where $\alpha^2 = (2m/\hbar^2)E$ and $\beta^2 = (2m/\hbar^2)(V_0 - E)$. In the limit as $V_0 \to \infty$ and $b \to 0$ in such a way that $\beta b \to 0$ and $\beta^2 ab \to C$ (a finite constant), take the first terms in the series expansions for $\cosh \beta b$ and $\sinh \beta b$, and show that the equation reduces to $\cos kL = \cos \alpha L + (C/2\alpha L) \sin \alpha L$. (This is the **Kronig–Penney model**.)

7.14 (a) Convince yourself that (7.40) is correct, and calculate all the allowed values of k in the first Brillouin zone in a finite crystal containing (i) 2, (ii) 5 and (iii) 10 unit cells. Repeat your calculation for values of k in the second Brillouin zone.

(b) Given that E is an even function of k, show that all levels, except for those at the edge of a Brillouin zone or with $k = 0$, are doubly degenerate. Calculate all the allowed values of kL and $\cos kL$ in each of the cases $N = 2, 3, 4, 5, 6, 7$ and 10.

(c) Calculate the wavelengths of all possible standing waves in a unit cell of width L. Show that the corresponding values of k $(= 2\pi/\lambda)$ are at a Brillouin zone edge.

*7.15 Use the computer program *The Kronig–Penney Model* to investigate the energy levels in a one dimensional periodic lattice composed of square wells and barriers in the limit as $V_0 \to \infty$, $b \to 0$, $\beta^2 ab \to C$ mentioned in Problem 7.13. The program plots the right-hand side of the equation $\cos kL = \cos \alpha L + (C/2\alpha L) \sin \alpha L$ as a function of αL, and, if the number of unit cells $N > 10$, draws the horizontal lines $\cos kL = \pm 1$. The edges of the allowed energy bands are determined by the points where these straight lines intersect the curve, and the program prints the energy range (in eV) of each band that appears on the graph. If $N \leqslant 10$, the program draws a set of horizontal lines corresponding to all allowed values of $\cos kL$ (cf. Problem 7.14b). In these cases the program prints the energy of each allowed level (corresponding to each allowed value of k) in the first five 'bands'. For the choices $C = 2, 5, 10$ and 50, find the allowed energies in the lowest four bands:

(a) when there is such a large number of unit cells in the lattice that the spectrum of k is essentially continuous;

(b) when there are only two unit cells in the lattice;

(c) when there are five unit cells in the lattice.

7.16 The principal axes of a three-dimensional cubic lattice are the x, y and z axes, and the lattice spacings along these directions are l_x, l_y and l_z. Which of the following vectors are lattice vectors?

$$r_1 = l_x x - l_y y, \qquad r_2 = 1.5 l_x x + 1.5 l_y y + 1.5 l_z z$$

$$r_3 = -2 l_x x + 3 l_y y + l_z z$$

$$r_4 = l_x x + l_y y + 0.5 l_z z$$

(x, y and z are unit vectors along the Cartesian axes). Which of the following are reciprocal lattice vectors?

$$k_1 = 2\pi \left(\frac{1}{l_x} x + \frac{1}{l_y} y + \frac{1}{l_z} z \right)$$

$$k_2 = 2\pi \left(\frac{1}{3 l_x} x + \frac{1}{l_z} z \right)$$

$$k_3 = \pi \left(\frac{1}{l_x} x + \frac{1}{l_y} y + \frac{1}{l_z} z \right)$$

$$k_4 = 2\pi \left(\frac{3}{l_x} x + \frac{2}{l_y} y - \frac{4}{l_z} z \right)$$

Verify that each of the lattice vectors and reciprocal lattice vectors satisfy (7.43).

7.17 (a) Calculate the density of states, as a function of energy, for a free electron in one dimension by using the periodic boundary condition (7.39) in a region of width l. Find the corresponding densities of states in two and in three dimensions.

(b) Assume that there is a fixed density n_c of conduction electrons in a metal, and that they can be treated approximately as free. Find expressions for the *total* number of electron states per unit volume between energy 0 and E, and the energy E_F of the highest occupied level at $0\,K$ (the Fermi energy), in the case where the electrons can move in three dimensions and in the case where they are confined to a two-dimensional plane. In silver $n_c = 5.85 \times 10^{28}\,m^{-3}$ and $E_F = 5.48\,eV$. Calculate the effective mass of the conduction electrons in a three-dimensional sample in a free electron approximation.

7.18 (a) Calculate the maximum wavelength of the light that can excite an electron from the top of the valence band into the conduction band in each of the following cases (the transition is not accompanied by a change in crystal momentum):

(i)   GaAs, $E_G = 1.52\,eV$

(ii) PbS (lead sulphide), $E_G = 0.29\,eV$,

(iii) CdS (cadmium sulphide), $E_G = 2.58\,eV$

(iv) InAs (indium arsenide), $E_G = 0.36\,eV$

(b) At each of the temperatures $T = 5$, 50, 300 and 3000 K calculate the probability of an electron being thermally excited to the conduction band in each of the materials mentioned in (a). (Use the values quoted for the band gaps, though in practice these vary slightly with temperature.)

CHAPTER 8

Magnetic Effects in Quantum Systems

8.1 The Hamiltonian for a charged particle in an electromagnetic field

We know from classical electromagnetic theory that the force on a particle with charge q and mass m moving with velocity $\boldsymbol{v}$ in an electric field $\mathcal{E}$ and a magnetic field $\mathcal{B}$ is the Lorentz force:

$$\mathcal{F} = m\,\frac{\mathrm{d}\boldsymbol{v}}{\mathrm{d}t} = q(\mathcal{E} + \boldsymbol{v} \times \mathcal{B}) \qquad (8.1)$$

The electric force produces an acceleration in the direction of $\mathcal{E}$, but the magnetic force is perpendicular to the velocity, and therefore does no work on the particle. If $\mathcal{E} = 0$ and the charge moves in a purely magnetic field, the kinetic energy remains constant, and the velocity changes in direction but not in magnitude. In this case, if the initial velocity is in the plane perpendicular to $\mathcal{B}$, the path of the particle is a circle of constant radius $mv/q\mathcal{B}$, and the angular frequency of the particle in this circular path is $q\mathcal{B}/m$ [1]. This is called the **cyclotron frequency** ω_c:

$$\omega_c = \frac{|q|\mathcal{B}}{m} \qquad (8.2)$$

If the particle has a component of velocity parallel to $\mathcal{B}$, either due to the initial conditions or due to the acceleration produced by an electric field in that direction, the path will become a circular spiral (or helix) with the centre of the circle moving parallel to $\mathcal{B}$.

To treat the motion of a charged particle in an electromagnetic field quantum mechanically, we must write down a Hamiltonian to represent its total energy, rather than using the Lorentz force law (8.1) directly. We can then find its possible energy eigenstates by solving the Schrödinger equation. In an electrostatic field $\mathcal{E}$ this is easily done by introducing the

Coulomb potential $\mathcal{V}$ through $\mathcal{E} = -\nabla\mathcal{V}$. The potential energy of the charge q in this field is then

$$V = q\mathcal{V} = -q\int\mathcal{E}\cdot d\boldsymbol{r} \tag{8.3}$$

The potential energy that we used for an electron in a hydrogen atom, (7.8), is of this type. Similarly, a constant electric field $\mathcal{E}$ in the x direction acting on an electron is represented by the potential $V = e\mathcal{E}x$. (The change in the energy eigenvalues and eigenstates of a bound electron due to the effect of an externally applied electric field is called the **Stark effect**.)

Since a magnetic field does not change the kinetic energy of a particle, we should not expect to represent its effect by a potential energy. However, in some particular situations where the magnetic field is static it is possible, and convenient, to introduce an effective magnetic potential energy. In particular, we shall be interested in a charged particle bound in an orbit, like an electron in an atom. Such a bound charge behaves like a current loop, which, according to classical electromagnetic theory, has a magnetic dipole moment $\boldsymbol{\mu}$. In a magnetic field $\mathcal{B}$ a magnetic dipole has potential energy [2]

$$V_{\text{mag}} = -\mathcal{B}\cdot\boldsymbol{\mu} \tag{8.4}$$

If the magnetic moment is not parallel to the field, its direction precesses about the direction of the field (as shown in Figure 8.1a) with the Larmor frequency ω_L [3]:

$$\omega_L = \frac{|q|\mathcal{B}}{2m} \tag{8.5}$$

The magnetic dipole moment of a current loop [2] has a magnitude equal to the product of the current with the area of the loop, and is in a direction perpendicular to the loop, as shown in Figure 8.1(b). This means that the magnetic dipole moment associated with an orbiting charge is proportional to its orbital angular momentum. To see this, consider a particle of mass m and charge q in a circular orbit of radius r. The equivalent current $\mathcal{I}$ is the charge divided by the time for one complete revolution: $\mathcal{I} = qv/2\pi r$, where v is the speed of the particle. Multiplying this by the area of the orbit, and using Figure 8.1(b) to identify the direction of the dipole moment, we obtain

$$\boldsymbol{\mu} = \frac{q}{2m}(m\boldsymbol{r}\times\boldsymbol{v}) \tag{8.6}$$

The term in parentheses is just the orbital angular momentum of the charge. The quantum mechanical operator $\hat{\boldsymbol{\mu}}_l$ representing the dipole moment of a charge due to its orbital motion can be obtained from the correspondence principle: we substitute the orbital angular momentum

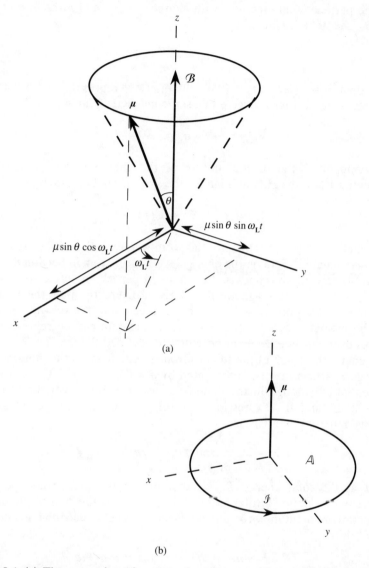

Figure 8.1 (a) The precession of a magnetic moment μ (due to the rotation of a negatively charged particle) about the direction of the magnetic field $\mathcal{B}$, with angular frequency ω_L. (b) The magnetic dipole moment μ of a loop of area A carrying a current $\mathcal{I}$. The angular momentum of the current carriers is in the opposite direction to μ if they are negatively charged.

operator $\hbar\hat{L}$ for the classical angular momentum in (8.6):

$$\hat{\mu}_l = \frac{q\hbar}{2m}\,\hat{L} \tag{8.7}$$

If the particle is an electron, with charge $q = -e$ and mass $m = m_e$, we define the **Bohr magneton** by

$$\mu_B = \frac{e\hbar}{2m_e} \tag{8.8}$$

and then $\hat{\boldsymbol{\mu}}_l = -\mu_B \hat{\boldsymbol{L}}$. From (8.4), the operator representing the potential energy of an orbiting charge q in a static magnetic field $\mathcal{B}$ is

$$\hat{V}_{\text{mag}} = -\mathcal{B} \cdot \hat{\boldsymbol{\mu}}_l = -\frac{q\hbar}{2m} \mathcal{B} \cdot \hat{\boldsymbol{L}} \tag{8.9}$$

Assuming that the orbital motion of the charge is due to a central potential $\hat{V}(r)$, the Hamiltonian for the system can be written as

$$\hat{H} = -\frac{\hbar^2}{2m} \nabla^2 + \hat{V}(r) + \hat{V}_{\text{mag}} \tag{8.10}$$

We shall need a Hamiltonian of this form when we discuss the effect of an applied magnetic field on the energy levels of an atom in Section 8.2.

In the case of an otherwise-free charged particle in an electromagnetic field, although the magnetic field does not directly alter the particle's energy, it does produce a change in the direction of the velocity, and thus of the momentum. This effect on the momentum can be represented in terms of the magnetic vector potential $\mathcal{A}$, defined though $\mathcal{B} = \nabla \times \mathcal{A}$ [4]. In general the effect of the full electromagnetic field on the motion of the charged particle may be represented by a scalar potential $\mathcal{V}$ and a vector potential $\mathcal{A}$ (though in the non-static case $\mathcal{V}$ is no longer the Coulomb potential), and the expressions for the electric and magnetic fields in terms of these potentials are

$$\mathcal{E} = -\frac{\partial \mathcal{A}}{\partial t} - \nabla \mathcal{V}, \qquad \mathcal{B} = \nabla \times \mathcal{A} \tag{8.11}$$

Just as the total energy E_{tot} is the sum of the kinetic and potential energies, we define a total momentum $\boldsymbol{p}_{\text{tot}}$ that is the sum of the normal (kinetic) momentum $m\boldsymbol{v}$ and an electromagnetic **potential momentum** $q\mathcal{A}$ [5]:

$$E_{\text{tot}} = \tfrac{1}{2}mv^2 + q\mathcal{V}, \qquad \boldsymbol{p}_{\text{tot}} = m\boldsymbol{v} + q\mathcal{A} \tag{8.12}$$

As usual, to transform the classical expressions (8.12) into quantum mechanical ones, we use the correspondence principle, and replace the classical variables by appropriate operators. The vector potential is a function of position, like the scalar potential, and since the components of the position operator are simply the algebraic variables x, y and z, $q\mathcal{V}$ and $q\mathcal{A}$ are represented as respectively scalar and vector functions of x, y and z. It turns out that it is the total momentum operator $\hat{\boldsymbol{P}}_{\text{tot}}$ that must be represented by the differential operator (3.9) (just as the total, rather than the kinetic, energy is represented by the differential operator (3.7)).

The expression for the total energy in terms of the kinetic and potential energies determines the Hamiltonian operator, so the quantum mechanical equations corresponding to (8.12) are

$$\hat{H} = \frac{1}{2m}\,\hat{P}_{\text{kin}}^2 + q\mathcal{V}(x, y, z) \left.\vphantom{\begin{array}{c}a\\b\end{array}}\right\}$$
$$\hat{P}_{\text{tot}} = -i\hbar\nabla = \hat{P}_{\text{kin}} + q\mathcal{A}(x, y, z) \quad (8.13)$$

where $\hat{P}_{\text{kin}}$ is the operator representing the kinetic momentum $m\boldsymbol{v}$. The full expression for the Hamiltonian of a (non-relativistic) charged particle in an electromagnetic field derivable from the potentials $\mathcal{V}$ and $\mathcal{A}$ is therefore

$$\hat{H} = \frac{1}{2m}\,[-i\hbar\nabla - q\mathcal{A}(x, y, z)]^2 + q\mathcal{V}(x, y, z) \quad (8.14)$$

8.2 The effects of applied magnetic fields on atoms

A magnetic field $\mathcal{B}$ applied to an atom interacts with the magnetic dipole moment produced by the orbital motion of the electrons. This affects the energy levels through the effective magnetic potential energy (8.9). For a single electron in a constant uniform field $\mathcal{B}$ in the z direction, (8.9) becomes

$$\hat{V}_{\text{mag}} = \mathcal{B}\mu_{\text{B}}\hat{L}_z = \hbar\omega_{\text{L}}\hat{L}_z \quad (8.15)$$

where μ_{B} is the Bohr magneton, (8.8), and ω_{L} is the Larmor frequency, (8.5). The Hamiltonian for a hydrogen atom in this magnetic field is given by (8.10), with $\hat{V}(r)$ identified as the appropriate Coulomb potential:

$$\hat{H} = -\frac{\hbar^2}{2m}\nabla^2 - \frac{e^2}{4\pi\varepsilon_0 r} + \hbar\omega_{\text{L}}\hat{L}_z = \hat{H}_0 + \hbar\omega_{\text{L}}\hat{L}_z \quad (8.16)$$

$\hat{H}_0$ is the Hamiltonian for the free hydrogen atom, and we have already found its eigenvalues E_n, (7.13), and eigenfunctions, (7.16). So the eigenstates of $\hat{H}_0$ can be classified by the set of quantum numbers (n, l, m_l).

The eigenfunctions of $\hat{H}_0$ given by (7.16) are eigenfunctions of $\hat{L}_z$, and so are also eigenfunctions of $\hat{V}_{\text{mag}}$, and thus of the complete Hamiltonian $\hat{H}$. The only effect of the additional potential is to add an energy proportional to the eigenvalue of $\hat{L}_z$. In Section 7.1 we saw that the energy of a particle in a spherically symmetric potential cannot depend on the eigenvalue of $\hat{L}_z$, but the potential $\hat{V}_{\text{mag}}$ of (8.15) is not spherically symmetric – an applied magnetic field breaks the spherical symmetry of the free atom by defining a particular direction in space (which we label the z axis), and the energy of the electron then depends on $\hat{L}_z$. So states

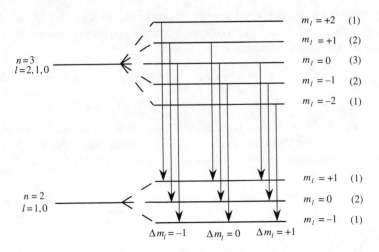

Figure 8.2 The Zeeman splitting of the $n = 2$ and $n = 3$ levels of hydrogen predicted by (8.17). The degeneracies of the magnetic sublevels are indicated in parentheses. The transitions from the $n = 3$ to the $n = 2$ level that are responsible for lines in the electric dipole emission spectrum are indicated, together with the associated change in m_l.

with the same n but different m_l no longer have the same energy, and each energy level of the free hydrogen atom is split into a number of sublevels:

$$E_n \rightarrow E_{n,m_l} = E_n + \hbar\omega_L m_l \tag{8.17}$$

The splitting of the $n = 2$ and $n = 3$ levels of hydrogen predicted by this equation is illustrated in Figure 8.2.

The splitting of the energy levels produces a corresponding splitting of the atomic spectral lines when atoms emit or absorb radiation in a region where there is a magnetic field. In optical spectroscopy we normally observe electric dipole radiation, which involves a change in the orbital angular momentum quantum numbers from $\{l, m_l\}$ to $\{l', m_l'\}$, where

$$\Delta l = l' - l = \pm 1, \qquad \Delta m_l = m_l' - m_l = \pm 1, 0 \tag{8.18}$$

(These selection rules are derived in Section 12.4.) The energy absorbed by a hydrogen atom in a magnetic field during a transition from a state with quantum numbers $\{n, l, m_l\}$ to one with quantum numbers $\{n', l', m_l'\}$ should therefore be, from (8.17) and (8.18),

$$E_{n',m_l'} - E_{n,m_l} = E_{n'} - E_n + \hbar\omega_L(m_l' - m_l)$$

$$= h\nu + \begin{cases} \hbar\omega_L \\ 0 \\ -\hbar\omega_L \end{cases} \tag{8.19}$$

where v is the frequency of the spectral line in the absence of the magnetic field. This means that, however large the number of different m_l values in the levels concerned, only three different frequencies of light should be observed in the spectrum, as indicated in Figure 8.2. This is true for hydrogen in strong magnetic fields, above approximately 0.1 T.

An atom with more than one electron has an orbital magnetic moment that is the resultant magnetic moment due to the orbital motion of all the electrons. This changes the total energy of the atom when it is placed in a uniform magnetic field, through a potential $\hat{V}_{\text{mag}}$ of the form given in (8.15), but with $\hat{L}_z$ interpreted as the z component of the resultant atomic orbital angular momentum operator, which is the (vector) sum of the orbital angular momentum operators of the individual electrons. In Section 7.3 we introduced the notation $\{L, M_L\}$ for the quantum numbers describing the resultant orbital angular momentum state of an atom, and in the absence of a magnetic field the atomic energy levels depend on L, but not on M_L. A magnetic field breaks the spherical symmetry of the atom and splits each atomic energy level into sublevels whose energy is proportional to M_L. When electric dipole radiation is emitted or absorbed by an atom containing more than one electron the selection rules (8.18) apply to the quantum numbers $\{L, M_L\}$, instead of to the single-electron quantum numbers $\{l, m_l\}$. So the spectral lines of an atom would be expected to split into three components in a magnetic field, in a similar way to that indicated in (8.19).

The splitting of an atomic spectral line into three components when the atom is placed in a magnetic field is known as the **normal Zeeman effect**. However, the spectra of most atoms exhibit much more complicated patterns of splitting, referred to as the **anomalous Zeeman effect**. Both the normal and anomalous Zeeman effects are illustrated in Figure 8.3. It is the anomalous effect that is usually observed, and the normal effect is only seen in exceptional cases, but the Zeeman splittings in any atom are reduced to three if sufficiently high magnetic fields are used. (We shall return to this at the end of Section 8.4.) Our discussion in this section has ignored the intrinsic magnetic moment of the electron, which we introduced in Section 6.2, and it is this, in addition to the magnetic moment due to its orbital motion, that explains the anomalous Zeeman effect.

8.3 The Stern–Gerlach experiment and electron spin

The idea that the electron has an intrinsic magnetic moment was first introduced to explain the anomalous Zeeman effect, but direct experimental evidence is provided by the Stern–Gerlach experiment. This was originally designed to test whether or not the angle between the angular momentum vector and the magnetic field is quantized. As we saw in

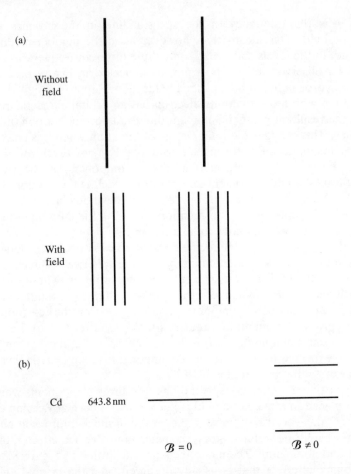

Figure 8.3 The anomalous and normal Zeeman effects: (a) the anomalous effect in the sodium D lines, which have wavelengths 589.0 nm and 589.6 nm in the absence of a magnetic field [6]; (b) the normal effect in the transition in cadmium at 643.8 nm, between the levels $n = 4$, $L = 2$, $S = 0$ and $n = 5$, $L = 1$, $S = 0$.

Section 4.5, the orientation depends on the z component of the angular momentum, which, according to quantum theory, can take $2l + 1$ different values, corresponding to all the possible values of m_l for a fixed l. The angular momentum of an atomic electron can be investigated through its magnetic effect, defined by (8.7), as discussed in Section 8.2. In their original experiment Stern and Gerlach used atoms of silver, which has only one valence electron outside a completely filled core (see Table 7.2), so that the resultant magnetic moment of each atom is due to the single valence electron. An atom passing through a static but non-uniform magnetic field will experience a force $-\nabla V_{\mathrm{mag}}$, which, from (8.4), is proportional to the component of its magnetic dipole moment

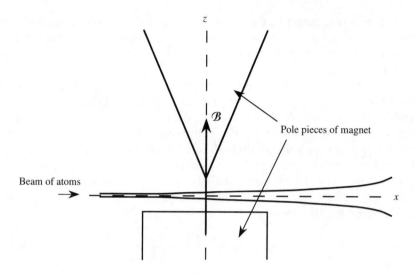

Figure 8.4 The experimental arrangement for the Stern–Gerlach experiment.

parallel to the field, and this is proportional to the corresponding component of the angular momentum. So atoms in a beam in which the magnetic moments are initially randomly oriented will experience different forces when they enter a region containing a non-uniform magnetic field. Classically the beam would be expected to spread out in a continuous manner, while quantum mechanically it should split into $2l + 1$ discrete sub-beams.

A Stern–Gerlach magnet has pole pieces shaped so that the field $\mathcal{B}$ is in the z direction and varies rapidly with z, but is almost constant in the (x, y) plane. The experimental arrangement is shown schematically in Figure 8.4. A one-electron atom passing between the pole pieces, with its momentum in the x direction, experiences a force that is only appreciable in the z direction, represented by

$$\hat{F}_z = - \frac{\partial \hat{V}_{\text{mag}}}{\partial z} = \mu_{\text{B}} \frac{\mathrm{d}\mathcal{B}}{\mathrm{d}z} \hat{L}_z \tag{8.20}$$

So the atom is deflected in the z direction by an amount proportional to the z component of angular momentum. In a Stern–Gerlach experiment the incident beam of atoms is parallel (so that, to a very good approximation, all atoms have momenta in the same direction before entering the field) and the magnetic dipole moments of the atoms are initially randomly orientated. As the beam travels between the poles of the magnet each atom experiences a force that depends on the orientation of its dipole moment, and the beam is split into a discrete number of sub-beams. This confirms that the angle between an atomic magnetic moment and the external magnetic field is restricted to a discrete set of values, and that the angular momentum is quantized. But Stern and Gerlach found that the beam of silver atoms split into two, apparently implying $2l + 1 = 2$, and $l = \frac{1}{2}$.

The restriction of orbital angular momentum quantum numbers to zero or integer values is essential if the corresponding eigenfunction is to satisfy the condition (4.36) that it should be single-valued. So the magnetic moment responsible for the splitting of the beam of silver atoms cannot possibly be due to the orbital motion of the valence electron. Instead it is attributed to an intrinsic magnetic dipole moment of the electron, $\hat{\boldsymbol{\mu}}_s$, and we introduce an intrinsic angular momentum $\hbar\hat{S}$ (the spin of the electron), which we discussed in Section 6.2. In analogy with (8.7), we define the operator

$$\hat{\boldsymbol{\mu}}_s = -g\mu_B\hat{S} = -g\,\frac{e\hbar}{2m}\,\hat{S} \qquad (8.21)$$

It is convenient to measure $\hat{\boldsymbol{\mu}}_s$ in units of the Bohr magneton (8.8), but (8.21) cannot be derived in a similar way to (8.7) for $\hat{\boldsymbol{\mu}}_l$, since $\hat{\boldsymbol{\mu}}_s$ is an intrinsic property of the electron and is not due to any sort of rotational motion. So we must introduce a dimensionless constant of proportionality g, the **g-factor** of the electron, whose value is close to 2, but which cannot be predicted by our theory. (However, Dirac's relativistic treatment of the electron gives $g = 2$, and quantum field theory gives a value just over 0.1% greater than 2, which agrees with the observed value to within the limits of experimental accuracy.) As we saw in Section 6.2, the number of possible orientations of $\hat{\boldsymbol{\mu}}_s$ is correctly represented if the spin quantum number is $s = \frac{1}{2}$, so that there are two possible eigenvalues of $\hat{S}_z$, given by $m_s = \pm s = \pm\frac{1}{2}$.

8.4 Spin–orbit coupling

If the electron has an intrinsic magnetic moment, the Hamiltonian that we used in the Schrödinger equation for the hydrogen atom, (7.10), is incomplete. In the rest frame of the electron the proton appears to orbit the electron (just as, in our rest frame, the Sun appears to orbit the Earth), and this moving positive charge produces a magnetic field, which interacts with the intrinsic magnetic moment of the electron. This adds a term of the form $V_{\text{mag}} = -\mathcal{B}\cdot\boldsymbol{\mu}$, (8.4), to the potential energy, where the magnetic moment is the intrinsic magnetic moment of the electron and the magnetic field is due to the orbital motion of the proton relative to the electron. This field is that of a magnetic dipole whose moment is of the form (8.7), proportional to the relative orbital angular momentum of the proton and electron, which is represented by $\hbar\hat{L}$. Since the intrinsic magnetic moment of the electron is proportional to its spin $\hat{S}$, $\hat{V}_{mag}$ is proportional to $\hat{L}\cdot\hat{S}$, and it is called the potential energy due to **spin–orbit coupling**. In the usual frame of reference, in which the atom as a whole is at rest, the spin–orbit coupling potential (which can only be obtained correctly by a relativistic calculation) is

$$\hat{V}_{so}(r)\hat{L}\cdot\hat{S} = \frac{\hbar^2}{2m^2c^2}\frac{1}{r}\frac{\partial V(r)}{\partial r}\hat{L}\cdot\hat{S} \qquad (8.22)$$

where $V(r)$ is the usual Coulomb potential for the hydrogen atom, (7.8). This is also a reasonable approximation for many multi-electron atoms if an appropriate form is used for $V(r)$, and $\hbar\hat{L}$ and $\hbar\hat{S}$ respectively are interpreted as the resultant orbital angular momentum and resultant spin of all the electrons.

Adding the spin–orbit coupling to the Hamiltonian for the hydrogen atom, (7.10), we obtain

$$\hat{H} = -\frac{\hbar^2}{2m}\nabla^2 - \frac{e^2}{4\pi\varepsilon_0 r} + \hat{V}_{so}(r)\hat{L}\cdot\hat{S} = \hat{H}_0 + \hat{V}_{so}(r)\hat{L}\cdot\hat{S} \quad (8.23)$$

We know from Section 7.2 that when electron spin is included the eigenstates of $\hat{H}_0$ can be labelled by the quantum numbers $\{n, l, m_l, m_s\}$, since $\hat{H}_0$ commutes with $\hat{L}^2$, $\hat{L}_z$ and $\hat{S}_z$ (and with $\hat{S}^2$, but we do not include the spin eigenvalue s explicitly in the label because it is always equal to $\frac{1}{2}$ for a single electron). Using the form given for $\hat{L}\cdot\hat{S}$ in (6.35), we may write the spin–orbit potential as

$$\hat{V}_{so}(r)\hat{L}\cdot\hat{S} = \hat{V}_{so}(r)\tfrac{1}{2}(\hat{J}^2 - \hat{L}^2 - \hat{S}^2) \qquad (8.24)$$

where $\hat{J}$ is the total angular momentum operator for the electron (defined by (6.27) as the sum of its orbital and spin angular momentum operators). Equations (6.30), (6.36) and (6.37) show that $\hat{L}\cdot\hat{S}$ commutes with $\hat{L}^2$, $\hat{S}^2$, $\hat{J}^2$ and $\hat{J}_z$, but (6.40) indicates that it does not commute with $\hat{L}_z$ or $\hat{S}_z$. So an eigenstate of the Hamiltonian $\hat{H}$ (which includes the spin–orbit potential) cannot be an eigenstate of $\hat{L}_z$ or of $\hat{S}_z$. The Hamiltonian $\hat{H}_0$ commutes with $\hat{J}^2$ and $\hat{J}_z$, so that, as mentioned in Section 7.2, its eigenstates may be labelled by the set of quantum numbers $\{n, l, j, j_z\}$ instead of the set $\{n, l, m_l, m_s\}$. An eigenstate labelled by $\{n, l, j, j_z\}$ is simultaneously an eigenstate of the operator $\hat{L}\cdot\hat{S}$, and the corresponding eigenvalue of $\hat{L}\cdot\hat{S}$ is $\frac{1}{2}[j(j+1) - l(l+1) - s(s+1)]$.

The spin–orbit coupling splits each of the sublevels (n, l) into two, one with $j = l + \frac{1}{2}$ and the other with $j = l - \frac{1}{2}$, and this gives rise to the **fine structure** observed in the optical spectrum of atomic hydrogen. However, we cannot immediately deduce the magnitude of the splitting from the eigenvalues of $\hat{H}_0$ and (8.24) because of the r dependence of the spin–orbit potential. Rather than attempting to solve the new radial equation that results from $\hat{H}$ (which would be very difficult), we shall (in Chapter

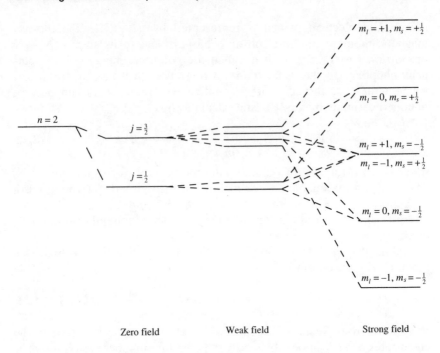

Figure 8.5 The $n = 2$ energy level in hydrogen. The single level of energy E_2 shown on the left is split into a doublet by spin–orbit coupling in zero magnetic field. This doublet is further split in a magnetic field, the character of the splitting changing as the strength of the field is increased.

11) use perturbation theory, which is a method of successive approxima-tions, to calculate quantitatively how spin–orbit coupling changes the eigenvalues E_n of $\hat{H}_0$. This approximation method is possible because the changes are very small – typically the corrections are of the order of $\alpha^2 E_R$ in hydrogen, where E_R is the Rydberg energy and $\alpha = e^2/4\pi\varepsilon_0\hbar c \approx \frac{1}{137}$ is the **fine structure constant**.

The spin–orbit potential is one of three terms that arise as relativistic corrections to the non-relativistic hydrogen atom Hamiltonian. (The other two are corrections to the kinetic energy operator and the so-called Darwin term [7].) In the complete relativistic expression for the energy of the hydrogen atom the l dependence introduced by the spin–orbit coup-ling is cancelled, and the energy depends only on n and j. The final expression, correct to first order in v/c (where v is the average relative velocity of electron and proton) is [7]

$$E_{nj} = E_n\left[1 + \frac{\alpha^2}{n^2}\left(\frac{n}{j + \frac{1}{2}} - \frac{3}{4}\right)\right] \qquad \textbf{(8.25)}$$

The fine structure of the $n = 2$ level of hydrogen is shown in Figure 8.5.

Spin–orbit coupling produces similar fine structure in the spectra of multi-electron atoms. In the case of light atoms, where L–S coupling applies, the energy eigenvalues already depend on the atomic orbital angular momentum quantum number L and the total spin S (as well as on the principal quantum number n), even before spin–orbit coupling is included. Spin–orbit coupling splits each sublevel into levels corresponding to different values of the total angular momentum quantum number J. This gives rise to the fine structure of atomic spectral lines, such as the doublet structure of the sodium line illustrated in Figure 8.3(a). Other relativistic corrections only produce a shift in the energy levels rather than a splitting, and so spin–orbit coupling is the most significant relativistic correction.

The spin–orbit splitting of atomic energy levels does not depend on the eigenvalue of the z component of any of the angular momenta, because an internal magnetic field does not break the spherical symmetry of the atom as a whole – its direction is randomly oriented relative to any external direction in space. But an external magnetic field produces further splitting, through a potential of the form (8.4), where now we must use the total magnetic dipole moment, which is the sum of the orbital and spin magnetic moments:

$$\hat{V}_{\text{mag}} = -\hat{\boldsymbol{\mu}} \cdot \mathcal{B} = -(\hat{\boldsymbol{\mu}}_l + \hat{\boldsymbol{\mu}}_s) \cdot \mathcal{B}$$
$$= \mu_{\text{B}}(\hat{\boldsymbol{L}} + 2\hat{\boldsymbol{S}}) \cdot \mathcal{B} = \hbar\omega_{\text{L}}(\hat{L}_z + 2\hat{S}_z) \qquad \textbf{(8.26)}$$

(We have used (8.7) for $\hat{\boldsymbol{\mu}}_l$ and (8.21) with $g = 2$ for $\hat{\boldsymbol{\mu}}_s$, and ω_{L} is the Larmor frequency, (8.5).) Equation (8.26) can be applied to multi-electron atoms, with the usual proviso that all angular momenta are interpreted as the resultant angular momenta of the atom as a whole. $\hat{V}_{\text{mag}}$ is proportional to $\hat{J}_z + \hat{S}_z$ rather than to $\hat{J}_z$, and this means that (because of (6.41)) $\hat{J}^2$ does not commute with $\hat{V}_{\text{mag}}$, and the eigenstates of the complete Hamiltonian are no longer eigenstates of $\hat{J}^2$. However, it turns out that the number of eigenvalues of $\hat{J}_z$ determines the number of magnetic sublevels into which each energy level is split.

If the applied magnetic field is strong enough, the spin–orbit coupling is much weaker than the magnetic potential (8.26), and can be neglected to a first approximation. The magnetic splitting of the energy levels is then similar to that given in (8.17), except that m_l (or M_L for a multi-electron atom) is replaced by $m_l + 2m_s$ (or $M_L + 2M_S$). This can be seen in the case of the $n = 2$ level in hydrogen in Figure 8.5. Since the spin quantum number is unchanged in an electric dipole transition, the spectral lines are split into three, as indicated in (8.19). The spin–orbit coupling can then be included as a perturbation, and results in fine structure of the three spectral lines.

If, on the other hand, the applied field is weak, so that the spin–orbit coupling is larger than the magnetic interaction, the magnetic potential must be treated as a perturbation that further splits the already complicated pattern of splittings due to spin–orbit coupling. The anomalous Zeeman effect (mentioned at the end of Section 8.2) is satisfactorily explained by such calculations. Figure 8.5 compares the splitting of the $n = 2$ level in hydrogen in weak and strong magnetic fields, and Figure 8.3(a) shows the anomalous Zeeman splitting of the sodium D lines. The normal Zeeman effect occurs, even in weak fields, if the transition concerned is between two states with zero spin, as illustrated by Figure 8.3(b) for a transition between two levels in cadmium. For intermediate magnetic field strengths the magnetic potential is of the same order of magnitude as the spin–orbit coupling, and both terms must be included simultaneously in the perturbation calculation. Experimentally the pattern of splittings characteristic of the anomalous Zeeman effect changes continuously as the field is increased, and the spectral lines gradually coalesce into three lines (each with fine structure) at sufficiently high fields.

8.5 The motion of free electrons in a uniform magnetic field: Landau levels

The effects of a magnetic field on the properties of electrical conductors and semiconductors can give useful information about the energy states of the conduction electrons. Some very dramatic experimental results, such as the de Haas–van Alphen effect and the quantum Hall effect, can only be explained in terms of the quantum nature of the electrons [8]. So we shall examine the energy eigenvalue problem for a conduction electron in a specimen of conducting material placed in a constant uniform magnetic field. The general nature of the quantum effects that arise can be demonstrated by using the free electron model. According to this picture, an electron in the conduction band moves freely through the specimen, but the free mass is replaced by an effective mass m^*, since the effective inertia of the electron is changed by interactions with the lattice and with other electrons.

Suppose that a conductor is placed in a constant uniform magnetic field $\mathcal{B}$ in the z direction. Equation (8.11) shows that this field can be represented by the vector potential

$$\mathcal{A}_x = 0, \qquad \mathcal{A}_y = \mathcal{B}x, \qquad \mathcal{A}_z = 0 \qquad \textbf{(8.27)}$$

(Other choices can also give the same field $\mathcal{B}$, for example $\{\mathcal{A}_x = -\mathcal{B}y,$ $\mathcal{A}_y = 0, \mathcal{A}_z = 0\}$, or $\{\mathcal{A}_x = \frac{1}{2}\mathcal{B}y, \mathcal{A}_y = \frac{1}{2}\mathcal{B}x, \mathcal{A}_z = 0\}$. It does not matter that the choice of the vector potential is not unique, since the observed effects depend on the magnetic field but not on the particular form of the potential from which we derive it. Two forms of the potential that give the same magnetic field are said to differ by a **gauge transformation**.) We are interested in the behaviour of an electron described by a Hamiltonian of the form (8.14), with $\mathcal{V} = 0$ (assuming there is no applied electric field) and the vector potential given in (8.27). The Schrödinger equation is

$$-\frac{\hbar^2}{2m^*}\left[\frac{\partial^2}{\partial z^2} + \frac{\partial^2}{\partial x^2} + \left(\frac{\partial}{\partial y} + \frac{ie\mathcal{B}}{\hbar}x\right)^2\right]\psi(x, y, z) = E\psi(x, y, z)$$

(8.28)

The y and z components of the momentum operator commute with the Hamiltonian $\hat{H}$ of (8.28):

$$\left[-i\hbar\frac{\partial}{\partial y}, \hat{H}\right] = 0, \qquad \left[-i\hbar\frac{\partial}{\partial z}, \hat{H}\right] = 0$$

Therefore the energy eigenfunctions can be simultaneously eigenfunctions of $\hat{P}_y$ and $\hat{P}_z$, with y and z dependence of the form given in (3.10), and we can write

$$\psi(x, y, z) = \psi(x)e^{ik_y y}e^{ik_z z}$$

(8.29)

where $\hbar k_y$ and $\hbar k_z$ are the eigenvalues of $\hat{P}_y$ and $\hat{P}_z$ respectively. Substitution of (8.29) into the Schrödinger equation (8.28) gives a differential equation for the x dependence of the eigenfunctions:

$$\left[-\frac{\hbar^2}{2m^*}\frac{d^2}{dx^2} + \tfrac{1}{2}m^*\omega_c^2(x - x_0)^2\right]\psi(x) = E'\psi(x)$$

(8.30)

where we have defined the constants

$$\omega_c = \frac{e\mathcal{B}}{m^*}, \qquad x_0 = -\frac{\hbar k_y}{e\mathcal{B}}, \qquad E' = E - \frac{\hbar^2}{2m^*}k_z^2$$

(8.31)

Equation (8.30) is of the same form as the one-dimensional simple harmonic oscillator equation (5.1), with the motion centred at the point x_0, and with natural frequency ω_c (which is the cyclotron frequency (8.2) for an electron with effective mass m^*). The eigenvalues E' are therefore given by (5.15):

$$E' = E_n = (n + \tfrac{1}{2})\hbar\omega_c \quad (n = 0, 1, 2, \ldots)$$

(8.32)

So E', the total energy associated with motion of the electron in the (x, y) plane, is quantized, and these oscillator-like energy levels are known as **Landau levels**. The total energy eigenvalue is

$$E = \frac{\hbar^2}{2m^*}k_z^2 + (n + \tfrac{1}{2})\hbar\omega_c \quad (n = 0, 1, 2, \ldots)$$

(8.33)

In a macroscopic specimen this spectrum is not completely discrete, since the first term has an effectively continuous range of eigenvalues.

Equation (8.32) gives the energy of the motion corresponding to the circular motion at the cyclotron frequency, in a plane perpendicular to the magnetic field, which we should expect from a classical calculation. The additional term in (8.33) corresponds to the classical kinetic energy of free motion along the z axis, which makes the classical trajectory a helix. Although at first sight it seems that the quantum particle is in a state with constant velocity along both the y and the z directions, we must remember that $\hbar k_y$ and $\hbar k_z$ are the eigenvalues of the total momentum defined in (8.13) and not of the usual kinetic momentum. Since the vector potential $\mathcal{A}$ that we are using, (8.27), has no z component, there is no distinction between total and kinetic momentum along this axis, and an eigenstate of $\hat{P}_z$ represents motion with constant velocity along the z direction. But an eigenstate of $\hat{P}_y$ corresponds to a classical state in which the y component of *total* momentum is constant, and because $\mathcal{A}$ has a y component this means that the velocity along the y axis is *not* constant. In fact, for a classical particle a constant value $\hbar k_y$ of the y component of p_{tot} implies circular motion in the (x, y) plane (see Problem 8.9).

The eigenfunctions of (8.30) are of the form given in (5.26), so that (8.29) becomes

$$\psi(x, y, z) = C_n e^{ik_y y} e^{ik_z z} \exp\left[-\tfrac{1}{2}(\xi - \xi_0)^2\right] H_n(\xi - \xi_0) \qquad \textbf{(8.34)}$$

where we have introduced $\xi = (m^*\omega_c/\hbar)^{1/2}x$, corresponding to the definition of ξ in (5.2). The eigenfunctions associated with the Landau levels represent the state of motion in the (x, y) plane, and so they are independent of z:

$$\psi_{n\xi_0}(x, y) = C_n e^{ik_y y} \exp\left[\tfrac{1}{2}(\xi - \xi_0)^2\right] H_n(\xi - \xi_0) \qquad \textbf{(8.35)}$$

These eigenfunctions depend on the values of ξ_0 and k_y as well as the Landau level number n, but ξ_0 and k_y are related through (8.31), so the state represented by (8.35) can be characterized by the two quantum numbers n and $x_0 = (\hbar/m^*\omega_c)^{1/2}\xi_0$. However, the Landau energy given in (8.32) depends only on n, so that the Landau levels are degenerate, and states with the same n but different x_0 have the same energy. Each of the states in one Landau level corresponds to a classical orbit about a point in the sample with x coordinate x_0. However, the y coordinate of the centre of the orbit cannot be well defined at the same time as the x component in the quantum case, because x_0 is proportional to $\hbar k_y$, the eigenvalue of the operator $\hat{P}_y$ which does not commute with y.

The degree of degeneracy of a Landau level is the number of possible values of x_0, and this depends on the number of different values of k_y. In a sample of finite width $\mathbb{L}_y$, (7.40) shows that allowed values of k_y are separated by $2\pi/\mathbb{L}_y$. (Note that $\mathbb{L}_y$ is the total length of the specimen, which corresponds to $N\mathbb{L}$ in (7.40).) So allowed values of x_0 are separated by $\Delta x_0 = h/e\mathcal{B}\mathbb{L}_y$, which means that the number of values of x_0 in a sample of length $\mathbb{L}_x$ is the nearest integer less than

$$\frac{\mathbb{L}_x}{\Delta x_0} = \frac{e\mathcal{B}\mathbb{L}_x\mathbb{L}_y}{h} = \frac{\Phi}{h/e} \tag{8.36}$$

where $\Phi = \mathcal{B}\mathbb{L}_x\mathbb{L}_y$ is the total magnetic flux through the sample. So each Landau level contains the same number, approximately equal to $e\Phi/h$, of different states with the same energy. In a semiclassical picture this number $e\Phi/h$ is the number of orbits in the sample with the same energy but centred on different points, and since the total flux through these orbits is Φ, each classical orbit encloses one unit of magnetic flux of magnitude h/e. This unit is called the **flux quantum**.

8.6 Periodic effects in two-dimensional conductors

Many of the effects on the properties of conductors produced by a magnetic field are associated with the periodic emptying of the highest occupied Landau level as the field is increased. To understand a little of how this comes about, we shall examine the motion of a fixed number N of conduction electrons in a two-dimensional conductor placed perpendicular to a constant, uniform magnetic field at zero kelvin. (This is not very realistic, since the number of electrons in a two-dimensional slice of a three-dimensional material is unlikely to remain constant.) At 0 K, in the absence of a magnetic field, all levels up to the Fermi level are filled (as we explained in Section 7.6). In two dimensions the density of electron states is independent of energy (Problem 7.17a), so the number of filled electron states in the conduction band at 0 K varies with energy in the way shown at the top of Figure 8.6(a). When a magnetic field is applied, the almost-continuous distribution of energy levels changes to the set of discrete Landau levels, and so electrons from a very large number of energy levels at zero field must occupy a much smaller number of Landau levels. This is possible because the Landau levels are degenerate, so that many electrons can occupy the same energy level. In Figure 8.6(a) light lines indicate the range of energy levels at zero field that contribute electrons to each Landau level in the field $\mathcal{B}_1$. Although some electrons gain energy, others lose energy, and for the field $\mathcal{B}_1$ the energy of the highest occupied Landau level is less than E_F.

Equation (8.36) shows that the degeneracy of each Landau level increases linearly with magnetic field. Suppose that at some field $\mathcal{B}$ the number of conduction electrons is an integer multiple of the number of

200

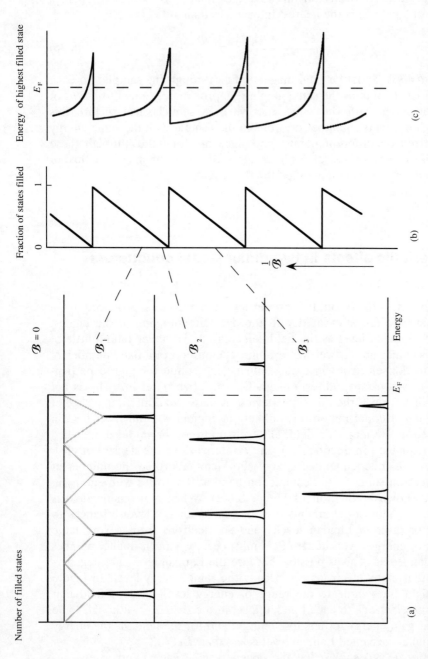

Figure 8.6 (a) The distribution of a fixed number of conduction electrons as a function of energy in a two-dimensional conductor at 0 K in the absence of a magnetic field, and when uniform magnetic fields $\mathcal{B}_1$, $\mathcal{B}_2$ and $\mathcal{B}_3$ are applied perpendicular to the sample. The light lines indicate the energy levels at zero field from which the Landau levels in the field $\mathcal{B}_1$ are filled. (b) The fraction of filled states in the highest occupied Landau level as a function of $1/\mathcal{B}$. Dashed lines indicate the values of $1/\mathcal{B}$ corresponding to the fields $\mathcal{B}_1$, $\mathcal{B}_2$ and $\mathcal{B}_3$. (c) The energy of the highest occupied Landau level as a function of $1/\mathcal{B}$.

states in each Landau level:

$$N = j \frac{\Phi}{h/e} = j \frac{e\mathcal{B}\mathcal{L}_x\mathcal{L}_y}{h} \quad (j \text{ an integer}) \qquad (8.37)$$

Then all the states in the j lowest Landau levels are completely filled, and since the lowest level has $n = 0$, the highest filled level is that with $n = j - 1$. If the magnetic field is increased, the number of states in each Landau level is increased, while the number of electrons does not change. The new empty states in the lower levels will be filled by electrons from higher levels, with the result that all the levels below $n = j - 1$ are filled, but the highest occupied level, $n = j - 1$, is only partially filled. As the field is increased further, the fraction of filled states in the $j - 1$ level decreases, until it reaches zero, at which point all the lowest levels from $n = 0$ to $j - 2$ are completely filled. A further increase in the field decreases the fraction of filled states in what is now the highest occupied level (with $n = j - 2$), until that level too is emptied, and the highest occupied level is that with $n = j - 3$, and so on. Equation (8.37) shows that a Landau level empties each time $1/\Phi$ changes by e/Nh, or $1/\mathcal{B}$ changes by $e\mathcal{L}_x\mathcal{L}_y/Nh$. So the fraction of filled states in the highest occupied Landau level fluctuates periodically between 1 and 0 as a function of $1/\mathcal{B}$, as shown in Figure 8.6(b).

Not only the degeneracy but also the energy of each Landau level, (8.32), increases linearly with the magnetic field, as can be seen from the definition of the cyclotron frequency in (8.31). Suppose that the lowest j levels are completely filled; then the highest electron energy at this field is $(j - 1 + \frac{1}{2})\hbar\omega_c$. As the field is increased, the highest electron energy remains that of the $n = j - 1$ level as long as it contains electrons, and this energy increases with the field. When the population of the level drops to zero, the highest electron energy decreases suddenly by $\hbar\omega_c$ to the energy of the $n = j - 2$ level, and then increases with field again until that level too is empty. As shown in Figure 8.6(c), the highest electron energy varies periodically about the zero-field Fermi energy E_F as a function of $1/\mathcal{B}$, with the same period as the fluctuation in the fraction of filled states in the highest occupied level.

Any property of a conductor that depends on either the population of the highest electron energy level or on the maximum energy of the electrons would be expected to oscillate as the magnetic field is changed, in the way just described. Although our treatment is too simplified to be a good model, and though we have considered only motion in a plane perpendicular to the magnetic field, such oscillations *are* observed, even in three-dimensional samples. Low temperatures and high magnetic fields are necessary, so that $\hbar\omega_c \gg k_B T$, and thermal excitation of electrons to Landau levels above the highest occupied at $0\,K$ is negligibly small. Under such conditions the **de Haas–Shubnikov effect** is observed. This is an oscillation of electrical resistance as the magnetic field is changed. It

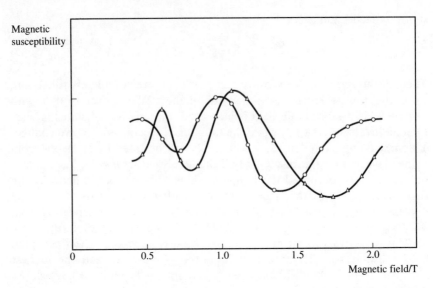

Figure 8.7 The original observation of the de Haas–van Alphen effect [9]: the oscillating dependence of the magnetic susceptibility of bismuth on the magnetic field, at a temperature of 14.2 K. The two sets of data correspond to two different orientations of the magnetic field relative to the axes of the crystal.

has a constant period when plotted as a function of $1/\mathcal{B}$, just like the population of the highest Landau level in our two-dimensional model. A similar oscillation in the magnetization produced in a conductor by a magnetic field is known as the **de Haas–van Alphen effect**, and this is illustrated in Figure 8.7.

8.7 The quantum Hall effect

The **Hall effect** is the appearance of a potential difference across the width of a conductor or semiconductor carrying a current when a magnetic field is applied perpendicular to the current flow. It can be predicted from the Lorentz force acting on the charges carrying the current (without invoking quantum mechanics), and was the first means of demonstrating that in some conductors the carriers are electrons while in others they are holes. The Hall resistance is the ratio of the electric field associated with the transverse potential difference to the current density, and the classical calculation indicates that this should increase linearly with magnetic field, as we shall show below. But in a semiconductor, if the current flow is confined to two dimensions at low temperatures, an interesting departure from this behaviour is observed: the (transverse) Hall resistance changes in discrete steps, as shown in Figure 8.8(a). This is the **quantum Hall effect**, and is associated with the existence of discrete Landau levels. (de Haas–Shubnikov oscillations in the longitudinal electrical resistance can also be seen in the figure.) Although we shall not be able to give a satisfactory explanation of the effect here, it must be mentioned briefly, since it has led to the establishment of a fundamental unit of electrical resistance.

The basic arrangement for observing the Hall effect is shown in Figure 8.8(b). A conductor carrying a current of density $\mathcal{J}_x$ in the x direction is placed in a magnetic field $\mathcal{B}$ in the z direction. The Lorentz force on the carriers of the current tends to deflect them in the negative y direction, whether the current is due to negatively charged carriers travelling in the $-x$ direction or positively charged ones travelling in the $+x$ direction, as can be seen from (8.1). Under equilibrium conditions this displacement of charge produces a voltage across the conductor, equivalent to an electric field $\mathcal{E}_y$ in the $-y$ direction for negative carriers or the $+y$ direction for positive ones. This is the **Hall field**, and the **Hall resistance** is defined as

$$\rho_{xy} = \mathcal{E}_y/\mathcal{J}_x = R_H\mathcal{B} \tag{8.38}$$

where R_H is the **Hall coefficient**, which has the same sign as the charge of the carriers. If the density of carriers of charge $-e$ is n_c and their velocity is v in the negative x direction then $\mathcal{J}_x = n_c e v$. In equilibrium the electric and magnetic parts of the Lorentz force in (8.1) cancel each other, so that $\mathcal{E}_y = v\mathcal{B}$, and

$$R_H = 1/n_c e \tag{8.39}$$

If the density of carriers is constant, as might be expected, the Hall coefficient is constant and the Hall resistance (8.38) increases linearly with $\mathcal{B}$. Even if n_c is not constant, a classical theory can only predict a continuous variation of the Hall resistance with magnetic field, totally unlike the stepped behaviour evident in Figure 8.8(a).

The quantum Hall effect appears in semiconductors (at temperatures sufficiently low that thermal excitation between Landau levels is negligible) only if the motion of the conduction electrons is confined to a plane. This is possible, for example, in a quantum well structure such as described in Section 3.3. An electron in the middle layer of the sandwich of two semiconductors illustrated in Figure 3.4 is confined in the z direction, but free to move in the (x, y) plane. If a magnetic field is applied along the z axis, the energy of such an electron in the conduction band is given by (8.33) with $(\hbar^2/2m^*)k_z^2$ replaced by the energy E_i of one of the discrete levels in the one-dimensional square well. The magnetic field does not change the energy associated with motion in its own direction, so, assuming no other fields are applied, and provided the

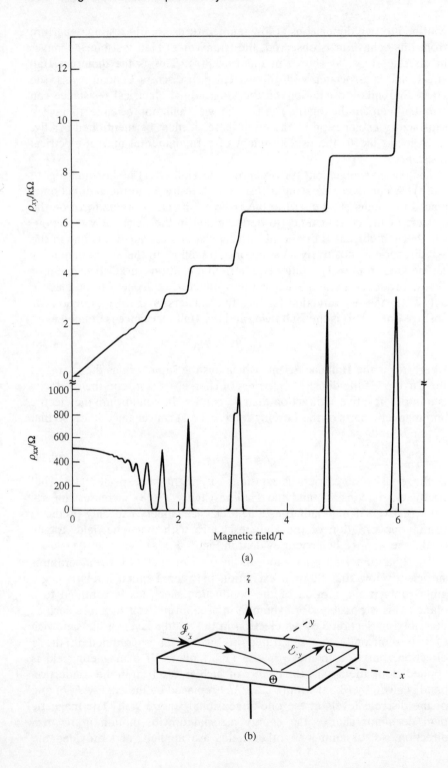

(a)

(b)

thermal energy is too small to excite the electron to the next level in the square well, E_i is fixed at a constant value (usually that of the lowest state, with $i = 1$). The motion is then effectively two-dimensional, with the Landau spectrum (8.32).

The experimental results (Figure 8.8a) show plateaux in ρ_{xy} at values given very accurately by the relation

$$\rho_{xy} = \frac{h}{je^2} \quad (j \text{ integer}) \tag{8.40}$$

That is, the Hall resistance remains constant at a value satisfying (8.40) over a relatively large range of fields, and then changes rather abruptly to the value corresponding to the next integer down, $j - 1$, as the field is increased. If we substitute the value of the Hall resistance given in (8.40) into (8.38), and compare the resulting expression for the Hall coefficient with (8.39), we find the density of conduction electrons to be

$$n_c = j\,\frac{e\mathcal{B}}{h} \quad (j \text{ integer}) \tag{8.41}$$

This implies that the total number of conduction electrons in the specimen is $\mathcal{L}_x\mathcal{L}_y n_c = j(e\mathcal{B}\mathcal{L}_x\mathcal{L}_y/h)$, just the right number to fill j Landau levels completely! So apparently each plateau in the Hall resistance corresponds to a completely filled set of Landau levels. (A **fractional quantum Hall effect (FQHE)** has also been observed, where the plateaux in the Hall resistance correspond to fractional values of j. However, this is due to the collective behaviour of the conduction electrons, and cannot be discussed in terms of an independent electron model and single-electron energy levels.)

It is not possible to explain in simple terms why the Hall resistance remains constant in the plateaux regions, nor the even more surprising fact that these constant values satisfy (8.40) to a very high degree of accuracy. Indeed, the accuracy is such that values of the Hall resistance in the plateaux regions are used to define a fundamental unit of resistance, h/e^2, and measurements give $h/e^2 = 2.591\,228\ \text{M}\Omega$ accurate to three parts per million [10]. This number is also used, with a value for the velocity of light, to obtain the most accurate value of the dimensionless fine structure constant, $\alpha = e^2/4\pi\varepsilon_0\hbar c$, which we introduced in Section 8.4.

Figure 8.8 (a) The quantum Hall effect (adapted from Figure 14 of [10]). The longitudinal resistance ρ_{xx} shows evidence of Shubnikov–de Haas oscillations, and vanishes at fields where there are plateaux in the Hall resistance ρ_{xy}. (b) The experimental arrangement for observing the Hall effect. The deflection of positively charged current carriers, giving rise to a Hall field $\mathcal{E}_y$ in the positive y direction, is shown.

References

[1] Grant I.S. and Phillips W.R. (1990). *Electromagnetism* 2nd edn, Section 4.7. London: Wiley.

[2] Ref [1], Section 4.3.

[3] Konopinski E.J. (1981). *Electromagnetic Fields and Relativistic Particles*, Section 5.5, pp. 140–2. New York: McGraw-Hill.

[4] Feynman R.P., Leighton R.B. and Sands M. (1964). *The Feynman Lectures on Physics*, Vol. II, Sections 14-1 and 18-6. Reading MA: Addison-Wesley.

[5] Ref. [3], Chap. 5, p. 123 and Chap. 6.2, pp. 158–60.

[6] Back E. and Landé A. (1925). *Zeeman-Effekt und Multiplettstruktur der Spektrallinien*. Berlin: Springer.

[7] Cohen-Tannoudji C., Diu B. and Laloë F. (1977). *Quantum Mechanics*, Vol. 2, Chap. XII B-1. New York: Wiley.

[8] Shoenberg D. (1984). *Magnetic Oscillations in Metals*. Cambridge University Press.

[9] de Haas W.J. and van Alphen P.M. (1930). *Proceedings of the Netherlands Royal Academy of Sciences* **33**, 1106.

[10] A transcript of the lecture given by von Klitzing on the presentation to him of the Nobel prize, for the discovery of the quantum Hall effect, is to be found in von Klitzing K. (1986). *Reviews of Modern Physics* **58**, 519. A more recent general discussion of the effect appears in Challis L.J. (1992). Physics in less than three dimensions. *Contemporary Physics* **33**, 111.

Problems

8.1 Repeat the calculation of the energy eigenvalues of a hydrogen atom in a magnetic field, neglecting spin–orbit coupling but including the interaction of the intrinsic magnetic moment by using (8.26) instead of (8.15) for $\hat{V}_{\text{mag}}$. Show that the prediction for the Zeeman splitting of the spectral lines, (8.19), is unaffected by the change in $\hat{V}_{\text{mag}}$ if the spin quantum number m_s does not change in the transition. (This is true for an electric dipole transition.)

8.2 (a) A beam of atoms with total angular momentum quantum number $j = 1$ and randomly oriented magnetic moments enters a Stern–Gerlach apparatus with its magnetic field in the z direction. How many beams will emerge from the apparatus, and what are their distinguishing quantum numbers? What are the allowed angles between the magnetic moment of an atom and the magnetic field?

   (b) If the beams emerging from the apparatus are blocked, except for one that is passed into a second Stern–Gerlach apparatus, also with its magnetic field in the z direction, what effects will be observed? What would happen if the magnetic field in the second apparatus were at some angle to the z direction?

8.3 (a) By putting $r \approx r_B$ (the Bohr radius) in (8.22), show that the spin–orbit potential $V_{so}(r)$ is of the order of magnitude of $\alpha^2 E_R$, where E_R is the Rydberg energy and α is the fine structure constant.

(b) The sodium D line is due to electric dipole transitions of the single valence electron between the levels $n = 3$, $l = 1$ and $n = 3$, $l = 0$. Find, qualitatively, how these levels are split by spin–orbit coupling, giving in each case the quantum numbers of each component of the split level. Deduce the fine structure of the spectral line.

(c) Show that the energy levels of a multi-electron atom that have total orbital angular momentum quantum number L and total spin quantum number $S = 0$ are not split by spin–orbit coupling.

(d) Find the number of levels into which an atomic energy level characterized by the quantum numbers (L, S) is split by spin–orbit coupling, and show that the splitting between adjacent levels is proportional to the J value of the higher level. (*Hint*: See Problem 6.6.)

(e) The fine structure in the electric dipole spectrum of calcium (which has two valence electrons) shows triplets of lines. Identify the L, S and J values of the energy levels involved in the transitions responsible for the two triplets of spectral lines whose spacing is shown below (the frequencies of the lines increase from the bottom of the diagram towards the top):

$$5.6\,\text{cm}^{-1}$$

$$105.8\,\text{cm}^{-1}$$

$$3.7\,\text{cm}^{-1}$$

$$52.2\,\text{cm}^{-1}$$

8.4 For the potential $\hat{V}_{mag}$ of (8.26) show that the commutator $[\hat{J}^2, \hat{V}_{mag}]$ is not equal to zero. Find the commutator of $\hat{V}_{mag}$ with each of the following operators: $\hat{L}^2$, $\hat{S}^2$, $\hat{J}_z$, $\hat{L}_z$, $\hat{S}_z$, $\hat{L} \cdot \hat{S}$. What can you deduce about the eigenstates of $\hat{V}_{mag}$?

8.5 (a) Find the precession frequencies of (i) the orbital and (ii) the intrinsic magnetic moment of an electron in a magnetic field of $1\,\text{T}$. Use the order of magnitude of the spin–orbit coupling from Problem 8.3(a) to estimate the minimum magnitude (in T) of a 'strong' magnetic field (that is, one that will split a spectral line into three main components).

(b) Sketch the splitting of the $n = 3$ level of hydrogen in (i) a weak and (ii) a strong magnetic field.

(c) The fine structure (Problem 8.3b) and Zeeman splitting of the sodium D line are shown in Figure 8.3(a). Identify the quantum numbers of the energy levels involved in the electric dipole transitions that produce the six lines in a magnetic field. Can you explain why the outer two lines in the latter case should be weaker than the other four?

(d) Explain why the spectral line associated with the transition in cadmium illustrated in Figure 8.3(b) shows the normal Zeeman effect.

8.6   The Hamiltonian for a quantum particle behaving like a rigid rotator is given in (4.47). If the particle also has spin half, show that spin–orbit coupling would split the energy level with quantum number l into two, unless $l = 0$. Give the quantum numbers characterizing the two sublevels, and show that their separation is proportional to $l + \frac{1}{2}$. Discuss the splitting of these levels in a magnetic field.

8.7   (a) Show that the vector potentials $\{\mathcal{A}_x = -\mathcal{B}y, \mathcal{A}_y = 0, \mathcal{A}_z = 0\}$ and $\{\mathcal{A}_x = -\frac{1}{2}\mathcal{B}y, \mathcal{A}_y = \frac{1}{2}\mathcal{B}x, \mathcal{A}_z = 0\}$ both represent the same magnetic field as the potential given in (8.27).

(b) Derive the Hamiltonian in (8.28) from the general form (8.14) using the potential given in (8.27). Prove that it commutes with $\hat{P}_y$ and $\hat{P}_z$.

(c) Substitute the form for the wave function given in (8.29) into (8.28), and obtain the Schrödinger equation (8.30).

(d) Find the Hamiltonian appropriate in the case of each of the two forms for the vector potential mentioned in (a). Investigate the commutation properties of each Hamiltonian with all three components of the momentum operator, and explain the significance of your results.

8.8   Outside an infinitely long solenoid the magnetic field is zero, but the vector potential does not vanish. At a point outside the solenoid, a perpendicular distance r from its axis, $\mathcal{A}$ is tangential to the solenoid, and of magnitude $(n\mathcal{I}\mathcal{A}/2\pi\varepsilon_0 c^2)/r$, where n is the number of turns per unit length of the solenoid, $\mathcal{I}$ the current it carries, and $\mathbb{A}$ its cross-sectional area. A free electron with constant speed v travels in a semicircular path of (constant) radius r about the solenoid, from point A to point B, as shown in Figure 8.9. Show that its wave function is of the form

$$C \exp[i(m\boldsymbol{v} \cdot \boldsymbol{r} - e\mathcal{A} \cdot \boldsymbol{r} - Et)/\hbar],$$

and that the phase at B depends on whether the electron has travelled from A by the upper or lower path in the figure. Show that the phase difference between the two paths is $h\Phi/e$, where Φ is the total magnetic flux through the solenoid. (This is the basis of the **Aharanov–Bohm effect**, in which an interference pattern is observed when a beam of electrons is split to pass on either side of a very long solenoid, and then recombined.)

8.9   The equations of motion in classical Hamiltonian mechanics are

$$\frac{\partial H}{\partial p_x} = \frac{dx}{dt}, \qquad \frac{\partial H}{\partial x} = -\frac{dp_x}{dt}$$

with similar expressions involving the y and z components. For a charged particle in a magnetic field represented by the vector potential $\mathcal{A}$, the Hamiltonian function is $H = (2m)^{-1}(\boldsymbol{p} - q\mathcal{A})^2$, where $\boldsymbol{p}$ is the total

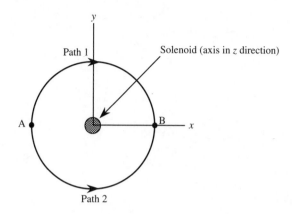

Figure 8.9 Diagram for Problem 8.8.

momentum (compare this with the quantum mechanical version (8.14)). Show that, for $\mathcal{A}$ given by (8.27), a classical calculation finds that the total momenta p_y and p_z are constants, and that the motion along the x axis is a simple harmonic oscillation at the cyclotron frequency. Calculate the x, y and z components of the velocities, and prove that if the particle is initially moving in the (x, y), plane its trajectory is a circle.

8.10   The density of conduction electrons in a conducting surface is 2×10^{19} m^{-2}. Estimate the number of completely filled Landau levels when a magnetic field of 0.01 T is applied perpendicular to the surface. What is the degeneracy (approximately) of each level in a square sample with sides of length 0.01 m? If the uncertainty in x_0, the x coordinate of the centre of a Landau orbit, were $0.01(h/e\mathcal{B}L_y)$ in this sample, what would be the minimum uncertainty in y_0, the y coordinate of the centre?

CHAPTER 9

The Superposition Principle

9.1 The prediction of the results of experiments on quantum systems

Postulate 2 (Section 3.1) enables us to predict all the possible results of a measurement of a dynamical variable: we have to identify a suitable Hermitian operator $\hat{A}$ to represent the variable concerned, and solve its eigenvalue equation. Then the set of eigenvalues $\{a_n\}$ is the set of all values that could result from a measurement of that variable. If the particle is in an eigenstate of the variable to be measured then there is only one possible result of such a measurement – the eigenvalue corresponding to that particular eigenstate. For example, if a hydrogen atom is in its ground state then the result of measuring the energy is certainly $E_1 = -E_R$, (7.13). However, we need to be able to make predictions about the result of a measurement when the particle is **not** in an eigenstate of the variable to be measured.

The value of any variable in a state that is not one of the corresponding eigenstates is indeterminate, in the sense discussed in Section 1.5. For example, (4.6) shows that an energy eigenstate of a particle bound in a potential well is not simultaneously a momentum eigenstate, so that for a particle in such an energy eigenstate the momentum value is indeterminate. This does not preclude a measurement of the particle's momentum, but it means that there is no unique prediction for the result that would be obtained. So, in general, a series of identical measurements on a set of quantum particles in identical states will not give identical results, as we discussed in Section 2.3. If a large number of particles are prepared in the same state, represented by the wave function $\Psi(x, y, z, t)$, then, in general, measurements of the variable represented by $\hat{A}$ will yield a range of values. Each number is one of the eigenvalues a_n, and the mean $\langle a \rangle$ and the standard deviation Δa of these results are given by (2.7) and (2.8), in terms of the probability $\mathbb{P}_n$ of obtaining a particular result a_n. Only in the special case where $\Psi(x, y, z, t)$ is an eigenfunction of $\hat{A}$, say

$\Psi(x, y, z, t) \equiv \psi_k(x, y, z, t)$, will the measurements all yield the same value, which will be the particular eigenvalue a_k corresponding to the eigenfunction $\psi_k(x, y, z, t)$. In this case the probability $\mathbb{P}_k$ will be equal to 1, while $\mathbb{P}_n = 0$ for $n \neq k$, but in general there will be a set of non-zero probabilities $\mathbb{P}_n$ with different values. Postulate 2 tells us how to predict the *possible* outcome of any measurement, but clearly we also need to find out how to calculate the *probability* $\mathbb{P}_n$ of obtaining any particular result.

In Chapter 3 we saw that an energy eigenfunction of a particle in a bounded region of constant potential has components corresponding to states of equal but opposite momentum. For example, the energy eigenfunction for a particle in Region I of Figure 3.5 is given by (3.45):

$$\psi_0(z) = A_0 e^{ik_0 z} + B_0 e^{-ik_0 z}, \qquad k_0^2 = \frac{2m(E - V_0)}{\hbar^2}$$

This is a linear combination of two momentum eigenfunctions, one corresponding to a momentum in the $+z$ direction, and the other to a momentum in the $-z$ direction. In Section 3.4 we identified the coefficients A_0 and B_0 as probability amplitudes for finding the particle travelling to the right and left respectively if a momentum measurement is performed. This is a particular example of a general assumption that allows us to predict the probability of obtaining any particular result when a measurement is made on a particle in an arbitrary state. This assumption is our third postulate.

Postulate 3

A normalized wave function $\Psi(x, y, z, t)$, representing an arbitrary state of a quantum particle, can be expanded as a unique linear superposition of the complete set of normalized eigenfunctions $\{\psi_n(x, y, z, t)\}$ of any operator $\hat{A}$ representing a dynamical variable. The probability of obtaining the particular eigenvalue a_n of $\hat{A}$ when the corresponding variable is measured on the particle in the state $\Psi(x, y, z, t)$ is $\mathbb{P}_n = |c_n|^2$, where c_n is the coefficient of the particular eigenfunction $\psi_n(x, y, z, t)$ in the linear superposition.

Postulate 3 is the **superposition principle**. This states first that any wave function $\Psi(x, y, z, t)$ describing the state of a particle can be expressed as a linear superposition of the set of eigenfunctions of any particular operator $\hat{A}$ representing a dynamical variable:

$$\Psi(x, y, z, t) = \sum_n c_n \psi_n(x, y, z, t) \tag{9.1}$$

Here $\{\psi_n(x, y, z, t)\}$ is the set of normalized eigenfunctions of the operator $\hat{A}$, and for each normalized wave function $\Psi(x, y, z, t)$ the expansion is unique. Secondly, the postulate provides a physical interpretation of this superposition by identifying the expansion coefficient c_n as the **probability amplitude** for finding the particular eigenvalue a_n if the variable represented by $\hat{A}$ is measured when the particle is in the state represented by the normalized wave function $\Psi(x, y, z, t)$.

Thus the results of a set of measurements of a variable, represented by operator $\hat{A}$, on a number of quantum particles all in the same state, represented by $\Psi(x, y, z, t)$, are predicted as follows.

(1) The eigenvalue equation for $\hat{A}$ is solved, giving the set of eigenvalues $\{a_n\}$, which is the set of all possible results. This also gives the set of (normalized) eigenfunctions $\{\psi_n(x, y, z, t)\}$.

(2) The wave function $\Psi(x, y, z, t)$ is expanded as in (9.1), using the eigenfunctions of $\hat{A}$ found in Step 1. Provided the wave function is normalized, the probability of finding any particular value a_n is $\mathbb{P}_n \equiv |c_n|^2$, where c_n is the appropriate expansion coefficient.

(3) The mean and standard deviation of the results can be predicted in terms of the eigenvalues $\{a_n\}$ and the probabilities $\{\mathbb{P}_n\}$ by the statistical equations (2.7) and (2.8).

9.2 The superposition expansion

To understand how to use Postulate 3, let us look at a particle in an infinitely deep one-dimensional square well. Equation (3.40) gives the spatial dependence of the energy eigenfunctions, and there is an exponential time dependence of the form given in (3.5), so the eigenfunctions are

$$\psi_n(z, t) = \sqrt{\left(\frac{2}{L}\right)} \sin \frac{n\pi z}{L} \exp\left(-\,\mathrm{i}\,\frac{E_n t}{\hbar}\right) \qquad (9.2)$$

Mathematically, a function of z defined in the region $0 \leqslant z \leqslant L$ and vanishing at $z = 0$ and $z = L$ can be expanded as a Fourier sine series in that interval [1]:

$$f(z) = \sum_{n=1}^{\infty} A_n \sin \frac{n\pi z}{L} = \sum_{n=1}^{\infty} c_n \sqrt{\left(\frac{2}{L}\right)} \sin \frac{n\pi z}{L} \qquad (9.3)$$

This looks like a linear superposition of the form (9.1) in terms of the complete set of normalized energy eigenfunctions at time $t = 0$. (The Fourier series actually extends over all space, and $f(z)$ is antisymmetric about $z = 0$, vanishes at $z = 0$ and $z = L$, and repeats itself with a period length $2\mathcal{L}$. However, we may write an arbitrary wave function at $t = 0$ for a particle in the well as $\Psi(z, 0) = f(z)$ for $0 \leqslant z \leqslant \mathcal{L}$, and zero elsewhere.) According to Postulate 2, an energy measurement on the particle in the well must give one of the energy eigenvalues $E_n = h^2 n^2 / 8m\mathcal{L}^2$, (3.39). Postulate 3 indicates that if the particle is in the state described by $f(z)$ at $t = 0$ then an energy measurement at that instant has probability $|c_n|^2$ of giving the result E_n.

To calculate the Fourier coefficients A_n, and hence the probability amplitudes c_n, for some particular function $f(z)$, we need to invert (9.3). This is done by making use of the fact that different sine functions are orthogonal (in the sense defined by (2.5)):

$$\frac{2}{\mathcal{L}} \int_0^{\mathcal{L}} dz \sin \frac{n\pi z}{\mathcal{L}} \sin \frac{m\pi z}{\mathcal{L}} = \delta_{nm} \qquad (9.4)$$

Multiplying (9.3) by $\sqrt{(2/\mathcal{L})} \sin(m\pi z/\mathcal{L})$ and integrating from 0 to $\mathcal{L}$, we get

$$c_n = \int_0^{\mathcal{L}} dz \, f(z) \sqrt{\left(\frac{2}{\mathcal{L}}\right)} \sin \frac{n\pi z}{\mathcal{L}} \qquad (9.5)$$

For example, suppose that the state of the particle at time $t = 0$ is represented by

$$\Psi(z, 0) = \begin{cases} Cz(\mathcal{L} - z) & (0 \leqslant z \leqslant \mathcal{L}) \\ 0 & (z < 0, z > \mathcal{L}) \end{cases} \qquad (9.6)$$

where $\Psi(z, 0)$ is normalized if $C = \sqrt{(30/\mathcal{L}^5)}$ (Problem 2.1). Then (9.5) gives

$$c_n = \begin{cases} \dfrac{8\sqrt{15}}{\pi^3 n^3} & (n \text{ odd}) \\ 0 & (n \text{ even}) \end{cases} \qquad (9.7)$$

This means that there is no possibility of measuring the energy to be any of the eigenvalues with even n, E_2, E_4, ..., and the probability of finding it to be E_n with n odd is $|c_n|^2 = 960/(n\pi)^6$.

In fact we can see that $c_n = 0$ for even n from the symmetry properties of the wave function, without evaluating the integral in (9.5). Figure 9.1 shows that $\Psi(z, 0)$ is symmetric about the centre of the well, $z = \frac{1}{2}\mathcal{L}$, while $\sin(n\pi z/\mathcal{L})$ is symmetric for $n = 1, 3, 5, \ldots$ and antisymmetric for $n = 2, 4, 6, \ldots$. (Look back at Figure 3.3, which shows some of these eigenfunctions.) The expansion for a symmetric wave function cannot contain any antisymmetric components. This argument can be expressed in terms of parity. If we place the origin at the centre of the well, the energy eigenfunctions are simultaneously eigenfunctions of parity with eigenvalue $+1$ for $n = 1, 3, 5, \ldots$ and -1 for $n = 2, 4, 6, \ldots$. $\Psi(z, 0)$ is also an eigenfunction of parity with eigenvalue $+1$, and therefore can

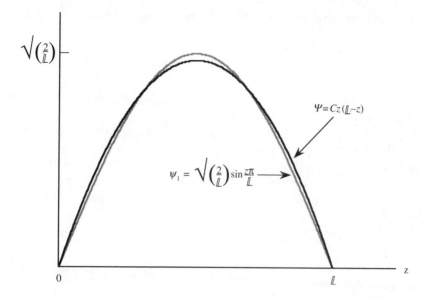

Figure 9.1 The wave functions $\Psi(z,0) = Cz(\mathbb{L} - z)$ and $\psi_1(z) = \sqrt{(2/\mathbb{L})} \sin(\pi z/\mathbb{L})$ representing states of a particle in an infinitely deep one-dimensional square well.

only be composed from the energy eigenfunctions with parity +1. Thus $c_n = 0$ if n is even.

The probability, given by (9.7), that a measurement at $t = 0$ will give the result E_n is largest for $n = 1$, and $\mathbb{P}_1 = |c_1|^2$ is just less than 0.999. (This is not surprising when we look at Figure 9.1 and see how similar the function $\Psi(z,0)$ is to the eigenfunction $\psi_1 = \sqrt{(2/\mathbb{L})} \sin(\pi z/\mathbb{L})$.) Clearly the probability of finding any of the higher energy eigenvalues is very small indeed. However, it is not exactly zero, which implies that the mean of a large number of identical measurements will be greater than E_1. Using the statistical equation (2.7) for the mean, with the values $\mathbb{P}_n = |c_n|^2$ from (9.7) and the energy eigenvalues given in (3.39), we find

$$\langle E \rangle = \sum_n |c_n|^2 E_n = \frac{10}{\pi^2} E_1 \approx 1.013 E_1 \tag{9.8}$$

Similarly,

$$\langle E^2 \rangle = \sum_n |c_n|^2 E_n^2 = \frac{120}{\pi^4} E_1^2 \tag{9.9}$$

(To obtain these values, the results

$$\sum_{n \text{ odd}} \frac{1}{n^4} = \tfrac{1}{96}\pi^4, \qquad \sum_{n \text{ odd}} \frac{1}{n^2} = \tfrac{1}{8}\pi^2$$

have been used.) Then, from the definition (2.8), the standard deviation is

$$\Delta E = E_1 \frac{2\sqrt{5}}{\pi^2} \approx 0.45 E_1$$

These predictions refer to an energy measurement at the time $t = 0$. What can we say about a measurement made at any later time? Provided the particle is not disturbed by any external interaction (which would alter the Hamiltonian, and thus the energy eigenvalues and eigenfunctions), the coefficients c_n will be constant, with values given by (9.7). The mean $\langle E \rangle = \sum_n |c_n|^2 E_n$ and the standard deviation ΔE are constant, but the probability distribution $|\Psi(z, t)|^2$ changes with time. This is because the wave function at any time t is given by (9.1), which includes the time dependence of the energy eigenfunctions:

$$\Psi(z, t) = \sum_n c_n \sqrt{\left(\frac{2}{\mathbb{L}}\right)} \sin \frac{n\pi z}{\mathbb{L}} \exp\left(-\frac{iE_n t}{\hbar}\right) \qquad \textbf{(9.10)}$$

The computer program *Wave packets* can be used to construct a superposition of energy eigenstates for a particle in an infinitely deep one-dimensional square well, and to examine the time dependence of the associated probability density. But note that (9.10) does not apply if the measurement at $t = 0$ is actually performed! The measurement constitutes an external disturbance, and immediately after an energy measurement that records a particular value E_n the particle must be in the corresponding energy eigenstate. So we usually assume that a measurement produces an abrupt change from the initial state to an eigenstate of the variable concerned, and avoid the question (which is the subject of **quantum measurement theory**) of exactly how the measurement produces this effect. Another type of external interaction, which we shall discuss in Chapter 12, changes the energy of the particle in a way that can be represented by adding a small time-dependent potential to the Hamiltonian. In that case it is possible to represent the state of the particle at an arbitrary time by an expansion of the form (9.1), but with time-dependent coefficients $\{c_n(t)\}$.

The sine and cosine functions are not the only sets of functions that can be used to express an arbitrary function as a linear superposition. In classical physics waves on a string are expressed in terms of trigonometric functions, but in cases of cylindrical symmetry, such as the vibrations of a circular membrane, the relevant functions are Bessel functions [2]. As another example, the angular dependence of the electrostatic potential due to a charge distribution that is symmetric about an axis can be expanded in terms of Legendre polynomials [3]. Mathematically, an arbitrary function has an expansion of the form (9.1) if the set of functions $\{\psi_n(x, y, z, t)\}$ is a **complete mutually orthogonal set**. Furthermore, the function $\Psi(x, y, z, t)$ must be sufficiently smooth, which in practice is always the case for any function suitable for use as a wave function.

We shall not discuss the mathematical meaning of completeness, but merely note that the set of eigenfunctions of any linear Hermitian operator that represents a dynamical variable is mathematically complete, and (as we shall prove in Section 10.3) eigenfunctions corresponding to different eigenvalues are always orthogonal. If the eigenfunctions are also normalized, they form an **orthonormal set**, and satisfy an equation similar to (9.4):

$$\int_{\text{all space}} \mathrm{d}^3 r\, \psi_m^*(x, y, z, t)\psi_n(x, y, z, t) = \delta_{mn} \qquad \textbf{(9.11)}$$

The range of integration is over all the space on which the wave function and the eigenfunctions are defined. This property is used to evaluate the expansion coefficients: we multiply (9.1) on the left by the complex conjugate of a particular eigenfunction $\psi_k(x, y, z, t)$, integrate over all space, and use the orthogonality relation (9.11). This gives

$$\int_{\text{all space}} \mathrm{d}^3 r\, \psi_k^*(x, y, z, t)\Psi(x, y, z, t)$$

$$= \sum_n c_n \int_{\text{all space}} \mathrm{d}^3 r\, \psi_k^*(x, y, z, t)\psi_n(x, y, z, t)$$

$$= \sum_n c_n \delta_{nk} = c_k$$

Thus the general expression for an expansion coefficient is

$$c_n = \int_{\text{all space}} \mathrm{d}^3 r\, \psi_n^*(x, y, z, t)\Psi(x, y, z, t) \qquad \textbf{(9.12)}$$

To summarize, whatever the measurement whose results we wish to predict, we can expand the wave function representing an arbitrary state of the quantum particle as a linear combination of the eigenfunctions of the operator representing the relevant variable (Postulate 3). Then (9.12) can be used to find the probability amplitude for obtaining a particular result.

9.3 Expectation values and uncertainties

In quantum mechanics the **expectation value** $\langle \hat{A} \rangle$ of an operator $\hat{A}$ in the state represented by $\Psi(x, y, z, t)$ is defined as

$$\langle \hat{A} \rangle = \frac{\displaystyle\int_{\text{all space}} \mathrm{d}^3 r \, \Psi^*(x, y, z, t) \hat{A} \, \Psi(x, y, z, t)}{\displaystyle\int_{\text{all space}} \mathrm{d}^3 r \, \Psi^*(x, y, z, t) \Psi(x, y, z, t)} \qquad (9.13)$$

The denominator is equal to 1 if the wave function $\Psi(x, y, z, t)$ is normalized.

The expectation value is identical to the predicted mean $\langle a \rangle$ of a set of measurements of the variable represented by $\hat{A}$ on particles in the state represented by $\Psi(x, y, z, t)$. This is easily demonstrated for the particle in an infinitely deep square well that we discussed in Section 9.2. In the region where the potential is finite the Hamiltonian is

$$\hat{H} = -\frac{\hbar^2}{2m} \frac{\mathrm{d}^2}{\mathrm{d}z^2} \quad (0 \leqslant z \leqslant \mathbb{L}) \qquad (9.14)$$

The wave function $\Psi(z, 0)$ in (9.6) (normalized by $C = \sqrt{(30/\mathbb{L}^5)}$ and the definition (9.13) then give

$$\begin{aligned}
\langle \hat{H} \rangle &= \int_0^{\mathbb{L}} \mathrm{d}z \, [Cz(\mathbb{L} - z)]\left(-\frac{\hbar^2}{2m} \frac{\mathrm{d}^2}{\mathrm{d}z^2}\right)[Cz(\mathbb{L} - z)] \\
&= -\frac{\hbar^2}{2m} \int_0^{\mathbb{L}} \mathrm{d}z \, [Cz(\mathbb{L} - z)][-2C] \\
&= \frac{C^2 \hbar^2 \mathbb{L}^3}{6m} = \frac{10}{\pi^2} E_1 \qquad (9.15)
\end{aligned}$$

This is identical to the value $\langle E \rangle$ in (9.8), which was calculated from the statistical equation (2.7).

The equivalence of the expectation value of an operator (defined by (9.13)) and the statistical mean predicted for a set of measurements of the corresponding variable can be demonstrated, in the general case, using the superposition principle. First we consider the denominator of (9.13), and expand $\Psi(x, y, z, t)$ in terms of the normalized eigenfunctions of $\hat{A}$, as in (9.1):

$$\int_{\text{all space}} \mathrm{d}^3 r \, \Psi^*(x, y, z, t) \Psi(x, y, z, t)$$

$$= \int_{\text{all space}} \mathrm{d}^3 r \left\{\sum_n c_n^* \psi_n^*(x, y, z, t)\right\}\left\{\sum_k c_k \psi_k(x, y, z, t)\right\}$$

$$= \sum_{nk} c_n^* c_k \int_{\text{all space}} \mathrm{d}^3 r \, \psi_n^*(x, y, z, t) \psi_k(x, y, z, t) \qquad (9.16)$$

The eigenfunctions satisfy the orthonormality condition (9.11). This means that, for each value of n, the summation over k contains only one non-zero term, namely the one for which $k = n$. So (9.16) becomes

$$\int_{\text{all space}} d^3r \, \Psi^*(x, y, z, t) \Psi(x, y, z, t) = \sum_n |c_n|^2 \qquad (9.17)$$

If $\Psi(x, y, z, t)$ is normalized, the integral in this equation must be equal to 1, and the expansion coefficients satisfy

$$\sum_n |c_n|^2 = 1 \qquad (9.18)$$

This is equivalent to (2.6), and is a necessary condition if $|c_n|^2$ is to be given a direct probability interpretation.

Now the numerator of (9.13) can be evaluated in a similar way, using the superposition (9.1). Assuming (9.18) is satisfied, we have

$$\langle \hat{A} \rangle = \sum_{nk} c_n^* c_k \int_{\text{all space}} d^3r \, \psi_n^*(x, y, z, t) \hat{A} \psi_k(x, y, z, t) \qquad (9.19)$$

Remember that $\psi_k(x, y, z, t)$ is an eigenfunction of $\hat{A}$:

$$\hat{A} \psi_k(x, y, z, t) = a_k \psi_k(x, y, z, t) \qquad (9.20)$$

Therefore

$$\langle \hat{A} \rangle = \sum_{nk} c_n^* c_k a_k \int_{\text{all space}} d^3r \, \psi_n^*(x, y, z, t) \psi_k(x, y, z, t)$$

$$= \sum_n |c_n|^2 a_n = \langle a \rangle \qquad (9.21)$$

where we have again used the orthonormality condition (9.11). Since $|c_n|^2 = \mathbb{P}_n$ (Postulate 3), (9.21) is identical to the statistical definition (2.7) of the mean value of the set of results.

The uncertainty in the value of the variable represented by $\hat{A}$ in a particular quantum state is defined by

$$\Delta \hat{A} = \sqrt{(\langle \hat{A}^2 \rangle - \langle \hat{A} \rangle^2)} \qquad (9.22)$$

Because the expectation value of an operator is equivalent to the mean of a set of measurements, (9.22) is equivalent to the standard deviation defined in the statistical equation (2.8).

Expectation values for the position and momentum of an oscillator

Suppose that a momentum or a position measurement is made on a one-dimensional simple harmonic oscillator in one of its energy eigenstates. To predict the results of a set of momentum measurements, we need first of all to know the momentum eigenvalues. The momentum eigenvalue equation for a particle moving along the x axis has solutions $\psi(x) = Ce^{ipx/\hbar}$, (3.10), and these functions are finite everywhere for any real value of p. The momentum spectrum for a free particle is therefore the continuous range of real values from $-\infty$ to $+\infty$, and correspondingly there is a continuous rather than a discrete set of momentum eigenfunctions. The spectrum of the position operator is also continuous, since the probability density for a position measurement, $|\psi(x)|^2$, is a continuous function of position.

To find the probability of any particular result for a momentum measurement when the particle is in an energy eigenstate, we should expand the energy eigenfunction as a linear superposition of momentum eigenfunctions. However, the superposition (9.1) is an expansion in terms of a discrete, not a continuous, set. Similarly, the expression (9.21) for the expectation value of an operator applies only if the eigenvalue spectrum is discrete. But the definition (9.13) does not depend on whether or not the variable to be measured has a discrete spectrum, and so it can be used to predict the expectation value of momentum or position.

Consider an oscillator in its nth energy eigenstate, represented by the eigenfunction $\psi_n(x)$, which is given in (5.26) as a function of $\xi = \sqrt{(m\omega/\hbar)}x$. Neither the momentum operator nor the position operator commutes with the simple harmonic oscillator Hamiltonian, so the energy eigenstates are not momentum or position eigenstates, and the momentum and position are indeterminate when the oscillator has a precise energy.

Provided we use the normalized energy eigenfunctions, the expectation values of momentum and position are given, according to (9.13), by

$$\langle \hat{P}_x \rangle = \int_{-\infty}^{\infty} \mathrm{d}x \; \psi_n^*(x) \left(-\mathrm{i}\hbar \, \frac{\mathrm{d}}{\mathrm{d}x} \right) \psi_n(x) \qquad (9.23)$$

$$\langle \hat{X} \rangle = \int_{-\infty}^{\infty} \mathrm{d}x \; \psi_n^*(x) x \, \psi_n(x) \qquad (9.24)$$

These can easily be seen to vanish from considerations of symmetry: the energy eigenfunction is a parity eigenfunction corresponding to the eigenvalue $(-1)^n$, while the operators $\hat{P}_x$ and $\hat{X}$ have negative parity. Each integrand has the parity eigenvalue $(-1)^n(-1)(-1)^n = -1$, so that it is an odd function of x, and the integral vanishes. The expectation values are therefore

$$\langle \hat{P}_x \rangle = 0, \qquad \langle \hat{X} \rangle = 0 \qquad (9.25)$$

(Remember that these are identical to the mean values $\langle p_x \rangle$ and $\langle x \rangle$ that would be obtained using the statistical definition of the mean, (2.7).)

Another method of calculating these integrals uses the properties of the ladder operators $\hat{A}$ and $\hat{A}^{\dagger}$, defined in (5.4) in terms of ξ and $\mathrm{d}/\mathrm{d}\xi$. Inverting these expressions, we find

$$\left. \begin{array}{l} \hat{P}_x = -\mathrm{i}\sqrt{(\tfrac{1}{2}\hbar m\omega)}(\hat{A} - \hat{A}^{\dagger}) \\[2mm] x = \hat{X} = \sqrt{\left(\dfrac{\hbar}{2m\omega} \right)}(\hat{A} + \hat{A}^{\dagger}) \end{array} \right\} \qquad (9.26)$$

But the ladder operators act on an energy eigenfunction to produce the eigenfunction of an adjacent energy level. To be precise, (5.29) and (5.30) state that

$$\hat{A}\psi_n(x) = \sqrt{(n)}\psi_{n-1}(x)$$
$$\hat{A}^\dagger\psi_n(x) = \sqrt{(n+1)}\psi_{n+1}(x)$$

Since the eigenfunctions $\{\psi_n(x)\}$ form an orthogonal set, (5.27), the integrals in (9.23) and (9.24) can be seen to vanish, and (9.25) is verified.

Equations (9.26) are useful in calculating the expectation values of $\hat{P}_x^2$ and $\hat{X}^2$. Squaring these expressions, we get

$$\hat{P}_x^2 = -\tfrac{1}{2}\hbar m\omega(\hat{A}^2 - \hat{A}\hat{A}^\dagger - \hat{A}^\dagger\hat{A} + \hat{A}^{\dagger 2}) \tag{9.27}$$

$$\hat{X}^2 = \frac{\hbar}{2m\omega}(\hat{A}^2 + \hat{A}\hat{A}^\dagger + \hat{A}^\dagger\hat{A} + \hat{A}^{\dagger 2}) \tag{9.28}$$

The expectation values of $\hat{A}^2$ and $\hat{A}^{\dagger 2}$ must be zero, in the same way as the expectation values of $\hat{A}$ and $\hat{A}^\dagger$, since the squared operators act on $\psi_n(x)$ to give functions proportional to $\psi_{n-2}(x)$ and $\psi_{n+2}(x)$ respectively, and these are orthogonal to $\psi_n(x)$. To evaluate the expectation values of $\hat{A}\hat{A}^\dagger$ and $\hat{A}^\dagger\hat{A}$, remember that they are related to the simple harmonic oscillator Hamiltonian through (5.5), which can be re-expressed as

$$\hat{H} = (\hat{A}^\dagger\hat{A} + \tfrac{1}{2})\hbar\omega = (\hat{A}\hat{A}^\dagger - \tfrac{1}{2})\hbar\omega$$

The expectation value of $\hat{H}$ in the nth energy eigenstate is the eigenvalue $E_n = (n + \tfrac{1}{2})\hbar\omega$, so the expectation values of the operators in (9.27) and (9.28) are

$$\langle\hat{P}_x^2\rangle = \hbar m\omega(n + \tfrac{1}{2}) \tag{9.29}$$

$$\langle\hat{X}^2\rangle = \frac{\hbar}{m\omega}(n + \tfrac{1}{2}) \tag{9.30}$$

Then from (9.22) the uncertainties in momentum and position are

$$\Delta\hat{P}_x = \Delta p_x = \sqrt{[\hbar m\omega(n + \tfrac{1}{2})]} \tag{9.31}$$

$$\Delta\hat{X} = \Delta x = \sqrt{\left[\frac{\hbar}{m\omega}(n + \tfrac{1}{2})\right]} \tag{9.32}$$

The product of these satisfies Heisenberg's uncertainty principle, as it should:

$$\Delta\hat{P}_x\,\Delta\hat{X} = \hbar(n + \tfrac{1}{2}) \geqslant \tfrac{1}{2}\hbar \tag{9.33}$$

Note that in the ground state, $n = 0$, the product of the uncertainties takes the minimum possible value. This is a peculiarity of the simple harmonic oscillator, and for energy eigenstates of particles in other potentials the minimum value of $\Delta\hat{P}_x\,\Delta\hat{X} = \Delta p_x\,\Delta x$ allowed by the uncertainty principle is always exceeded, even in the ground state.

9.4 Superpositions of momentum eigenfunctions

The expansion of an arbitrary wave function as a superposition of momentum eigenfunctions is not of the form (9.1), since they form a continuous rather than a discrete set. So to predict the mean of a set of momentum measurements in Section 9.3 we used the definition (9.13), which does not depend on whether the eigenvalue spectrum is continuous or discrete. We now wish to find how to expand an arbitrary wave function as a superposition of momentum eigenfunctions.

Just as a continuous periodic function is represented as a discrete Fourier series, a non-periodic function that is sufficiently smooth and that is negligible outside some finite region can be represented by a Fourier integral expansion [4]:

$$f(x) = \frac{1}{\sqrt{(2\pi)}} \int_{-\infty}^{\infty} \mathrm{d}k \, g(k) \mathrm{e}^{ikx} \qquad (9.34)$$

For example, the wave function representing an energy eigenstate of the one-dimensional simple harmonic oscillator could be expanded in this way, since it is localized in a region centred on the origin, of half-width somewhat greater than the amplitude of the corresponding classical oscillator as shown in Figure 5.2, and the probability amplitude outside this region is negligibly small, though not exactly zero. In (9.34) $g(k)$ is the Fourier transform of $f(x)$, which is given by

$$g(k) = \frac{1}{\sqrt{(2\pi)}} \int_{-\infty}^{\infty} \mathrm{d}x \, f(x) \mathrm{e}^{-ikx} \qquad (9.35)$$

Equation (9.34) looks like a superposition of the continuous set of momentum eigenfunctions

$$\psi_p(x) = C\mathrm{e}^{ikx}, \qquad p = \hbar k \qquad (9.36)$$

We shall therefore assume that at any instant t an arbitrary wave function $\Psi(x, t)$ can be expanded in terms of the momentum eigenfunctions by writing a Fourier integral expansion similar to that of (9.34):

$$\Psi(x, t) = \frac{1}{\sqrt{(2\pi)}} \int_{-\infty}^{\infty} \frac{\mathrm{d}p}{\hbar} \frac{1}{C} g(p, t) C\mathrm{e}^{ipx/\hbar}$$

$$= \frac{1}{\sqrt{(2\pi)}} \int_{-\infty}^{\infty} \mathrm{d}p \, \frac{1}{C\hbar} g(p, t)\psi_p(x)$$

(The Fourier transform has been written as a function of time to allow for a different expansion at each instant.) In this context it is conventional to choose the normalization constant to be

$$C = 1/\sqrt{(2\pi\hbar)} \qquad (9.37)$$

and to define

$$\phi(p, t) = \frac{1}{\sqrt{h}} \, g(p, t)$$

Then the expansion of $\Psi(x, t)$ in terms of momentun eigenfunctions is

$$\Psi(x, t) = \int_{-\infty}^{\infty} \mathrm{d}p \, \phi(p, t) \psi_p(x) \qquad (9.38)$$

and (9.35) for the Fourier transform gives

$$\phi(p, t) = \int_{-\infty}^{\infty} \mathrm{d}x \, \psi_p^*(x) \Psi(x, t) \qquad (9.39)$$

In three dimensions (9.38) and (9.39) become

$$\Psi(x, y, z, t) = \int_{\text{all momenta}} \mathrm{d}^3 p \, \phi(p_x, p_y, p_z, t) \psi_p(x, y, z) \qquad (9.40)$$

$$\phi(p_x, p_x, p_x, t) = \int_{\text{all space}} \mathrm{d}^3 r \, \psi_p^*(x, y, z) \Psi(x, y, z, t) \qquad (9.41)$$

where $\psi_p(x, y, z) = (2\pi h)^{-3/2} e^{i k \cdot r}$, $p = \hbar k$, and the integral in (9.40) is over all values of the momentum – that is, over all momentum space.

Equation (9.40) is analogous to the discrete expansion (9.1), with the integration over a continuous set of eigenfunctions replacing the summation over a discrete set, and the continuous function $\phi(p_x, p_y, p_z, t)$ replacing the set of discrete coefficients $\{c_n\}$. Equation (9.41), which gives $\phi(p_x, p_y, p_z, t)$ for any particular (p_x, p_y, p_z) at some instant, is of the same form as (9.12), which determines the expansion coefficient c_n for any particular n.

Returning to the momentum measurements on the one-dimensional simple harmonic oscillator discussed in Section 9.3, we can now calculate not only the expectation value and standard deviation but also the probability of obtaining any particular result. Following Postulate 3, the function $\phi(p, t)$ in the expansion (9.38) should be related to this probability. However, since the possible results cover a continuous range, we have to identify $|\phi(p, t)|^2 \, \mathrm{d}p$ as the probability of measuring the momentum to lie in the range between p and $p + \mathrm{d}p$. Then $|\phi(p, t)|^2$ is a probability density for the momentum measurement. For the simple harmonic oscillator in its nth energy cigenstate the probability density amplitude for the momentum measurement at the instant t can be found by substituting the energy eigenfunction (5.26) into (9.39). For the ground state the result is

$$\phi(p, t) = \frac{1}{(\hbar m \omega \pi)^{1/4}} \exp\left(-\frac{p^2}{2m\omega\hbar}\right) e^{-i\omega t/2} \qquad (9.42)$$

The momentum probability distribution $|\phi(p, t)|^2$ is shown in Figure 9.2 as a function of p. Note that it is independent of time, and this will be true for the momentum probability distribution in any energy eigenstate,

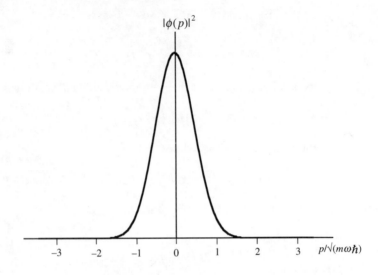

Figure 9.2 The momentum probability distribution for a one-dimensional simple harmonic oscillator in its ground state.

because $\phi(p, t)$ has the exponential time dependence of the energy eigenfunction.

Wave packets

A superposition of momentum eigenfunctions like that in (9.40) can describe the state of a free particle whose position is known to lie within some finite region of space at time t. The wave function describing such a state is a **wave packet**, with negligibly small probability amplitude outside the region in which the particle is localized. Consider the one-dimensional case, and suppose that at $t = 0$ the particle lies within a region of width Δx. The uncertainty principle indicates that at that instant its momentum is not determined to better than $\Delta p \geqslant \hbar/(2\,\Delta x)$, so that it cannot be in a momentum eigenstate. In fact Fourier theory [5] tells us that the integral expansion (9.34) of a wave pulse of width Δx contains contributions from all k values within a range $\Delta k \approx 1/(2\,\Delta x)$. So all momentum eigenfunctions corresponding to eigenvalues within the range $\hbar/(2\,\Delta x)$ contribute to the expansion (9.38), and the uncertainty principle is automatically satisfied.

For a free particle the momentum eigenfunctions are simultaneously eigenfunctions of the energy. So the appropriate time dependence $e^{-iEt/\hbar}$, with $E = p^2/2m$, should be associated with $\psi_p(x)$, and the wave packet describing a free particle localized in some region of space can be written, using (9.38), as

$$\Psi(x, t) = \int_{-\infty}^{\infty} dp \; \phi(p) \; \frac{1}{\sqrt{(2\pi\hbar)}} \; e^{i(px - Et)/\hbar} \qquad (9.43)$$

The function $\phi(p)$ is calculated for any given wave packet from (9.39) at $t = 0$, and is independent of time provided the particle remains free. (If the particle does not remain free, the momentum eigenfunctions are no longer energy eigenfunctions, and the new time dependence must be included by making $\phi(p)$ time-dependent.)

The wave packet (9.43) will change with time, because it is a super-position of waves with different phase velocities E/p. As in the case of classical wave pulses [6], this produces **dispersion**, or spreading, of the wave packet. The centre of the packet travels with the group velocity, which corresponds to the velocity of the physical particle. But as time goes on, the position of the particle becomes less well determined, since Δx increases with time. This does not mean that the particle itself spreads out – for example a measurement to determine the radius of a free proton gives the same value of roughly 10^{-15} m independent of time. What increases with time is the volume of space that a detector must cover to be sure of detecting the particle.

An ideal classical particle state has zero uncertainty in both position and momentum, and the state of a free quantum particle that is the closest approximation to this is the **minimum-uncertainty state**, in which the product $\Delta x \, \Delta p_x$ takes its minimum value $\frac{1}{2}\hbar$. The wave function at $t = 0$ is

$$\Psi(x, 0) = \frac{1}{(2\pi\sigma^2)^{1/4}} \exp\left[- \frac{(x - x_0)^2}{4\sigma^2}\right] e^{ip_0 x/\hbar} \qquad (9.44)$$

This is normalized, and gives a Gaussian probability distribution centred on $x = x_0$ and with width at half-height $\Delta(0) = \sigma\sqrt{(2\ln 2)}$, as shown in Figure 9.3. (This probability distribution has the same x dependence as the ground state of a simple harmonic oscillator, which, as we noted at the end of Section 9.3, is a minimum-uncertainty state.)

The expectation values of $\hat{X}$, $\hat{P}_x$, $\hat{X}^2$ and $\hat{P}_x^2$ for the wave function (9.44) can be calculated using the definition (9.13), and we find

$$\langle \hat{X} \rangle = \langle x \rangle = x_0, \qquad \langle \hat{P}_x \rangle = \langle p_x \rangle = p_0$$
$$\Delta x = \sigma, \qquad \Delta p_x = \hbar/2\sigma$$

The time dependence of the Gaussian wave packet can be found from (9.43), using the constant-momentum probability amplitude $\phi(p)$ calculated from (9.39). The packet spreads as it moves with the group velocity p_0/m, its width at time t being given by

$$\sigma^2(t) = \sigma^2(0) + \frac{\hbar^2 t^2}{4m^2\sigma^2(0)}$$

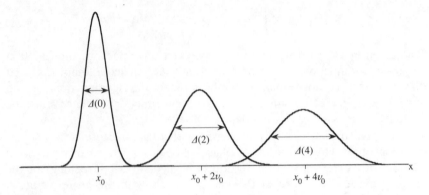

Figure 9.3 The spatial probability distribution for a Gaussian wave packet describing a free particle moving in one dimension, at times $t = 0$, $t = 2$ and $t = 4$. The group velocity of the packet is v_0, and the width at half-height, $\sigma(t)\sqrt{(2\ln 2)}$, is labelled $\Delta(t)$.

This spreading of the wave packet is shown in Figure 9.3, where the Gaussian packet centred on x_0 at $t = 0$ is also shown at later times $t = 2$ and $t = 4$.

The construction of a wave packet for a bound particle has been demonstrated experimentally for the single valence electron of an alkali metal atom. The wave packet is a superposition of different energy states, and since the bound states of the atom are discrete, the superposition is of the form (9.1), with $\psi_n(x, y, z, t) \propto e^{-iE_n t/\hbar}$. The energy levels of the valence electron in such an atom are similar to those of hydrogen, varying with the principal quantum number as $1/n^2$, and they get closer and closer for large n. The spread in frequency of the beam produced by a pulsed laser is sufficient to excite coherently several different energy levels at high n. The electron wave packet that results is localized radially, and oscillates backwards and forwards between a minimum and a maximum distance from the nucleus. These distances are the same as the minimum and maximum distances of an electron in the corresponding elliptical Bohr orbit, as shown in Figure 9.4, and the wave packet oscillates at exactly the classical orbital frequency expected in the Bohr model. The experiment has been performed with rubidium atoms excited to a superposition of energy states centred on values of n between 20 and 100 [7], and two complete oscillations of the wave packet could be observed before the packet dispersed. A wave packet that orbits the nucleus exactly like the electron in the Bohr model would have to be localized in terms of the angular variables as well as the radial distance. This would entail exciting simultaneously and coherently an appropriate superposition of eigenstates with different angular momentum quantum numbers (l, m_l) as well as different energies.

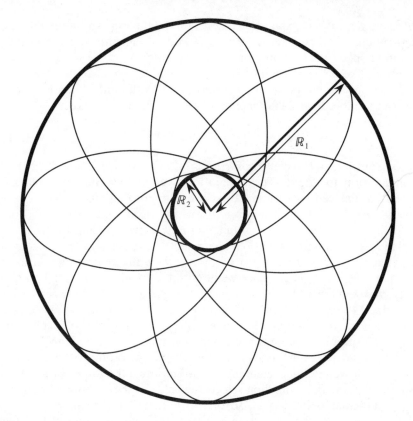

Figure 9.4 Bohr orbits for an electron in an atom, with differently oriented symmetry axes. The radially localized wave packet oscillates between the maximum and minimum radial distances $\mathbb{R}_1$ and $\mathbb{R}_2$.

9.5 Position eigenstates and the Dirac delta function

According to our discussion in Section 9.4, the function $|\phi(p_x, p_y, p_z, t)|^2$, where $\phi(p_x, p_y, p_z, t)$ is defined by (9.41), is the probability density for a momentum measurement at time t. This is reminiscent of the interpretation of the wave function $\Psi(x, y, z, t)$ as the probability density amplitude for a position measurement at the instant t, given by Postulate 1. Indeed, $\phi(p_x, p_y, p_z, t)$ plays an equivalent role in predicting the result of a momentum measurement to $\Psi(x, y, z, t)$ in the case of a position measurement, and $\phi(p_x, p_y, p_z, t)$ is referred to as the **momentum space wave function**. According to (9.40), $\phi(p_x, p_y, p_z, t)$ is essentially the expansion coefficient in a description of the state as a superposition of momentum eigenfunctions. So we can think of the wave function

$\Psi(x, y, z, t)$ as an expansion coefficient in the superposition that describes an arbitrary state in terms of position eigenfunctions. Since the position spectrum is continuous, this expansion must be an integral expansion, like the expansion (9.40).

But what are the position eigenfunctions? In the one-dimensional case we can use the **Dirac delta function** $\delta(x - x')$, defined to satisfy

$$\Psi(x, t) = \int_{-\infty}^{\infty} dx' \, \Psi(x', t)\delta(x - x') \tag{9.45}$$

This is an integral expansion for an arbitrary wave function $\Psi(x, t)$, of similar form to (9.38). The integration over momentum eigenvalues in (9.38) is replaced by an integration over the position eigenvalues x', the eigenfunction of momentum is replaced by the delta function $\delta(x - x')$ (which is to be regarded as a function of x corresponding to the eigenvalue x'), and the expansion coefficient for a particular momentum eigenvalue p, $\phi(p, t)$, is replaced by the wave function at the particular point x', $\Psi(x', t)$. Then, since we interpreted $|\phi(p, t)|^2 \, dp$ as the probability that a momentum measurement will yield a value between p and $p + dp$, $|\Psi(x', t)|^2 \, dx'$ can be interpreted as the probability that a position measurement will yield a value between x' and $x' + dx'$. This is the same interpretation of the wave function as that given in Section 2.1.

The Dirac delta function is not a function in the ordinary sense, since it only has a well-defined meaning within an integral over x. From (9.45) we see that the delta function selects the value of the rest of the integrand at a particular point $x' = x$, so that in a sense we can think of it as a function that vanishes everywhere except at the unique point $x' = x$. One useful way of representing the delta function is [8]

$$\delta(x - x') = \frac{1}{2\pi\hbar} \int_{-\infty}^{\infty} dp \, e^{ip(x-x')/\hbar} \tag{9.46}$$

To see that this form leads to the expansion (9.45), we substitute it in the right-hand side of that equation, and remember that, according to (9.36) and (9.37), $(2\pi\hbar)^{-1/2}e^{ipx}$ is a momentum eigenfunction:

$$\int_{-\infty}^{\infty} dx' \, \Psi(x', t)\delta(x - x') = \frac{1}{2\pi\hbar} \int_{-\infty}^{\infty} dx' \int_{-\infty}^{\infty} dp \, e^{ip(x-x')/\hbar} \Psi(x', t)$$

If we exchange the order of the p and x' integrations, this becomes

$$\int_{-\infty}^{\infty} dp \left[\int_{-\infty}^{\infty} dx' \, \psi_p^*(x')\Psi(x', t) \right] \psi_p(x) = \int_{-\infty}^{\infty} dp \, \phi(p, t)\psi_p(x)$$

$$= \Psi(x, t)$$

where we have used (9.38) and (9.39).

Note that (9.46) is an expansion of the delta function in terms of the momentum eigenfunctions $\psi_p(x)$, of the same form as (9.38), with expansion coefficients $\phi(p, t) = (2\pi\hbar)^{-1/2}e^{-ipx'/\hbar}$. Since this gives $|\phi(p, t)|^2 = (2\pi\hbar)^{-1}$, the momentum probability density is independent of p, and a momentum measurement on a particle in a position eigenstate is as likely to result in one particular value of p as in any other. In other words, the momentum is completely indeterminate, as we should expect.

References

[1] Boas M.L. (1983). *Mathematical Methods in the Physical Sciences* 2nd edn, Chap. 7. New York: Wiley.
[2] See Ref. [1], Sections 12.12–12.20 and 13.6.
[3] See Ref. [1], Section 12.5, pp. 492–5.
[4] See Ref. [1], Section 15.4.
[5] Main I.G. (1984). *Vibrations and Waves in Physics* 2nd edn, Section 11.3. Cambridge University Press.
[6] French A.P. (1971). *Vibrations and Waves*, Chap. 7, pp. 230–4. London: Thomas Nelson.
[7] ten Wolde A., Noordam L.D., Lagendijk A. and van Linden van den Heuvell H.B. (1988). *Physical Review Letters* **61**, 2099.
[8] See Ref. [1], Section 15.7.

Problems

9.1 Equations (3.44) and (3.45) determine the energy eigenfunction for a particle in the potential of Figure 3.5 with energy eigenvalue $E > V_0, V_1$. Use the superposition principle to find an expression for the probability of each of the following results of a momentum measurement.

(a) the particle is travelling towards $z = 0$ in Region I;
(b) the particle is travelling towards $z = 0$ in Region II;
(c) the particle is travelling away from $z = 0$ in Region I.

Hence justify the expressions for the reflection and transmission coefficients given in (3.46).

9.2 (a) Substitute the wave function (9.6) into (9.5) to verify the form of the expansion coefficients given in (9.7).

(b) Perform the appropriate integration to verify (9.15).

9.3 At time $t = 0$ a particle in an infinitely deep one-dimensional square well is in the state described by the wave function $\Phi(z, 0)$, which is not an energy eigenfunction. Expand this function in terms of the energy eigenfunctions of the particle in each of the following cases:

(a)
$$\Phi(z,0) = \begin{cases} Cz & (0 \leqslant z \leqslant \frac{1}{2}L) \\ C(L-z) & (\frac{1}{2}L \leqslant z \leqslant L) \end{cases}$$

(b)
$$\Phi(z,0) = C'z(z - \tfrac{1}{2}L)(L-z) \quad (0 \leqslant z \leqslant L)$$

(In each case the wave function is zero outside the region $0 \leqslant z \leqslant L$.) For each wave function find the probability of obtaining the results E_1, E_2 and E_3 if an energy measurement is performed (E_n is the nth energy eigenvalue in the well). Find also the expectation value of the energy in each case. (The necessary integration is easier if the sine function is expanded in terms of exponentials and the recurrence relation

$$\int_0^L dz\, z^n e^{ikz} = \frac{L^n}{ik} e^{ikL} - \frac{n}{ik} \int_0^L dz\, z^{n-1} e^{ikz}$$

is used. The results

$$\sum_{n \text{ odd}} \frac{1}{n^2} = \tfrac{1}{8}\pi^2, \qquad \sum_{n \text{ even}} \frac{1}{n^4} = \tfrac{1}{1440}\pi^4$$

will be needed.)

*9.4 Find the numerical values of the first 13 coefficients in your expansion for the wave function $\Phi(z,0)$ in each case in Problem 9.2, and use them in the computer program *Wave packets: infinite square well* to construct (approximately) the wave packet representing the particle at time $t = 0$. Write down the expansion for the wave function at $t \neq 0$ in each case. Observe the variation of the probability density with time in each case, by selecting the appropriate option in the program, and compare it with that of the probability density for an energy eigenfunction (for the level $n = 1$ in case (a) and the level $n = 2$ in case (b)).

9.5 (a) At time $t = 0$ a one-dimensional simple harmonic oscillator is in a state represented by $\Psi(x,0) = \sqrt{(\tfrac{1}{2})}\psi_0(x) - \sqrt{(\tfrac{1}{3})}\psi_1(x) + \sqrt{(\tfrac{1}{6})}\psi_2(x)$, where $\psi_n(x)$ is a normalized simple harmonic oscillator energy eigenfunction, (5.26). If energy measurements are performed on a set of oscillators in this state, what is the probability of obtaining each of the following results: $\tfrac{1}{2}\hbar\omega$, $\tfrac{3}{2}\hbar\omega$, $\tfrac{7}{6}\hbar\omega$, $\tfrac{5}{2}\hbar\omega$ and $\tfrac{9}{2}\hbar\omega$? Find the expectation value of the energy, and its uncertainty.

* (b) Write down an expansion for the state of the oscillator at any time t, and examine the time dependence of the probability distribution both analytically and using the computer program *Wave packets: harmonic oscillator*.

9.6 An electron confined in a quantum well has four discrete energy levels $E_1 = 0.27\,\text{eV}$, $E_2 = 1.08\,\text{eV}$, $E_3 = 3.65\,\text{eV}$ and $E_4 = 4.06\,\text{eV}$. It is in a state in which the probabilities associated with these energies are $\tfrac{1}{2}$, $\tfrac{1}{4}$, $\tfrac{3}{16}$ and $\tfrac{1}{16}$ respectively.

(a) Find the expectation value of its energy $\langle \hat{H} \rangle$ and the corresponding uncertainty $\Delta \hat{H}$.

(b) Obtain an expression for the wave function describing the state of the particle in terms of its energy eigenfunctions $\psi_n(z)$ at time $t = 0$. Why is the expression not unique? Write down two different wave functions corresponding to the same values of $\langle \hat{H} \rangle$ and $\Delta \hat{H}$ that you found in (a).

9.7   A particle is in a state described by the wave function

$$\psi(z) = \tfrac{1}{2}e^{i\pi z/\mathbb{L}} - \sqrt{(\tfrac{1}{2})}e^{-i\pi z/\mathbb{L}} + \tfrac{1}{4}\sqrt{3}\, e^{i2\pi z/\mathbb{L}} - \tfrac{1}{4}e^{-i2\pi z/\mathbb{L}}$$

Find all possible results of a measurement of $\hat{P}_z$, and the corresponding probabilities. Find also the expectation value and uncertainty of the momentum.

9.8   A rigid rotator (with Hamiltonian given by (4.47)) is in a state described by the wave function $\psi(\theta) = 1 + 3\cos\theta + 4\cos^2\theta$. Expand this function in terms of the simultaneous eigenfunctions of energy and orbital angular momentum ((4.45) and Table 4.1). What are the possible results of measurements of the energy and of the z component and square of the orbital angular momentum? Find the probability of obtaining each, and calculate the expectation values of $\hbar^2 \hat{L}^2$ and $\hbar \hat{L}_z$, together with their uncertainties.

9.9   Prove that the expectation value of an operator $\hat{A}$ in one of its eigenstates ψ_n is the corresponding eigenvalue a_n.

9.10 (a) Prove that the probability distribution for a state that is a superposition of exactly two different energy eigenstates oscillates with a frequency that is determined by the difference in their energy eigenvalues.

*(b) Use the computer program *Wave packets* to demonstrate this in the case of a particle in an infinitely deep one-dimensional square well: examine the time dependence of the probability distribution when the first coefficient and the kth are $\sqrt{\tfrac{1}{2}}$ and all others arc 0, for a range of k. You should be able to make a rough estimate of the period of the oscillation for a given k by observing the time intervals displayed on the screen. Calculate the ratios of the periods when $k = 3, 4, 5, \ldots$ to that when $k = 2$, and compare the result with the values you would expect on the basis of your calculation in (a). Repeat your observations in the case of the one-dimensional simple harmonic oscillator.

9.11 (a) Show that the expectation value of the potential energy of a one-dimensional simple harmonic oscillator is equal to the expectation value of its kinetic energy, and find the uncertainties in each.

(b) Show that the integrals

$$\int_{-\infty}^{\infty} dx\, \psi_n^*(x) x \psi_k(x), \qquad \int_{-\infty}^{\infty} dx\, \psi_n^*(x) \hat{P}_x \psi_k(x)$$

(where $\psi_k(x)$ is an energy eigenfunction of the one-dimensional simple harmonic oscillator) vanish unless $n = k \pm 1$, and evaluate the integrals when this condition is satisfied. (*Hint*: Refer to Problem 5.4(c).)

9.12   Calculate the expectation values of the radial component of the position operator, $\hat{R} = r$, and the radial momentum operator $\hat{P}_r$, (4.22), in each of the states in the $n = 1$ and $n = 2$ levels of the hydrogen atom. Compare your result for $\langle \hat{R} \rangle$ with the most probable value of r in each case. Find the uncertainties in $\hat{R}$ and $\hat{P}_r$ in each state.

9.13   (a) Find the Fourier transform $\phi(p,0)$ of the wave function $\Psi(z,0)$ of (9.6). Deduce the probability density for the results of a momentum measurement performed on a particle in the state represented by $\Psi(z,0)$. Note that the probability density is an even function of p, and deduce the expectation value of the momentum. Check that this is the same as the expectation value obtained using the wave function $\Psi(z,0)$ and the definition (9.13) for the expectation value.

   (b) Find $\phi(p,0)$ and the probability density for the results of a momentum measurement for a particle in an energy eigenstate of the infinitely deep one-dimensional square well. Deduce the expectation value of the momentum.

9.14   Prove (9.42) for the amplitude of the momentum probability density for a one-dimensional simple harmonic oscillator in its ground state. (The results

$$\int_{-\infty}^{\infty} dx \exp(-\alpha x^2 + i\beta x) = \int_{-\infty}^{\infty} dx \exp\left[-\alpha\left(x - \frac{i\beta}{2\alpha}\right)^2\right] \exp\left(-\frac{\beta^2}{4\alpha}\right)$$

$$= \int_{-\infty}^{\infty} dx' \exp(-\alpha x'^2) \exp\left(-\frac{\beta^2}{4\alpha}\right)$$

$$\int_{-\infty}^{\infty} dx \exp(-\alpha x^2) = \sqrt{\left(\frac{\pi}{\alpha}\right)}.$$

can be used.)

9.15   (a) Find the momentum probability distribution for the Gaussian wave packet (9.44).

   (b) Is the wave packet (9.44) an energy eigenfunction?

   (c) Calculate the expectation values $\langle \hat{X} \rangle$, $\langle \hat{P}_x \rangle$, $\langle \hat{X}^2 \rangle$ and $\langle \hat{P}_x^2 \rangle$ for a particle in the state described by the Gaussian wave packet (9.44). (You will need the integrals given at the end of Problem 9.14.)

 * (d) Using the computer program *Transmission: Gaussian wave packet*, investigate the behaviour of a Gaussian wave packet incident on a square barrier in one dimension. Using the default barrier width (0.02 in arbitrary units), watch the behaviour of the wave packet for a range of values of V_0/E (ratio of barrier height to energy). For example, try $V_0/E = 0.2$, 0.8, 1.0, 1.3, 1.5, 2 and 3. Note the proportion of the packet that is transmitted each time, and compare what happens with what you would expect from a classical point of view. Repeat your

investigation using different barrier widths, for example 0.06, 0.1 and 0.2.

9.16 (a) Evaluate the following integrals involving the Dirac delta function, which was introduced in (9.45):

(i) $\int_{-\infty}^{\infty} dx \, (x^3 - 3x + 2)\delta(x - 1)$

(ii) $\int_{-\infty}^{0} dx \, (x^3 - 3x + 2)\delta(x - 1)$

(iii) $\int_{0}^{2} dx \, (x^3 - 3x + 2)\delta(x - 1)$

(iv) $\int_{1-\epsilon}^{1+\epsilon} dx \, (x^3 - 3x + 2)\delta(x - 1)$, where ϵ is any small positive number.

(b) One way of representing the Dirac delta function was shown in (9.46). Another is the following:

$$\delta(x - x_0) = \lim_{\epsilon \to 0} \frac{\Theta(x - x_0 + \epsilon) - \Theta(x - x_0)}{\epsilon} \quad (\epsilon > 0)$$

where $\Theta(x)$ is the step function defined by

$$\Theta(x) = \begin{cases} 1 & (x > 0) \\ -1 & (x < 0) \end{cases}$$

(Note that this definition implies that $\delta(x)$ is the first derivative of the step function.) To help you to visualize $\delta(x - x_0)$, sketch roughly the functions $\Theta(x)$ and $[\Theta(x + \epsilon) - \Theta(x)]/\epsilon$ for each of the values $\epsilon = 0.1$, 0.01 and 0.001.

CHAPTER 10

The Matrix Formulation of Quantum Mechanics

10.1 Alternatives to Schrödinger's wave mechanics

Chapter 9 has given us a new view of the significance of a wave function – it is, for a position measurement, the equivalent of the set of expansion coefficients $\{c_n\}$ that are relevant in the case of the measurement of a variable with a discrete spectrum. Just as $|c_n|^2$ is the probability that the result of a measurement of the variable concerned will be its nth eigenvalue, $|\Psi(x, y, z, t)|^2$ is the probability density for finding the particle at the point (x, y, z) if a position measurement is performed at time t. In Schrödinger's wave mechanics a state is described by the wave function $\Psi(x, y, z, t)$, which emphasizes the role of position in defining the state of a particle, and allows the direct use of the classical potential function $V(x, y, z)$ in setting up the energy eigenvalue equation.

If we had initially defined the state of a particle in terms of momentum instead of position, we would have represented the momentum components by algebraic variables p_x, p_y, p_z, and introduced a momentum-space wave function, $\phi(p_x, p_y, p_z, t)$. Again the state would be represented by a continuous function, but this time of momentum rather than position. Postulate 1 (Section 2.2) would be replaced by an interpretation of $|\phi(p_x, p_y, p_z, t)|^2$ as a probability density for the results of a momentum measurement. As we saw in Section 4.1, the incompatibility of position and momentum measurements (which is a fundamental ingredient of quantum theory) is represented by the non-commutation of the corresponding operators, (4.3). This means that if we represent momentum components by algebraic variables, we cannot also use algebraic variables to represent the position, since all algebraic variables commute. The commutation relations of (4.3) are preserved if we use p_x, p_y and p_z instead of the differential operators $\hat{P}_x$, $\hat{P}_y$ and $\hat{P}_z$ defined in (3.9), and replace the algebraic variables x, y and z by differential operators

$\hat{X} = i\hbar\, \mathrm{d}/\mathrm{d}p_x$, $\hat{Y} = i\hbar\, \mathrm{d}/\mathrm{d}p_y$ and $\hat{Z} = i\hbar\, \mathrm{d}/\mathrm{d}p_z$. Other operators, such as those for the kinetic and potential energies, have to be redefined correspondingly.

But now suppose we were to make the measurement of a variable with a discrete spectrum the starting point for describing a quantum state. After all, the energy levels of molecules, atoms and nuclei are of fundamental importance, and we might well prefer to specify the state of a bound quantum particle in terms of its energy content, rather than its position or momentum distribution. In this case the state would be represented by the set of coefficients $\{c_n\}$ that determine the probability of finding the particle in a particular energy state. These discrete coefficients can be thought of as the components of a vector in some abstract N-dimensional space, where N is the number of energy eigenstates. Since the coefficients are, in general, complex, the vectors constructed from them will be complex vectors. Many bound systems, like the simple harmonic oscillator or the hydrogen atom, have an infinite number of discrete energy levels, so that the vector with components $(c_1, c_2, c_3, \ldots)$ will be infinite-dimensional! However, much of the mathematics appropriate to finite-dimensional vectors also applies to infinite-dimensional ones, and we shall not be concerned in this book with the difficulties that may arise when $N \rightarrow \infty$. In any case, there are many situations of interest where we can restrict our attention to a finite number of energy levels. For example, we might be interested in a physical situation where the probability of all transitions except those between some finite set of levels is negligible, so that it is a good approximation to include only these particular levels when calculating how the state of the system develops in time. Such a situation arises in electron spin resonance or nuclear magnetic resonance experiments. In that case a magnetic field splits each energy eigenstate into components corresponding to states that differ in the orientation of their magnetic moments, and thus of their total angular momentum. (Remember (8.26).) A second, oscillating, magnetic field induces transitions between states with different eigenvalues of $\hat{J}_z$ but the same eigenvalue $j(j + 1)$ of $\hat{J}^2$. To predict how the state of a particle evolves, it is possible to confine our interest to a set of states with the same angular momentum quantum number j. So in this case an arbitrary state can be expanded as a superposition of the $2j + 1$ states with the same j but different eigenvalues of $\hat{J}_z$, and can therefore be represented by a $(2j + 1)$-dimensional vector. (We shall discuss the particular case of magnetic resonance in a two-level system, $j = \frac{1}{2}$, in Section 12.2.)

In general, we can define the state of a quantum particle in terms of the complete set of eigenstates of any variable we like, by making use of the superposition principle. The **basis set** is the set of eigenfunctions of the operator representing the chosen variable, and the wave function representing an arbitrary state is expanded in the form (9.1) if the basis set is discrete, or in a form similar to (9.40) if it is continuous. For example in

Schrödinger's wave mechanics the basis is the continuous set of position eigenstates, and an arbitrary state is represented by the continuous set of expansion coefficients that appear in (9.45) – that is, by the wave function $\Psi(x, y, z, t)$. We shall refer to this as the **Schrödinger basis**. Note that we did not have to use the expansion (9.45), or the explicit representation of the position eigenfunctions as Dirac delta functions, to set up the formalism of wave mechanics – Postulate 1 defined the significance of the wave function without the necessity of defining a position eigenfunction. Similarly, if we use the continuous basis of momentum eigenstates (the **momentum basis**), an arbitrary state is described by the momentum space wave function $\phi(p_x, p_y, p_z, t)$ that appears in the expansion (9.40). If we choose a discrete basis, such as the set of energy eigenstates of a bound particle, an arbitrary state will be represented by a vector whose components are the discrete set of coefficients $\{c_n\}$ in an expansion of the form (9.1). This type of basis leads to the **matrix formulation** of quantum mechanics (originated by Heisenberg at the same time that Schrödinger developed his wave formulation). It does not matter which basis we choose, and any quantum mechanical operator or equation can be translated into the appropriate form for each basis. Quantum mechanics is the same theory whichever basis we choose to work in.

10.2 The representation of the state of a particle in a discrete basis

Suppose that the wave function Ψ representing the state of a particle is expressed in terms of a complete set of N eigenfunctions $\{\psi_n; n = 1, \ldots, N\}$ through an expansion of the form (9.1). The state represented by Ψ may equally well be described by the N coefficients $\{c_n; n = 1, \ldots, N\}$ in the expansion, and these can be arrayed as an N-dimensional column vector $\boldsymbol{c}$:

$$\boldsymbol{c} = \begin{pmatrix} c_1 \\ c_2 \\ c_3 \\ \vdots \\ c_N \end{pmatrix} = c_1 \begin{pmatrix} 1 \\ 0 \\ 0 \\ \vdots \\ 0 \end{pmatrix} + c_2 \begin{pmatrix} 0 \\ 1 \\ 0 \\ \vdots \\ 0 \end{pmatrix} + c_3 \begin{pmatrix} 0 \\ 0 \\ 1 \\ \vdots \\ 0 \end{pmatrix} + \ldots + c_N \begin{pmatrix} 0 \\ 0 \\ 0 \\ \vdots \\ 1 \end{pmatrix}$$

$$(10.1)$$

$\boldsymbol{c}$ is a (complex) **state vector**. The expression on the right of (10.1) is an expansion in terms of N unit **basis vectors** $\{\boldsymbol{b}_n; n = 1, \ldots, N\}$, each of which has only one non-zero component:

$$
\boldsymbol{b}_1 = \begin{pmatrix} 1 \\ 0 \\ 0 \\ \vdots \\ 0 \end{pmatrix}, \quad
\boldsymbol{b}_2 = \begin{pmatrix} 0 \\ 1 \\ 0 \\ \vdots \\ 0 \end{pmatrix}, \quad \ldots, \quad
\boldsymbol{b}_{N-1} = \begin{pmatrix} 0 \\ \vdots \\ 0 \\ 1 \\ 0 \end{pmatrix}, \quad
\boldsymbol{b}_N = \begin{pmatrix} 0 \\ 0 \\ \vdots \\ 0 \\ 1 \end{pmatrix}
$$

(10.2)

Each of the basis vectors represents one of the basis eigenstates. For example, the components of $\boldsymbol{b}_k$ are $b_n = \delta_{kn}$, so that in a state represented by this vector, $\boldsymbol{c} = \boldsymbol{b}_k$, the probability of finding the particle to be in the kth basis eigenstate is 1, while it is zero for all the others. Equation (10.1) is a form of the expansion (9.1), in which the unit vectors $\{\boldsymbol{b}_n\}$ representing the basis eigenstates take the place of the basis eigenfunctions.

In (10.2) we have chosen basis vectors that are constant, but the basis states in an expansion of the form (9.1) may vary with time. For example, if we use a discrete set of energy eigenstates as our basis states (the **energy basis**), each eigenfunction has a time dependence of the form given in (3.5):

$$
\psi_n(x, y, z, t) = \psi_n(x, y, z, 0)\mathrm{e}^{-\mathrm{i}E_n t/\hbar}
$$

We could define basis vectors with the appropriate time dependence, but it is easier to choose them to be constant, and the exponential time-dependent factor must then be included in the coefficients $\{c_n\}$ representing any state. The nth energy eigenstate is therefore represented by the time-dependent state vector

$$
\boldsymbol{c}_n(t) = \mathrm{e}^{-\mathrm{i}E_n t/\hbar}\,\boldsymbol{b}_n
$$

(10.3)

where $\boldsymbol{b}_n = \boldsymbol{c}_n(0)$ is the appropriate constant basis vector.

The usual rules of matrix algebra [1] can be applied to any column vector, which behaves like a matrix with only one column. The product of an $N' \times N$ matrix $\boldsymbol{A}$ with an $N \times N''$ matrix $\boldsymbol{B}$ is defined to be the $N' \times N''$ matrix $\boldsymbol{C}$, with elements

$$
C_{ij} = \sum_{k=1}^{N} A_{ik} B_{kj}
$$

(10.4)

The number N of columns of $\boldsymbol{A}$ must be identical to the number of rows of $\boldsymbol{B}$. The **transpose** $\boldsymbol{A}^{\mathrm{T}}$ of a matrix $\boldsymbol{A}$ is obtained from $\boldsymbol{A}$ by exchanging rows and columns:

$$
(\boldsymbol{A}^{\mathrm{T}})_{ij} = A_{ji}
$$

(10.5)

In particular, the transpose of an N-dimensional column vector is an

N-dimensional row vector. The **Hermitian conjugate** $\mathbf{A}^\dagger$ of a matrix $\mathbf{A}$ is obtained by taking the transpose, and then taking the complex conjugate of each element:

$$(\mathbf{A}^\dagger)_{ij} = A_{ji}^* \qquad (10.6)$$

$\mathbf{A}$ is a **Hermitian matrix** if $\mathbf{A}^\dagger = \mathbf{A}$, or, in terms of matrix elements,

$$(\mathbf{A}^\dagger)_{ij} = A_{ji}^* = A_{ij} \qquad (10.7)$$

According to (10.4), a scalar (1×1 matrix) can be formed by taking the product of an N-dimensional row vector with an N-dimensional column vector. So we can define the scalar product of two complex vectors $\mathbf{c}$ and $\mathbf{d}$, each with N components, to be

$$\mathbf{c}^\dagger \cdot \mathbf{d} = \sum_{n=1}^{N} c_n^* d_n = (\mathbf{d}^\dagger \cdot \mathbf{c})^* \qquad (10.8)$$

(We have treated $\mathbf{d}$ as a matrix with one column and elements $d_{n1} = d_n$, and $\mathbf{c}^\dagger$ as a matrix with one row and elements $c_{1n}^* = c_n^*$.) The square of the vector (equal to the square of its magnitude) is thus

$$\mathbf{c}^\dagger \cdot \mathbf{c} = \sum_{n=1}^{N} |c_n|^2$$

Looking back to (9.17), we see that if the components of the vector $\mathbf{c}$ are expansion coefficients for the wave function $\Psi(x, y, z, t)$ then the value of $\mathbf{c}^\dagger \cdot \mathbf{c}$ is identical to that of the normalization integral:

$$\mathbf{c}^\dagger \cdot \mathbf{c} = \int_{\text{all space}} d^3r \, \Psi^*(x, y, z, t) \Psi(x, y, z, t) \qquad (10.9)$$

If the wave function is normalized, $\mathbf{c}$ must be a vector with magnitude equal to 1 – that is, a unit vector. (Note that each of the basis vectors $\mathbf{b}_k$ is a unit vector, since $\mathbf{b}_k^\dagger \mathbf{b}_k = 1$. This is also true of the time-dependent vectors defined by (10.3) in the energy basis, since $\mathbf{c}_n^\dagger(t) = (e^{-iE_n t/\hbar})^* \mathbf{b}_n^\dagger = e^{iE_n t/\hbar} \mathbf{b}_n^\dagger$.)

In general, if one state of a particle is represented either by the wave function $\Psi(x, y, z, t)$ or by the state vector $\mathbf{c}$, while another state is represented either by $\Phi(x, y, z, t)$ or by $\mathbf{d}$, then we can show that

$$\mathbf{c}^\dagger \cdot \mathbf{d} = \int_{\text{all space}} d^3r \, \Psi^*(x, y, z, t) \Phi(x, y, z, t) \qquad (10.10)$$

That is, the scalar product of the two state vectors (expressed, of course, in the same basis) is equal to the integral over all space of the product of one wave function with the complex conjugate of the other. To prove this, it is easiest to start with the integral, and then to use (9.1) to expand the two wave functions in terms of the set of basis eigenfunctions

$\{\psi_n(x, y, z, t)\}$, which satisfy the orthonormality requirement (9.11):

$$
\left.
\begin{aligned}
\Psi(x, y, z, t) &= \sum_n c_n \psi_n(x, y, z, t) \\
\Phi(x, y, z, t) &= \sum_n d_n \psi_n(x, y, z, t)
\end{aligned}
\right\}
\tag{10.11}
$$

(The coefficients $\{c_n\}$ and $\{d_n\}$ are the components of the vectors $\boldsymbol{c}$ and $\boldsymbol{d}$ respectively.) The integral on the right-hand side of (10.10) can be written as

$$
\sum_{nk} c_n^* d_k \int_{\text{all space}} \mathrm{d}^3 r\, \psi_n^*(x, y, z, t) \psi_k(x, y, z, t) = \sum_{nk} c_n^* d_k \delta_{nk} = \sum_n c_n^* d_n
$$

(We have used (9.11).) Comparing this with (10.8), we see that it is equal to $\boldsymbol{c}^\dagger \cdot \boldsymbol{d}$, and (10.10) is verified. It follows that if two states are represented by wave functions that are orthogonal, as defined in (2.5), the scalar product of the vectors representing the states in any discrete basis must vanish:

$$
\boldsymbol{c}^\dagger \cdot \boldsymbol{d} = 0
\tag{10.12}
$$

This is exactly what we mean by orthogonality in ordinary vector algebra: if the scalar product of two vectors in ordinary three-dimensional space vanishes, the two vectors are orthogonal. It is easy to show that the basis vectors (10.2) satisfy $\boldsymbol{b}_k^\dagger \cdot \boldsymbol{b}_n = 0$ if $n \neq k$. Since they are also unit vectors, we can write

$$
\boldsymbol{b}_k^\dagger \cdot \boldsymbol{b}_n = \delta_{nk}
\tag{10.13}
$$

These basis vectors therefore form an orthonormal set.

10.3 The matrix representation for dynamical variables

As we pointed out in Section 10.1, if we change the way we represent the state of a particle, we must also change the way we represent dynamical variables. If we use a discrete basis and represent the state of a particle by the state vector $\boldsymbol{c}$ then we shall find that the operators representing dynamical variables are matrices. Suppose that in the Schrödinger basis a variable is represented by a differential operator $\hat{A}$, and consider a state represented by $\Psi(x, y, z, t)$, which is not an eigenfunction of $\hat{A}$. The effect of using the operator $\hat{A}$ on $\Psi(x, y, z, t)$ is to produce a new wave function, $\Phi(x, y, z, t)$ say:

$$
\hat{A} \Psi(x, y, z, t) = \Phi(x, y, z, t)
\tag{10.14}
$$

If we now transform to a discrete basis and expand these wave functions in terms of the (same) set of basis eigenfunctions, (10.11), then (10.14) can be written as

$$\sum_n \hat{A} c_n \psi_n(x, y, z, t) = \sum_n d_n \psi_n(x, y, z, t)$$

The effect of $\hat{A}$ on $\Psi(x, y, z, t)$ is determined by the coefficients $\{d_n\}$, which are obtained, as in (9.12), by multiplying the equation on the left by $\psi_k^*(x, y, z, t)$, integrating over all space, and using the orthonormality condition (9.11). This gives

$$\sum_n \hat{A}_{kn} c_n = d_k \tag{10.15}$$

where $\hat{A}_{kn}$ is defined as

$$\hat{A}_{kn} = \int_{\text{all space}} d^3 r \, \psi_k^*(x, y, z, t) \hat{A} \psi_n(x, y, z, t) \tag{10.16}$$

Equation (10.15) is equivalent to (10.14), and is just the multiplication of a column vector $\boldsymbol{c}$ by a matrix $\boldsymbol{A}$ with elements $A_{kn} = \hat{A}_{kn}$ to produce a new column vector $\boldsymbol{d}$. The matrix elements are related to the differential form of the operator by (10.16), and henceforth we shall use the symbol $\hat{A}$ to represent both the matrix $\boldsymbol{A}$ and the differential operator $\hat{A}$, since the appropriate form will always be obvious from the context: when we work in the Schrödinger basis we use the differential operator, and when we work in a discrete basis we use the corresponding matrix.

So when we use a discrete basis we must represent dynamical variables by matrices, and the elements of the matrix are related to the differential operator representing the variable in the Schrödinger basis by (10.16), where $\psi_n(x, y, z, t)$ is the wave function representing the nth state in the discrete basis. Note that, according to (9.13), a diagonal element of the matrix, $\hat{A}_{nn}$, is just the expectation value of the variable represented by $\hat{A}$ in the nth basis state. Similarly, the expectation value of $\hat{A}$ in an arbitrary state, described by the wave function $\Psi(x, y, z, t)$ or the vector $\boldsymbol{c}$, is given by (9.19), which can be written as

$$\langle \hat{A} \rangle = \int_{\text{all space}} d^3 r \, \Psi^*(x, y, z, t) \hat{A} \Psi(x, y, z, t) = \sum_{nk} c_n^* \hat{A}_{nk} c_k = \boldsymbol{c}^\dagger \hat{A} \boldsymbol{c}$$

$$\tag{10.17}$$

It is also useful to define a generalization of the expectation value by using in (10.17) vectors $\boldsymbol{c}$ and $\boldsymbol{c}'$ that represent *different* states:

$$\boldsymbol{c}^\dagger \hat{A} \boldsymbol{c}' = \sum_{nk} c_n^* \hat{A}_{nk} c_k' = \int_{\text{all space}} d^3 r \, \Psi^*(x, y, z, t) \hat{A} \Psi'(x, y, z, t)$$

$$\tag{10.18}$$

(c and c' represent the same states as $\Psi(x, y, z, t)$ and $\Psi'(x, y, z, t)$ respectively.)

According to Postulate 2 (Section 3.1), a dynamical variable must be represented by a linear Hermitian operator. Matrices are linear operators, and a Hermitian matrix $\hat{A}$ is identical to its Hermitian conjugate $\hat{A}^\dagger$, so that its elements satisfy (10.7). Written in terms of the Schrödinger basis (using (10.16)), this condition becomes

$$\left[\int_{\text{all space}} d^3r \, \psi_k^*(x, y, z, t) \hat{A} \psi_n(x, y, z, t) \right]^*$$

$$= \int_{\text{all space}} d^3r \, \psi_n^*(x, y, z, t) \hat{A} \psi_k(x, y, z, t) \quad (10.19)$$

More generally, we can define a Hermitian operator in the case where $\hat{A}$ is a differential operator or a function of the spatial variables by

$$\left[\int_{\text{all space}} d^3r \, \Psi^*(x, y, z, t) \hat{A} \Psi'(x, y, z, t) \right]^*$$

$$= \int_{\text{all space}} d^3r \, \Psi'^*(x, y, z, t) \hat{A} \Psi(x, y, z, t) \quad (10.20)$$

where $\Psi(x, y, z, t)$ and $\Psi'(x, y, z, t)$ represent two arbitrary states. This follows from (10.18) and (10.19), since for a Hermitian matrix

$$\left(\sum_{nk} c_n^* \hat{A}_{nk} c_k' \right)^* = \sum_{nk} c_n \hat{A}_{nk}^* c_k'^* = \sum_{nk} c_k'^* \hat{A}_{kn} c_n$$

We can now prove our assertion (in Section 3.1) that a Hermitian operator must have real eigenvalues. If $\{\psi_n\}$ is the set of eigenfunctions of $\hat{A}$ then (10.19) becomes

$$a_n^* \left[\int_{\text{all space}} d^3r \, \psi_k^*(x, y, z, t) \psi_n(x, y, z, t) \right]^*$$

$$= a_k \int_{\text{all space}} d^3r \, \psi_n^*(x, y, z, t) \psi_k(x, y, z, t) \quad (10.21)$$

where $\{a_n\}$ is the set of eigenvalues of $\hat{A}$. Now

$$\left[\int_{\text{all space}} d^3r \, \psi_k^*(x, y, z, t) \psi_n(x, y, z, t) \right]^*$$

$$= \int_{\text{all space}} d^3r \, \psi_k(x, y, z, t) \psi_n^*(x, y, z, t)$$

$$= \int_{\text{all space}} d^3r \, \psi_n^*(x, y, z, t) \psi_k(x, y, z, t)$$

so that (10.21) can be rewritten as

$$(a_n^* - a_k) \int_{\text{all space}} d^3r \, \psi_n^*(x, y, z, t) \psi_k(x, y, z, t) = 0 \quad (10.22)$$

This is true for all values of n and k, and in particular if $k = n$ then the integral is equal to 1 (for normalized eigenfunctions), implying that

$$a_n^* = a_n \tag{10.23}$$

That is, the eigenvalues are real. As a bonus we can also prove that eigenfunctions corresponding to different eigenvalues are orthogonal, as we assumed in writing (9.11): when $n \neq k$, provided $a_n \neq a_k$, (10.22) can only be satisfied if

$$\int_{\text{all space}} \mathrm{d}^3 r \, \psi_n^*(x, y, z, t)\psi_k(x, y, z, t) = 0 \quad (n \neq k)$$

Therefore eigenfunctions of a Hermitian operator that correspond to different eigenvalues are always orthogonal.

10.4 Eigenvalue equations in the matrix formulation

In the Schrödinger basis an eigenvalue equation of the general form shown in (3.4) is a differential equation, but if we are working in a discrete basis, it becomes a matrix equation. In matrix algebra the eigenvalues of a square matrix (if they exist) can be found by diagonalizing it [2]. This is achieved by solving an eigenvalue equation of the form

$$\hat{A}\boldsymbol{c} = \lambda \hat{I}\boldsymbol{c} = \lambda\boldsymbol{c} \tag{10.24}$$

where λ is an eigenvalue of the matrix $\hat{A}$, $\hat{I}$ is the unit matrix with the same dimensions as $\hat{A}$, and $\boldsymbol{c}$ is the eigenvector corresponding to the eigenvalue λ. When $\hat{A}$ is Hermitian this is of the same form as the general eigenvalue equation (3.4), but with a state vector replacing the wave function, and using the matrix representation of the operator. Equation (10.24) can only be satisfied if

$$\det\left(\hat{A} - \lambda\hat{I}\right) = 0 \tag{10.25}$$

If $\hat{A}$ is a Hermitian $N \times N$ matrix then in general there are N solutions to this equation, and N real eigenvalues. (These N eigenvalues may not all be different – if two or more of them are equal, they are said to be **degenerate**.) For each eigenvalue there is a corresponding eigenvector, which can be found by solving (10.24) using the appropriate value of λ. (If M of the N eigenvalues are degenerate, there are nevertheless N independent eigenvectors.) In the basis formed by its eigenvectors, the matrix $\hat{A}$ is diagonal: that is, all its elements are zero except for those on the diagonal, which are the eigenvalues. It is always possible to diagonalize a Hermitian matrix, and all its eigenvalues are real.

As an example of how to construct matrices to represent variables we shall look at the case of electron spin. It is particularly useful to find a matrix representation for the spin operators, since, as we explained in Section 6.2, they cannot be represented by differential operators in the same way that orbital angular momentum operators can.

The spin operator $\hat{S}$ was defined to be a Hermitian operator whose components satisfy the commutation relations (6.20), and for a particle with spin $\frac{1}{2}$ the spectrum is given by (6.22) and (6.23). The two eigenstates, represented by $\chi_{+1/2}$ and $\chi_{-1/2}$, satisfy the eigenvalue equations (6.24) and (6.25). Since $\hat{S}_z$ has two real eigenvalues, $m_s = \pm\frac{1}{2}$, we can represent it by the diagonal matrix

$$\hat{S}_z = \tfrac{1}{2}\begin{pmatrix} 1 & 0 \\ 0 & -1 \end{pmatrix} \qquad (10.26)$$

This satisfies (10.7), showing that it is a Hermitian matrix. The normalized eigenvectors of $\hat{S}_z$ are the two-dimensional unit column vectors: $\begin{pmatrix} 1 \\ 0 \end{pmatrix}$ for the positive eigenvalue and $\begin{pmatrix} 0 \\ 1 \end{pmatrix}$ for the negative one. We can therefore represent the eigenstates of $\hat{S}_z$ by these vectors:

$$\chi_{+1/2} = \begin{pmatrix} 1 \\ 0 \end{pmatrix}, \qquad \chi_{-1/2} = \begin{pmatrix} 0 \\ 1 \end{pmatrix} \qquad (10.27)$$

The Hermitian conjugates of these basis vectors, defined in accordance with (10.6), are

$$\chi_{+1/2}^\dagger = (1 \quad 0), \qquad \chi_{-1/2}^\dagger = (0 \quad 1) \qquad (10.28)$$

Note that these eigenvectors satisfy (6.26), which we quoted as the condition for $\chi_{+1/2}$ and $\chi_{-1/2}$ to form an orthonormal set. We can now recognize $\chi_{m_s}^\dagger \chi_{m_s'}$ as the scalar product, defined in (10.8), of two state vectors, and (6.26) as an orthonormality condition of the form (10.13).

It is easy to check that the eigenvalue equation (6.25) is automatically satisfied if we use the representations (10.26) and (10.27) for $\hat{S}_z$ and its eigenstates. (In fact, solving the eigenvalue equation (6.25) is equivalent to finding the diagonal form of the matrix representing $\hat{S}_z$.) Similarly (6.24) is satisfied if

$$\hat{S}^2 = \tfrac{3}{4}\begin{pmatrix} 1 & 0 \\ 0 & 1 \end{pmatrix} = \tfrac{3}{4}\hat{I} \qquad (10.29)$$

where $\hat{I}$ is the 2×2 unit matrix. In (6.21) $\hat{S}^2$ is defined in terms of the x, y and z components of the spin operator. We must find two more Hermitian matrices to represent $\hat{S}_x$ and $\hat{S}_y$, which, together with the matrix (10.26) for $\hat{S}_z$, satisfy the commutation relations (6.20) and give the form (10.29) for $\hat{S}^2$. The easiest way to proceed is to construct the matrices representing the ladder operators $\hat{S}_\pm = \hat{S}_x \pm i\hat{S}_y$, defined in accordance with (6.6). We know from Section 6.1 that the effect of $\hat{S}_+$ and $\hat{S}_-$ on a spin eigenstate is to transform it into an eigenstate corresponding to the next higher or lower eigenvalue of $\hat{S}_z$. However, we cannot assume that the eigenvector resulting from the action of $\hat{S}_-$ on $\chi_{+1/2}$ or $\hat{S}_+$ on $\chi_{-1/2}$ is normalized, so we write

$$\hat{S}_+\chi_{-1/2} = \hat{S}_+\begin{pmatrix} 0 \\ 1 \end{pmatrix} \propto \chi_{+1/2}, \qquad \hat{S}_-\chi_{+1/2} = \hat{S}_-\begin{pmatrix} 1 \\ 0 \end{pmatrix} \propto \chi_{-1/2} \qquad (10.30)$$

If we use $\hat{S}_+$ on the state with the highest $\hat{S}_z$ eigenvalue, or $\hat{S}_-$ on the state with the lowest $\hat{S}_z$ eigenvalue, we get zero, according to (6.13):

$$\hat{S}_+\chi_{+1/2} = \hat{S}_+\begin{pmatrix} 1 \\ 0 \end{pmatrix} = 0, \qquad \hat{S}_-\chi_{-1/2} = \hat{S}_-\begin{pmatrix} 0 \\ 1 \end{pmatrix} = 0 \qquad (10.31)$$

In order to satisfy (10.30) and (10.31), the 2×2 matrices for $\hat{S}_+$ and $\hat{S}_-$ must be of the form

$$\hat{S}_+ = \begin{pmatrix} 0 & \hat{S}_{12} \\ 0 & 0 \end{pmatrix}, \qquad \hat{S}_- = \begin{pmatrix} 0 & 0 \\ \hat{S}_{21} & 0 \end{pmatrix} \qquad (10.32)$$

From the definition of the ladder operators in terms of the operators $\hat{S}_x$ and $\hat{S}_y$, and from the requirement that $\hat{S}_x$ and $\hat{S}_y$ be Hermitian, we see that $\hat{S}_-$ is the Hermitian conjugate of $\hat{S}_+$:

$$\hat{S}_- = \hat{S}_x - i\hat{S}_y = (\hat{S}_x + i\hat{S}_y)^\dagger = (\hat{S}_+)^\dagger = \begin{pmatrix} 0 & 0 \\ \hat{S}_{12}^* & 0 \end{pmatrix} \qquad (10.33)$$

This means that $\hat{S}_{21} = \hat{S}_{12}^*$, and we can find its value by using (6.14):

$$\hat{S}_+\hat{S}_- = \hat{S}^2 - \hat{S}_z^2 + \hat{S}_z$$

If we multiply together the two matrices of (10.32), and use (10.26) and (10.29) for $\hat{S}_z$ and $\hat{S}^2$, this becomes

$$\begin{pmatrix} |\hat{S}_{12}|^2 & 0 \\ 0 & 0 \end{pmatrix} = \tfrac{3}{4}\begin{pmatrix} 1 & 0 \\ 0 & 1 \end{pmatrix} - \tfrac{1}{4}\begin{pmatrix} 1 & 0 \\ 0 & 1 \end{pmatrix} + \tfrac{1}{2}\begin{pmatrix} 1 & 0 \\ 0 & -1 \end{pmatrix} = \begin{pmatrix} 1 & 0 \\ 0 & 0 \end{pmatrix}$$

So $|\hat{S}_{12}|^2 = 1$, and if we choose $\hat{S}_{12}$ to be real then the matrices representing the ladder operators are

$$\hat{S}_+ = \begin{pmatrix} 0 & 1 \\ 0 & 0 \end{pmatrix}, \qquad \hat{S}_- = \begin{pmatrix} 0 & 0 \\ 1 & 0 \end{pmatrix} \qquad (10.34)$$

Equation (10.30) then becomes

$$\hat{S}_+\chi_{1/2} = \chi_{+1/2}, \qquad \hat{S}_-\chi_{+1/2} = \chi_{-1/2} \qquad (10.35)$$

(The only alternative is $\hat{S}_{12} = e^{i\gamma}$, where γ is a real constant phase, in which case the phase of the eigenvectors on the right-hand sides of (10.35) would be changed. However, as we explained in Section 2.2, the overall phase has no physical significance, so we can always choose $\gamma = 0$.) We can now invert the definitions of the ladder operators to find

$$\hat{S}_x = \tfrac{1}{2}(\hat{S}_+ + \hat{S}_-) = \tfrac{1}{2}\begin{pmatrix} 0 & 1 \\ 1 & 0 \end{pmatrix}, \qquad \hat{S}_y = \frac{1}{2i}(\hat{S}_+ - \hat{S}_-) = \tfrac{1}{2}\begin{pmatrix} 0 & -i \\ i & 0 \end{pmatrix}$$

The spin operators for $s = \frac{1}{2}$ are often written in terms of the **Pauli matrices** $\hat{\sigma}_x$, $\hat{\sigma}_y$ and $\hat{\sigma}_z$:

$$\hat{S}_i = \tfrac{1}{2}\hat{\sigma}_i \quad (i = x, y, z)$$

$$\hat{\sigma}_x = \begin{pmatrix} 0 & 1 \\ 1 & 0 \end{pmatrix}, \qquad \hat{\sigma}_y = \begin{pmatrix} 0 & -i \\ i & 0 \end{pmatrix}, \qquad \hat{\sigma}_z = \begin{pmatrix} 1 & 0 \\ 0 & -1 \end{pmatrix} \qquad \textbf{(10.36)}$$

Each of the Pauli matrices satisfies (10.7), showing that $\hat{S}_x$, $\hat{S}_y$ and $\hat{S}_z$ are Hermitian, as required. The rules of matrix multiplication show that they also satisfy commutation relations of the form (6.20):

$$\hat{S}_x \hat{S}_y - \hat{S}_y \hat{S}_x = [\hat{S}_x, \hat{S}_y] = i\hat{S}_z \qquad \text{and cyclic permutations}$$

Matrices representing the spin operator for any allowed value of the spin quantum number j can be constructed in a similar way [3].

10.5 A spin-half particle in a magnetic field

To see how to perform calculations using the matrix formulation of quantum mechanics, we shall look at the case of an electron moving in a constant uniform magnetic field $\mathcal{B}$ that makes an angle θ with the z axis, as shown in Figure 10.1. In terms of unit vectors x and z along the x and z directions in space, the magnetic field is

$$\mathcal{B} = \mathcal{B}(\sin\theta\, x + \cos\theta\, z)$$

The kinetic energy of a classical charged particle moving in the field is not changed, as we mentioned in Section 8.1. So we shall assume that the electron is in an eigenstate of kinetic energy. However, because the electron has an intrinsic magnetic moment, given by (8.21), its Hamiltonian contains a magnetic interaction term of the form (8.4):

$$\hat{V}_{\text{mag}} = -\hat{\boldsymbol{\mu}}_s \cdot \mathcal{B} = 2\mu_B \mathcal{B} \cdot \hat{S} = 2\mu_B \mathcal{B}(\sin\theta\, \hat{S}_x + \cos\theta\, \hat{S}_z)$$
$$= 2\hbar\omega_L(\sin\theta\, \hat{S}_x + \cos\theta\, \hat{S}_z)$$

Here we have put the g factor equal to 2 and used the definition of the Larmor frequency in (8.5). $\hat{V}_{\text{mag}}$ commutes with the kinetic energy operator, so that the energy eigenstate of the electron is simultaneously an eigenstate of the kinetic energy and of $\hat{V}_{\text{mag}}$. We can therefore find the magnetic contribution to the total energy by solving the eigenvalue equation for $\hat{V}_{\text{mag}}$:

$$2\hbar\omega_L(\sin\theta\, \hat{S}_x + \cos\theta\, \hat{S}_z)c = \hbar\omega_L \lambda c$$

Here $\hbar\omega_L \lambda$ is an eigenvalue, and c the corresponding eigenvector. Using

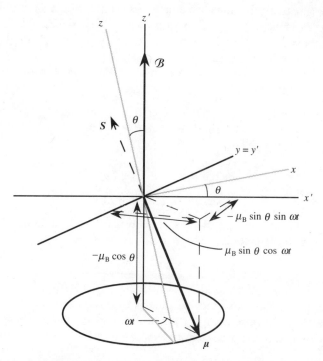

Figure 10.1 A semiclassical representation of the magnetic moment of an electron in a magnetic field $\mathcal{B}$ in the z' direction defined by angles $(\theta, 0)$. If the electron enters this field with spin up along the z axis, its magnetic moment (in the opposite direction) precesses about the z' axis with angular frequency $\omega = 2\omega_{\mathrm{L}}$.

the matrix representation (10.36) for the spin matrices, and writing **c** explicitly as a two-component vector, we obtain

$$\hbar\omega_{\mathrm{L}}\begin{pmatrix} \cos\theta & \sin\theta \\ \sin\theta & -\cos\theta \end{pmatrix}\begin{pmatrix} c_1 \\ c_2 \end{pmatrix} = \hbar\omega_{\mathrm{L}}\lambda\begin{pmatrix} c_1 \\ c_2 \end{pmatrix} \tag{10.37}$$

The eigenvalues of $\hat{V}_{\mathrm{mag}}$ are found from a determinantal equation of the form (10.25):

$$\det\begin{pmatrix} \cos\theta - \lambda & \sin\theta \\ \sin\theta & -\cos\theta - \lambda \end{pmatrix} = 0$$

which gives

$$\lambda^2 - \cos^2\theta - \sin^2\theta = \lambda^2 - 1 = 0$$

The solutions to this equation are $\lambda_{\pm} = \pm 1$, and

$$\hbar\omega_{\mathrm{L}}\lambda_{\pm} = \pm\hbar\omega_{\mathrm{L}} \tag{10.38}$$

Note that these eigenvalues are exactly the same as in the case where the magnetic field is in the z direction, when $\hat{V}_{\text{mag}} = 2\hbar\omega_L \hat{S}_z$. So, as we should expect, the magnetic energy of the particle does not depend on the direction of the field in space.

To find the eigenvectors we return to (10.37). Performing the matrix multiplication on the left-hand side, we get

$$\begin{pmatrix} c_1 \cos\theta + c_2 \sin\theta \\ c_1 \sin\theta - c_2 \cos\theta \end{pmatrix} = \lambda \begin{pmatrix} c_1 \\ c_2 \end{pmatrix}$$

or

$$c_1(\cos\theta - \lambda) + c_2 \sin\theta = 0$$

$$c_1 \sin\theta - c_2(\cos\theta + \lambda) = 0$$

We only need to use one of these equations (because the determinantal equation guarantees their consistency), and we find

$$\frac{c_2}{c_1} = \frac{\lambda - \cos\theta}{\sin\theta} = \frac{\pm 1 - \cos\theta}{\sin\theta}$$

If $\sin\theta$ and $\cos\theta$ are expanded in terms of $\cos\frac{1}{2}\theta$ and $\sin\frac{1}{2}\theta$, these ratios can be expressed more compactly, and the eigenvectors corresponding to the two eigenvalues can be written as

$$\boldsymbol{c}_+ = \begin{pmatrix} \cos\frac{1}{2}\theta \\ \sin\frac{1}{2}\theta \end{pmatrix}, \qquad \boldsymbol{c}_- = \begin{pmatrix} -\sin\frac{1}{2}\theta \\ \cos\frac{1}{2}\theta \end{pmatrix}$$

$\boldsymbol{c}_+$ corresponds to the eigenvalue λ_+, and $\boldsymbol{c}_-$ to λ_-, and we can interpret them as the eigenvectors representing states in which the spin is respectively 'up' or 'down' along the field direction. It is easy to check, using the definition of the scalar product in (10.8), that $\boldsymbol{c}_+$ and $\boldsymbol{c}_-$ are normalized and mutually orthogonal. The complete description of the magnetic energy eigenstates should include their time dependence, which is of the usual exponential form (3.5):

$$\boldsymbol{c}_+(t) = \begin{pmatrix} \cos\frac{1}{2}\theta \\ \sin\frac{1}{2}\theta \end{pmatrix} e^{-i\omega_L t}, \qquad \boldsymbol{c}_-(t) = \begin{pmatrix} -\sin\frac{1}{2}\theta \\ \cos\frac{1}{2}\theta \end{pmatrix} e^{i\omega_L t} \qquad \textbf{(10.39)}$$

Note that if $\theta = 0$ then $\boldsymbol{c}_+(0)$ and $\boldsymbol{c}_-(0)$ become the eigenvectors of $\hat{S}_z$ given in (10.27).

In Section 6.2 we emphasized that spin is not a physical rotation of the particle in space, and that the spin eigenstates are therefore not represented by functions of the spatial coordinates of the particle. The dependence of the eigenvectors $\boldsymbol{c}_+$ and $\boldsymbol{c}_-$ on the angle θ is not in conflict with this, since θ refers to the direction of the magnetic field, not to the position of the electron.

We have been working in a basis in which the matrix $\hat{S}_z$ is diagonal. To write $\hat{V}_{\text{mag}}$ in diagonal form, we must transform to a new basis, in which

$c_+(0)$ and $c_-(0)$ are the vectors $\boldsymbol{b}_+ = \begin{pmatrix} 1 \\ 0 \end{pmatrix}$ and $\boldsymbol{b}_- = \begin{pmatrix} 0 \\ 1 \end{pmatrix}$ respectively. This is achieved [4] by the transformation matrix $\hat{T}$, whose rows are formed by taking the transpose of the eigenvectors $c_+(0)$ and $c_-(0)$ in the old basis. So the first row of $\hat{T}$ is $c_+^{\mathrm{T}}(0)$ and the second is $c_-^{\mathrm{T}}(0)$, and, using the form of these vectors given by (10.39),

$$\hat{T} = \begin{pmatrix} \cos\frac{1}{2}\theta & \sin\frac{1}{2}\theta \\ -\sin\frac{1}{2}\theta & \cos\frac{1}{2}\theta \end{pmatrix} \tag{10.40}$$

It is easy to check that

$$\hat{T} \begin{pmatrix} \cos\frac{1}{2}\theta \\ \sin\frac{1}{2}\theta \end{pmatrix} = \begin{pmatrix} 1 \\ 0 \end{pmatrix} = \boldsymbol{b}_+, \qquad \hat{T} \begin{pmatrix} -\sin\frac{1}{2}\theta \\ \cos\frac{1}{2}\theta \end{pmatrix} = \begin{pmatrix} 0 \\ 1 \end{pmatrix} = \boldsymbol{b}_- \tag{10.41}$$

Any matrix $\hat{A}'$ representing a variable in this new basis is related to the matrix $\hat{A}$ representing the same variable in the old basis by

$$\hat{A}' = \hat{T}\hat{A}\hat{T}^\dagger \tag{10.42}$$

($\hat{T}^\dagger$ is the Hermitian conjugate of $\hat{T}$). Applying this to the matrix

$$\hat{V}_{\mathrm{mag}} = \hbar\omega_{\mathrm{L}} \begin{pmatrix} \cos\theta & \sin\theta \\ \sin\theta & -\cos\theta \end{pmatrix}$$

of (10.37) produces the diagonal form

$$\hat{V}'_{\mathrm{mag}} = \hat{T}\hat{V}_{\mathrm{mag}}\hat{T}^\dagger = \hbar\omega_{\mathrm{L}} \begin{pmatrix} 1 & 0 \\ 0 & -1 \end{pmatrix}$$

Note that this is $2\hbar\omega_{\mathrm{L}}$ times the matrix identified as $\hat{S}_z$ in (10.36). However, in the new basis this matrix represents the component of the spin operator in the direction of the magnetic field, which we have labelled z' in Figure 10.1. The components of the spin operator in the x' and y' directions shown in the figure are represented, in the new basis, by the matrices (10.36), which were identified as $\hat{S}_x$ and $\hat{S}_y$ in the old basis.

Suppose that a beam of electrons is prepared in an eigenstate of $\hat{S}_z$ corresponding to the eigenvalue $+\frac{1}{2}$ before it enters a region in which there is a constant uniform magnetic field $\mathcal{B}$ at angle θ to the z axis, and we wish to predict the results of measurements that might be made at some later time. In the basis where $\hat{S}_z$ is diagonal the initial state is represented by the vector $\boldsymbol{d}(0) = \begin{pmatrix} 1 \\ 0 \end{pmatrix}$ at $t = 0$. To calculate the probability of finding an electron in a particular energy eigenstate at some later time, we transform to the basis in which $\hat{V}_{\mathrm{mag}}$ is diagonal, using the transformation matrix (10.40):

$$\boldsymbol{d}'(0) = \hat{T}\begin{pmatrix} 1 \\ 0 \end{pmatrix} = \begin{pmatrix} \cos\frac{1}{2}\theta \\ -\sin\frac{1}{2}\theta \end{pmatrix} = \cos\frac{1}{2}\theta\,\boldsymbol{b}_+ - \sin\frac{1}{2}\theta\,\boldsymbol{b}_-$$

where $\boldsymbol{b}_+$ and $\boldsymbol{b}_-$ are the basis vectors defined in (10.41), which represent energy eigenstates of the electron. At any time t we expect

$$\boldsymbol{d}'(t) = \cos\tfrac{1}{2}\theta\,\mathrm{e}^{-\mathrm{i}\omega_\mathrm{L}t}\boldsymbol{b}_+ - \sin\tfrac{1}{2}\theta\,\mathrm{e}^{\mathrm{i}\omega_\mathrm{L}t}\boldsymbol{b}_- \qquad \textbf{(10.43)}$$

This indicates that the probability of finding the electron in the higher-energy level (with spin up along the direction of the field) is $|\cos\tfrac{1}{2}\theta\,\mathrm{e}^{-\mathrm{i}\omega_\mathrm{L}t}|^2 = \cos^2\tfrac{1}{2}\theta$, independent of the time, while the probability of finding it in the lower level (with spin down) is $\sin^2\tfrac{1}{2}\theta$.

The predicted mean of a set of measurements of a component of the spin is the expectation value, determined by (10.17). For electrons in the state represented by $\boldsymbol{d}'(t)$ the expectation values of $\hat{S}_{x'}$, $\hat{S}_{y'}$ and $\hat{S}_{z'}$ (given by (10.36) in the basis in which $\hat{V}_\mathrm{mag}$ is diagonal) are

$$\left.\begin{aligned}
\boldsymbol{d}'^\dagger(t)\hat{S}_{x'}\boldsymbol{d}'(t) &= -\tfrac{1}{2}\cos\tfrac{1}{2}\theta\sin\tfrac{1}{2}\theta\,(\mathrm{e}^{2\mathrm{i}\omega_\mathrm{L}t} + \mathrm{e}^{-2\mathrm{i}\omega_\mathrm{L}t}) \\
&= -\tfrac{1}{2}\sin\theta\cos2\omega_\mathrm{L}t \\
\boldsymbol{d}'^\dagger(t)\hat{S}_{y'}\boldsymbol{d}'(t) &= -\tfrac{1}{2}\mathrm{i}\cos\tfrac{1}{2}\theta\sin\tfrac{1}{2}\theta\,(\mathrm{e}^{2\mathrm{i}\omega_\mathrm{L}t} - \mathrm{e}^{-2\mathrm{i}\omega_\mathrm{L}t}) \\
&= \tfrac{1}{2}\sin\theta\sin2\omega_\mathrm{L}t \\
\boldsymbol{d}'^\dagger(t)\hat{S}_{z'}\boldsymbol{d}'(t) &= \tfrac{1}{2}(\cos^2\tfrac{1}{2}\theta - \sin^2\tfrac{1}{2}\theta) = \tfrac{1}{2}\cos\theta
\end{aligned}\right\} \qquad \textbf{(10.44)}$$

This illustrates the way in which quantum mechanical expectation values correspond to classically predicted values of the variables: classically, if a negatively charged particle with magnetic dipole moment $\boldsymbol{\mu}$ in the $-z$ direction were placed in a magnetic field in the z' direction, the magnetic moment would precess about the field direction as shown in Figure 10.1. (The precession frequency in the case of the intrinsic magnetic moment of the electron is g times the Larmor precession frequency, and we have used $g = 2$.) From the diagram we can see that the x', y' and z' components of the magnetic moment at any instant would be $\mu_\mathrm{B}\sin\theta\cos2\omega_\mathrm{L}t$, $-\mu_\mathrm{B}\sin\theta\sin2\omega_\mathrm{L}t$ and $-\mu_\mathrm{B}\cos\theta$ respectively. These are exactly the expectation values of the components of the operator $\hat{\boldsymbol{\mu}}_s = -2\mu_\mathrm{B}\hat{\boldsymbol{S}}$ we should obtain using (10.44). However, the magnitude of a classical component is not the result of any possible measurement on a quantum particle, but corresponds to the mean of a set of measurements, each of which gives either the result $+\mu_\mathrm{B}$ or $-\mu_\mathrm{B}$.

10.6 The Dirac notation

At this point it is convenient to introduce some new notation, which does not depend on which particular basis we are using: we introduce the **Dirac ket vector** $|\Psi\rangle$ to denote the state that is represented by the wave function $\Psi(x, y, z, t)$ in the Schrödinger basis. But $|\Psi\rangle$ represents the physical state, independent of the basis. $|\Psi\rangle$ therefore corresponds to a different vector or function for each basis, and, for example, in a discrete N-dimensional basis it would be a column vector $\boldsymbol{c}$ of the form given in (10.1). The vector **conjugate** to the ket vector $|\Psi\rangle$ is the **bra vector** $\langle\Psi|$ (Note that 'bra' + 'ket' = 'bracket'.) $\langle\Psi|$ is defined to be the Hermitian conjugate of $|\Psi\rangle$, and we define the scalar product of the state vector with itself, in analogy with (10.9), to be

$$\langle\Psi|\Psi\rangle = \int_{\text{all space}} d^3r\ \Psi^*(x, y, z, t)\Psi(x, y, z, t) \qquad (10.45)$$

Similarly, the scalar product of two different state vectors, $|\Psi\rangle$ and $|\Phi\rangle$, which may be complex, is defined in analogy with (10.10):

$$\langle\Psi|\Phi\rangle = \langle\Phi|\Psi\rangle^* = \int_{\text{all space}} d^3r\ \Psi^*(x, y, z, t)\Phi(x, y, z, t)$$

$$(10.46)$$

In Dirac notation the linear Hermitian operator representing a dynamical variable is denoted by the same symbol whether it is in differential, functional or matrix form. Thus the eigenvalue equation for the operator $\hat{A}$ is written as

$$\hat{A}|n\rangle = a_n|n\rangle$$

where $|n\rangle$ represents the nth eigenstate of $\hat{A}$. The matrix element defined in (10.16) is written as $\langle k|\hat{A}|n\rangle$:

$$\langle k|\hat{A}|n\rangle = \hat{A}_{kn} = \int_{\text{all space}} d^3r\ \psi_k^*(x, y, z, t)\hat{A}\psi_n(x, y, z, t)$$

$$(10.47)$$

where the differential form must obviously be used for the operator $\hat{A}$ on the right-hand side of the equation. In general, the expectation value of $\hat{A}$ in an arbitrary state $|\Psi\rangle$ is written as

$$\langle\hat{A}\rangle = \langle\Psi|\hat{A}|\Psi\rangle \qquad (10.48)$$

and the condition (10.20) that $\hat{A}$ be Hermitian is

$$\langle\Psi|\hat{A}|\Psi'\rangle^* = \langle\Psi'|\hat{A}|\Psi\rangle \qquad (10.49)$$

It is evident that the Dirac notation is much less cumbersome than the Schrödinger notation, in addition to the advantage of having a basis-independent way of denoting the state of a quantum particle. In order to keep the presentation consistent throughout this book, we shall continue to use the Schrödinger notation, with the exception of Chapter 14, where the more compact notation becomes essential for a clear discussion. The reader is encouraged to become familiar with the Dirac notation by rewriting some of the mathematical development in the following chapters in that form.

References

[1] Kreyszig E. (1988). *Advanced Engineering Mathematics* 6th edn, Chapter 7. New York: Wiley.

[2] See Ref. [1], Sections 7.12, 7.14 and 7.15.

[3] Edmonds A.R. (1974). *Angular Momentum in Quantum Mechanics* 2nd edn, Section 2.3. Princeton University Press.

[4] See Ref. [1], Section 7.15.

Problems

10.1 The potential for an electron moving in a uniform electric field $\mathcal{E}$ in the $-x$ direction is $V(x) = -e\mathcal{E}x$. Express the potential energy operator and the energy eigenvalue equation in the momentum basis. Find the general solution to this equation, and write down an integral expression for the corresponding eigenfunction in the Schrödinger basis. The Schrödinger equation for the particle in this triangular potential well in the usual Schrödinger formulation is not so easy to solve – try it! The solutions are called **Airy functions**, and their properties have been studied by mathematicians. The differential equation in the Schrödinger basis can be integrated numerically – the computer program *Triangular well* does this in the case where the potential becomes infinite for all $x \geq 0$.

10.2 (a) Show that the form of the Hamiltonian for the one-dimensional simple harmonic oscillator is invariant under the transformation from the Schrödinger to the momentum representation. Write down the energy eigenfunctions in the momentum basis, and explain what boundary conditions they satisfy.

(b) Use the results of Problem 9.11(b) to find the matrices representing the position and momentum operators for the one-dimensional simple harmonic oscillator in the energy basis.

10.3 Write down a vector representing the state mentioned in Problem 9.5(a), and check that it is normalized. Construct a matrix to represent the Hamiltonian of the particle, and repeat the calculations of the expectation value and uncertainty of the energy using matrix methods (see (10.17)).

Similarly, find vectors representing the states mentioned in Problems 9.6–9.8. Construct the matrix representing the Hamiltonian in the case of Problems 9.6 and 9.8, and the matrix representing the momentum in the case of Problem 9.7. In each case use matrix methods to calculate the expectation value and uncertainty of the relevant operator.

10.4 Show that each of the following matrices is Hermitian, and find its eigenvalues and eigenvectors:

(a) $\begin{pmatrix} 4 & 6-2i \\ 6+2i & 1 \end{pmatrix}$

(b) $\begin{pmatrix} 1 & -3 & -2\sqrt{6} \\ -3 & 5 & 0 \\ -2\sqrt{6} & 0 & 2 \end{pmatrix}$ (one eigenvalue is 4)

Normalize the eigenvectors in each case, and demonstrate explicitly that eigenvectors corresponding to different eigenvalues are orthogonal. In each case construct the matrix $\hat{T}$ that transforms the given basis to one in which the matrix is diagonal.

10.5   What are the conditions that make the following matrices Hermitian?

(a) $\begin{pmatrix} 2a & a & a \\ b & 2a & -a \\ c & d & 2a \end{pmatrix}$

(b) $\begin{pmatrix} 0 & -3e\mathcal{E}r_B & 0 & 0 \\ -3e\mathcal{E}r_B & 0 & 0 & 0 \\ 0 & 0 & 0 & 0 \\ 0 & 0 & 0 & 0 \end{pmatrix}$

If these conditions are satisfied, find the eigenvalues of each matrix. You will find that each matrix has a doubly degenerate eigenvalue. In each case find a set of orthogonal eigenvectors. Show that if v_1 and v_2 are orthogonal eigenvectors corresponding to the same degenerate eigenvalue λ then $c_1 v_1 + c_2 v_2$ (where c_1 and c_2 are completely arbitrary complex numbers) is also an eigenvector of the matrix corresponding to the same eigenvalue λ. Construct the matrix $\hat{T}$ that transforms the given basis to one in which the matrix is diagonal.

10.6   Use the results of Problem 6.9 to show that the matrix representation of $\hat{L} \cdot \hat{S}$ for states with orbital angular momentum quantum number $l = 1$ and spin $s = \frac{1}{2}$, in the basis in which $\hat{L}^2$, $\hat{L}_z$, $\hat{S}^2$ and $\hat{S}_z$ are diagonal, is

$$\hat{L} \cdot \hat{S} = \begin{pmatrix} \frac{1}{2} & 0 & 0 & 0 & 0 & 0 \\ 0 & -\frac{1}{2} & \sqrt{\frac{1}{2}} & 0 & 0 & 0 \\ 0 & \sqrt{\frac{1}{2}} & 0 & 0 & 0 & 0 \\ 0 & 0 & 0 & 0 & \sqrt{\frac{1}{2}} & 0 \\ 0 & 0 & 0 & \sqrt{\frac{1}{2}} & -\frac{1}{2} & 0 \\ 0 & 0 & 0 & 0 & 0 & \frac{1}{2} \end{pmatrix}$$

Diagonalize this matrix, and find its eigenvectors.

10.7   (a)   Prove that if $\hat{A}$ is Hermitian then

$$\int_{\text{all space}} d^3r \, (\hat{A}\Psi)^* \hat{B}\Psi = \int_{\text{all space}} d^3r \, \Psi^* \hat{A}\hat{B}\Psi$$

(*Hint*: Write the equation in matrix notation, and remember that $\hat{A}$ acting on Ψ produces a new wave function, Φ say.)

(b) Suppose that two of the eigenvalues of the operator $\hat{A}$ are degenerate. Then the orthogonality of the corresponding eigenstates, ψ_n and ψ_k say, does not follow from (10.22). Show that it is always possible to choose two linear combinations of ψ_n and ψ_k that are orthogonal. (*Hint*: What distinguishes these two states?)

10.8 Construct matrices representing the spin in the case where the spin quantum number is $S = 1$, following the procedure used in Section 10.4 to construct the $S = \frac{1}{2}$ matrices.

10.9 (a) Suppose that the Hamiltonian $\hat{H}$ representing the energy of a particle has only two eigenvalues. In a particular basis the diagonal elements of the matrix $\hat{H}$ are E_1 and E_2, and the off-diagonal elements satisfy $|\hat{V}| \leqslant |E_1 - E_2|$. Prove that the eigenvalues of $\hat{H}$ are

$$\lambda_1 = E_1 + \frac{2|V|^2}{|E_1 - E_2|}, \qquad \lambda_2 = E_2 - \frac{2|V|^2}{|E_1 - E_2|}$$

Find the corresponding normalized eigenvectors.

(b) In a particular basis the Hamiltonian matrix for a particle with only three energy eigenstates has all its diagonal elements equal, and only the off-diagonal elements connecting adjacent states are non-zero. Find the energy eigenvalues and eigenvectors.

10.10 Diagonalize the matrix $\hat{S} \cdot n$ for a spin-$\frac{1}{2}$ particle, where n is the unit vector $n = \sin\theta\cos\phi\, x + \sin\theta\sin\phi\, y + \cos\theta\, z$ (with x, y and z unit vectors parallel to the Cartesian axes). Write down expressions for the eigenvectors of $\hat{S} \cdot n$ in the basis in which $\hat{S}_z$ is diagonal. This is a particular example of the use of the superposition principle (9.1). If a particle is in the eigenstate of $\hat{S}_z$ with eigenvalue $+\frac{1}{2}$ at a time $t = 0$, and the Hamiltonian commutes with $\hat{S} \cdot n$, write down the eigenvector representing the state of the particle at any later time t. (Compare your answer with (10.43).) If, on the other hand, the Hamiltonian commutes with $\hat{S}_z$, what would the state be at time t?

10.11 (a) Rewrite the derivation of the angular momentum spectrum given in Section 6.1 in Dirac notation. (Define $|j\, j_z\rangle$ to be an eigenstate of $\hat{J}^2$ and $\hat{J}_z$.)

(b) Express (6.34) and (6.42) in Dirac notation.

(c) Express (10.43) and (10.44) in Dirac notation.

CHAPTER 11

Approximate Methods for Solving the Schrödinger Equation

11.1 Time-independent perturbation theory

We were able to solve the Schrödinger equation (7.10) for the hydrogen atom analytically (Appendix B3). The Hamiltonian used in this equation was the sum of a kinetic energy term and a Coulomb potential, but this is an approximation that neglects the spin–orbit coupling, and a better approximation is the Hamiltonian (8.23). However, the Schrödinger equation for this improved Hamiltonian cannot be solved analytically, and we must use a method of successive approximations. This is typical of the quantum mechanical treatment of the energy eigenvalue problem for any physical system: the first approximation to the Hamiltonian, $\hat{H}_0$, leads to a Schrödinger equation that can be solved relatively easily, and then further small terms are added to $\hat{H}_0$ to give a better description of the system. The new Schrödinger equation that results is then solved by successive approximations, starting from the eigenvalues and eigenfunctions of $\hat{H}_0$. The two most important approximate methods for solving energy eigenvalue problems are **time-independent perturbation theory** and the **variational method**. We shall discuss a widely used perturbation method in Sections 11.1–11.4, and a variational method in Section 11.5 [1].

Suppose that we have solved the Schrödinger equation

$$\hat{H}_0 \psi_n^{(0)} = E_n^{(0)} \psi_n^{(0)} \tag{11.1}$$

That is, we have found the set of energy eigenvalues $\{E_n^{(0)}\}$ and the corresponding eigenfunctions $\{\psi_n^{(0)}\}$ of a Hamiltonian $\hat{H}_0$. Now we wish to improve our description of the system by means of adding a small time-independent potential $\hat{V}'$ to $\hat{H}_0$, and finding the energy eigenvalues of the new Hamiltonian $\hat{H} = \hat{H}_0 + \hat{V}'$. That is, we wish to solve the Schrödinger equation

$$\hat{H}\psi_n = (\hat{H}_0 + \hat{V}')\psi_n = E_n\psi_n \qquad (11.2)$$

We shall assume that $\hat{V}'$ is small, in the sense that the eigenvalues and eigenfunctions of $\hat{H}$ do not differ very greatly from those of $\hat{H}_0$. In this case $\hat{V}'$ is a small perturbation on $\hat{H}_0$, and (11.2) can be solved by time-independent perturbation theory, using the known solutions to (11.1) as the **zeroth-order approximation**.

In order to derive corrections to the zeroth-order approximation in a systematic way, it is convenient to introduce a parameter λ that takes values between 0 and 1, and rewrite the Hamiltonian $\hat{H}$ as

$$\hat{H} = \hat{H}_0 + \lambda\hat{V}' \quad (0 \leqslant \lambda \leqslant 1) \qquad (11.3)$$

Then $\hat{H}$ reduces to the unperturbed Hamiltonian $\hat{H}_0$ as $\lambda \to 0$, and we shall assume that the eigenfunctions and eigenvalues of $\hat{H}$ reduce to those of $\hat{H}_0$ in the same limit. Each eigenvalue and eigenfunction of $\hat{H}$ is therefore expanded as a power series in λ:

$$\left. \begin{aligned} E_n &= E_n^{(0)} + \lambda E_n^{(1)} + \lambda^2 E_n^{(2)} + \lambda^3 E_n^{(3)} + \ldots \\ \psi_n &= \psi_n^{(0)} + \lambda\psi_n^{(1)} + \lambda^2\psi_n^{(2)} + \lambda^3\psi_n^{(3)} + \ldots \end{aligned} \right\} \qquad (11.4)$$

$E_n^{(0)}$ is the nth eigenvalue of $\hat{H}_0$ (and is the zeroth-order approximation to the eigenvalue E_n of $\hat{H}$). $E_n^{(1)}$, $E_n^{(2)}$, $E_n^{(3)}$, ... are successive corrections to this, and the superscript denotes the **order** of the correction. Similarly $\psi_n^{(1)}$, $\psi_n^{(2)}$, $\psi_n^{(3)}$, ... are first-, second-, third-, ... order corrections to the zeroth-order approximation $\psi_n^{(0)}$ for the eigenfunction of the nth eigenstate. Of course, the perturbation expansions (11.4) only make sense if they are convergent, which is not necessarily the case. So the method can only be used successfully if the corrections to the eigenvalues and the eigenfunctions decrease sufficiently rapidly as the order is increased.

To obtain expressions for the corrections to any desired order, we first substitute (11.3) and (11.4) in (11.2):

$$(\hat{H}_0 + \lambda\hat{V}')(\psi_n^{(0)} + \lambda\psi_n^{(1)} + \lambda^2\psi_n^{(2)} + \ldots)$$
$$= (E_n^{(0)} + \lambda E_n^{(1)} + \lambda^2 E_n^{(2)} + \ldots)(\psi_n^{(0)} + \lambda\psi_n^{(1)} + \lambda^2\psi_n^{(2)} + \ldots)$$

Each side of this equation can be arranged as a power series in λ:

$$\hat{H}_0\psi_n^{(0)} + \lambda(\hat{V}'\psi_n^{(0)} + \hat{H}_0\psi_n^{(1)}) + \lambda^2(\hat{V}'\psi_n^{(1)} + \hat{H}_0\psi_n^{(2)}) + \ldots =$$
$$E_n^{(0)}\psi_n^{(0)} + \lambda(E_n^{(1)}\psi_n^{(0)} + E_n^{(0)}\psi_n^{(1)}) + \lambda^2(E_n^{(2)}\psi_n^{(0)} + E_n^{(1)}\psi_n^{(1)} + E_n^{(0)}\psi_n^{(2)}) + \ldots$$

$$(11.5)$$

According to (11.1), the terms on either side of this equation that are independent of λ are equal, and can be cancelled. The equation from which the first-order corrections are calculated is then found by equating the coefficients of λ on either side:

$$\hat{V}'\psi_n^{(0)} + \hat{H}_0\psi_n^{(1)} = E_n^{(1)}\psi_n^{(0)} + E_n^{(0)}\psi_n^{(1)} \qquad (11.6)$$

Similarly, we obtain an equation from which the second-order corrections can be found (once the first-order corrections are known) by equating the coefficients of λ^2:

$$\hat{V}'\psi_n^{(1)} + \hat{H}_0\psi_n^{(2)} = E_n^{(2)}\psi_n^{(0)} + E_n^{(1)}\psi_n^{(1)} + E_n^{(0)}\psi_n^{(2)} \qquad (11.7)$$

Equations from which the third- and higher-order corrections may be determined result from equating the coefficients of the third and higher powers of λ in (11.5).

We now need to extract expressions for the first-order corrections $E_n^{(1)}$ and $\psi_n^{(1)}$, from (11.6). According to the ideas of Section 9.1, the function $\psi_n^{(1)}$ can be written in terms of the complete set of eigenfunctions $\{\psi_n^{(0)}\}$ of $\hat{H}_0$, using an expansion of the form (9.1):

$$\psi_n^{(1)} = \sum_k c_{nk}^{(1)}\psi_k^{(0)} \qquad (11.8)$$

This reduces the problem of finding the function $\psi_n^{(1)}$ for a given n to that of finding the expansion coefficients $c_{nk}^{(1)}$ for the given n and all k. (In the language of Chapter 10, we choose the eigenstates of $\hat{H}_0$ as our basis states, represented by an orthogonal set of unit vectors, and for each n the first-order correction to the nth eigenstate is represented by the vector with components $\{c_{nk}^{(1)}; k = 1, 2, \ldots\}$.) The zeroth-order eigenfunctions, assuming they have been normalized, must satisfy the orthonormality property (9.11):

$$\int_{\text{all space}} \mathrm{d}^3r \, \psi_k^{(0)*}\psi_n^{(0)} = \delta_{kn} \qquad (11.9)$$

Substituting the expansion (11.8) in (11.6), we obtain

$$(\hat{V}' - E_n^{(1)})\psi_n^{(0)} = (E_n^{(0)} - \hat{H}_0)\sum_k c_{nk}^{(1)}\psi_k^{(0)} = \sum_k c_{nk}^{(1)}(E_n^{(0)} - \hat{H}_0)\psi_k^{(0)}$$

But $\psi_k^{(0)}$ is an eigenfunction of $\hat{H}_0$ corresponding to the eigenvalue $E_k^{(0)}$, and so this equation becomes

$$(\hat{V}' - E_n^{(1)})\psi_n^{(0)} = \sum_k c_{nk}^{(1)}(E_n^{(0)} - E_k^{(0)})\psi_k^{(0)} \qquad (11.10)$$

To obtain expressions for $c_{nk}^{(1)}$ and $E_n^{(1)}$ from this equation, we need to eliminate the wave functions from it, and there is a standard technique for doing this: multiply the equation on the left by the complex conjugate of one of the zeroth-order eigenfunctions, $\psi_j^{(0)*}$ say, and integrate over all space, using the fact that the zeroth-order eigenfunctions form an orthogonal set. (We used this technique in Section 9.2 to extract an expression for each expansion coefficient from the superposition expansion. In terms of the matrix formulation of quantum mechanics, this means taking the scalar product of a basis vector with the equation, which is an equation between two vectors in this formulation.) Use of this procedure on (11.10) gives

$$
\int_{\text{all space}} d^3r \, \psi_j^{(0)*} \hat{V}' \psi_n^{(0)} - E_n^{(1)} \int_{\text{all space}} d^3r \, \psi_j^{(0)*} \psi_n^{(0)}
$$

$$
= \sum_k c_{nk}^{(1)} (E_n^{(0)} - E_k^{(0)}) \int_{\text{all space}} d^3r \, \psi_j^{(0)*} \psi_k^{(0)}
$$

($E_n^{(1)}$, $E_n^{(0)}$ and $c_{nk}^{(1)}$ are all independent of the spatial coordinates, and so can be taken outside the integrals.) Now the orthonormality condition (11.9) can be used, to obtain

$$
\int_{\text{all space}} d^3r \, \psi_j^{(0)*} \hat{V}' \psi_n^{(0)} - E_n^{(1)} \delta_{jn} = \sum_k c_{nk}^{(1)} (E_n^{(0)} - E_k^{(0)}) \delta_{jk}
$$

$$
= c_{nj}^{(1)} (E_n^{(0)} - E_j^{(0)}) \tag{11.11}
$$

Referring back to (10.16), we can identify the remaining integral as a matrix element of $\hat{V}'$ in the basis determined by the eigenstates of $\hat{H}_0$:

$$
\int_{\text{all space}} d^3r \, \psi_j^{(0)*} \hat{V}' \psi_n^{(0)} = \hat{V}'_{jn} \tag{11.12}
$$

If we choose $j = n$ then (11.11) gives

$$
E_n^{(1)} = \hat{V}'_{nn} = \int_{\text{all space}} d^3r \, \psi_n^{(0)*} \hat{V}' \psi_n^{(0)} \tag{11.13}
$$

That is, the change in the nth energy eigenvalue due to the potential $\hat{V}'$, to first order, is just the expectation value of $\hat{V}'$ in the uncorrected (zeroth-order) nth eigenstate. This result is intuitively reasonable: if we were asked by how much the energy of a particular state is altered by an additional potential $\hat{V}'$, the most obvious first guess would be the expectation (mean) value of $\hat{V}'$ in the state concerned.

On the other hand, if we put $j \neq n$ in (11.11), we find an expression for the coefficients that determine the first-order correction to the eigenfunction:

$$
c_{nj}^{(1)} = \frac{\hat{V}'_{jn}}{E_n^{(0)} - E_j^{(0)}} \quad (j \neq n) \tag{11.14}
$$

If any of the energy eigenstates of $\hat{H}_0$ are degenerate, so that $E_n^{(0)} - E_j^{(0)} = 0$ for some n and j, the expression (11.14) for the corresponding coefficient will diverge. For the moment we shall assume that the spectrum of $\hat{H}_0$ is completely non-degenerate, and return to the problem of how to deal with the degenerate case in Section 11.4.

We have not yet determined the value of $c_{nn}^{(1)}$. If $\psi_n \approx \psi_n^{(0)} + \lambda \psi_n^{(1)}$ is normalized, and powers of λ higher than the first are neglected, then (11.9) gives

$$
1 = \int_{\text{all space}} d^3r \, \psi_n^* \psi_n \approx 1 + \lambda (c_{nn}^{(1)*} + c_{nn}^{(1)}) \tag{11.15}
$$

This requires $c_{nn}^{(1)*} + c_{nn}^{(1)} = 0$, and if $c_{nn}^{(1)}$ is real then this implies that

$$c_{nn}^{(1)} = 0 \qquad (11.16)$$

In fact, it is always possible to choose $c_{nn}^{(1)}$ to be real: since the overall phase of a wave function has no significance (as explained in Section 2.2), the phase of each eigenfunction $\psi_k^{(0)}$ in the expansion (11.8) is arbitrary, and so there is no physical significance in the overall phase of $c_{nn}^{(1)}$.

Equations (11.14) and (11.16) determine the first-order correction to the nth eigenfunction. The eigenfunctions of (11.2), correct to first order, are obtained by neglecting all terms proportional to the second and higher powers of λ in (11.4), and then putting $\lambda = 1$. The result is

$$\psi_n \approx \psi_n^{(0)} + \psi_n^{(1)} = \psi_n^{(0)} + \sum_{k \neq n} c_{nk}^{(1)} \psi_k^{(0)}$$

$$= \psi_n^{(0)} + \sum_{k \neq n} \frac{\hat{V}_{kn}'}{E_n^{(0)} - E_k^{(0)}} \psi_k^{(0)} \qquad (11.17)$$

The perturbation can be thought of as mixing small amounts of the other unperturbed states with the nth, and $|c_{nk}^{(1)}|^2$ determines (to first order) the proportion of the kth unperturbed eigenfunction $\psi_k^{(0)}$ contributing to the nth eigenfunction ψ_n.

The second-order corrections to the eigenfunctions can be expanded in a similar way to the first-order corrections (11.8):

$$\psi_n^{(2)} = \sum_k c_{nk}^{(2)} \psi_k^{(0)} \qquad (11.18)$$

Putting these expansions in (11.7) and rearranging it, we get

$$E_n^{(2)} \psi_n^{(0)} = \sum_k c_{nk}^{(1)} (\hat{V}' - E_n^{(1)}) \psi_k^{(0)} + \sum_k c_{nk}^{(2)} (\hat{H}_0 - E_n^{(0)}) \psi_k^{(0)}$$

When this is multiplied on the left by $\psi_j^{(0)*}$ and integrated over all space, making use of (11.9) and (11.1), we obtain

$$E_n^{(2)} \delta_{jn} = \sum_k c_{nk}^{(1)} (\hat{V}_{jk}' - E_n^{(1)} \delta_{jk}) + \sum_k c_{nk}^{(2)} (E_k^{(0)} - E_n^{(0)}) \delta_{jk}$$

$$= \sum_k c_{nk}^{(1)} \hat{V}_{jk}' + c_{nj}^{(1)} E_n^{(1)} + c_{nj}^{(2)} (E_j^{(0)} - E_n^{(0)}) \qquad (11.19)$$

If we put $j = n$ and use (11.16), this gives the second-order correction to the energy eigenvalue:

$$E_n^{(2)} = \sum_k c_{nk}^{(1)} \hat{V}_{nk}' = \sum_{k \neq n} \frac{\hat{V}_{kn}' \hat{V}_{nk}'}{E_n^{(0)} - E_k^{(0)}} = \sum_{k \neq n} \frac{|\hat{V}_{kn}'|^2}{E_n^{(0)} - E_k^{(0)}} \qquad (11.20)$$

where we have used the fact that $\hat{V}'$ must be Hermitian and therefore satisfy (10.7). Note that to calculate the second-order correction to the eigen*value*, we need to know the first-order (but not the second-order) correction to the eigen*function*. We do not need to calculate the second-order correction to the eigenfunction unless we want to calculate the third-order correction to the energy eigenvalue, $E_n^{(3)}$.

As we have already indicated, it is necessary to check whether the expansions in (11.4) (with $\lambda = 1$) converge. The first-order corrections to the energy are small provided

$$\hat{V}'_{nn} \ll E_n^{(0)} \tag{11.21}$$

For the second-order corrections to the energy to be very small, (11.20) indicates that we need

$$|c_{nk}^{(1)}| \ll 1, \quad \text{that is } |\hat{V}'_{kn}| \ll E_n^{(0)} - E_k^{(0)} \quad \text{for all } k \neq n \tag{11.22}$$

That is, the first-order corrections to the eigenfunctions should also be very small.

11.2 First-order perturbations: a one-dimensional problem

As an example of the use of first-order perturbation theory, consider a perturbation $\hat{V}'$ applied to a particle of mass m moving in an infinitely deep one-dimensional potential well between $z = 0$ and $z = \mathbb{L}$. Suppose that the perturbation is parabolic and symmetric about the centre of the well, as shown in Figure 11.1:

$$\hat{V}' = Cz(z - \mathbb{L}) \tag{11.23}$$

where C is a small constant. The eigenvalue spectrum of the unperturbed particle (which is completely non-degenerate) is given by (3.39), and the eigenfunctions are given by (3.40), so we identify the zeroth-order approximations as

$$E_n^{(0)} = \frac{h^2}{8m\mathbb{L}^2} n^2 \quad (n = 1, 2, 3, \ldots)$$

$$\psi_n^{(0)} = \begin{cases} \sqrt{\left(\dfrac{2}{\mathbb{L}}\right)} \sin \dfrac{n\pi z}{\mathbb{L}} & (0 \leqslant z \leqslant \mathbb{L}) \\ 0 & (z \leqslant 0, z \geqslant \mathbb{L}) \end{cases} \tag{11.24}$$

Equation (11.13) identifies the first-order correction to the nth eigenvalue of the unperturbed Hamiltonian $\hat{H}_0$ as the expectation value of $\hat{V}'$

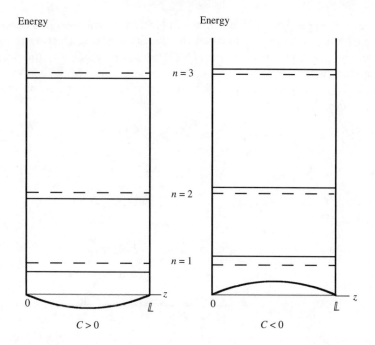

Figure 11.1 The first three energy levels of a particle in an infinitely deep one-dimensional square well perturbed by the potential $V = Cz(z - L)$: (a) for $C > 0$; (b) for $C < 0$. The base of the unperturbed well is shown as a light line, and the positions of the unperturbed levels are indicated by dashed lines.

in the nth eigenstate of $\hat{H}_0$:

$$E_n^{(1)} = \hat{V}'_{nn} = \int_0^{L} dz \, [\psi_n^{(0)}(z)]^* \hat{V}'(z) \psi_n^{(0)}(z)$$

$$= \frac{2}{L} \int_0^{L} dz \, Cz(z - L) \sin^2 \frac{n\pi z}{L}$$

Use of the identity $\sin^2(n\pi z/L) = \frac{1}{2}[1 - \cos(2n\pi z/L)]$ and integration by parts gives

$$E_n^{(1)} = -CL^2 \left(\frac{1}{6} + \frac{1}{2n^2\pi^2} \right) \tag{11.25}$$

The first-order changes in the energies of the first three levels are indicated in Figure 11.1, for both possible signs of the constant C. The first-order correction $E_n^{(1)}$ is small compared with the unperturbed energy $E_n^{(0)}$ given in (11.24), for any n, provided

$$|C| \ll 0.5 \, \frac{h^2}{mL^4} \tag{11.26}$$

(To get this upper limit, we put $n = 1$ in (11.25).) However, this condition is not necessarily sufficient in order for the first-order correction to give a reasonable approximation, since (11.22) must be satisfied if the second- and higher-order corrections are to be very small. To check this, we need to calculate the off-diagonal elements of the matrix representing the perturbation, $\hat{V}'_{kn}$ with $k \neq n$:

$$
\begin{aligned}
\hat{V}'_{kn} &= \frac{2}{L} \int_0^L \mathrm{d}z\, Cz(z - L) \sin \frac{k\pi z}{L} \sin \frac{n\pi z}{L} \\
&= \frac{C}{L} \int_0^L \mathrm{d}z\, z(z - L) \left[\cos \frac{(k - n)\pi z}{L} - \cos \frac{(k + n)\pi z}{L} \right] \\
&= \frac{CL^2}{\pi^2} \left[\frac{(-1)^{k-n} + 1}{(k - n)^2} - \frac{(-1)^{k+n} + 1}{(k + n)^2} \right]
\end{aligned}
$$

Since $k + n$ and $k - n$ are either both odd or both even, this gives

$$
\hat{V}'_{kn} = \begin{cases} 0 & (k + n \quad \text{odd}) \\[2mm] \dfrac{CL^2}{\pi^2} \dfrac{8kn}{(k^2 - n^2)^2} & (k + n \quad \text{even}) \end{cases} \tag{11.27}
$$

and we see that

$$
c^{(1)}_{nk} = \frac{\hat{V}'_{kn}}{E^{(0)}_n - E^{(0)}_k} = \begin{cases} 0 & (k + n \quad \text{odd}) \\[2mm] \dfrac{CmL^4}{h^2\pi^2} \dfrac{64kn}{(n^2 - k^2)^3} & (k + n \quad \text{even}) \end{cases}
$$

The maximum value of this occurs when $k = 1$ and $n = 3$, and all the coefficients $\{c^{(1)}_{nk}\}$ are very much less than 1 provided

$$
|C| \ll 26 \frac{h^2}{mL^4} \tag{11.28}
$$

This sets a higher limit on $|C|$ than (11.26), but in general the condition determined by (11.22) may be more restrictive than that given by (11.21).

If a Hamiltonian $\hat{H}_0$ commutes with the parity operator, its eigenstates are also parity eigenstates, as we saw in Section 4.4. This means that the square of the modulus of an eigenfunction of $\hat{H}_0$ must have even parity (since $|\psi^{(0)}_n(z)|^2$ is an even function of z whether $\psi^{(0)}_n(z)$ is even or odd), and consequently the integrand in (11.13) for the expectation value of $\hat{V}'$ has the same parity as $\hat{V}'$. If $\hat{V}'$ has odd parity, the integrand is odd, and the first-order correction to the energy of any level must vanish. For the one-dimensional case, if $\hat{V}'(z) = -\hat{V}'(-z)$ then

$$
E^{(1)}_n = \hat{V}'_{nn} = \int_{-\infty}^{\infty} \mathrm{d}z\, [\psi^{(0)}_n(z)]^* \hat{V}'(z) \psi^{(0)}_n(z) = 0
$$

In the case we have just examined the unperturbed Hamiltonian of the particle in an infinitely deep square well is symmetric about the centre of the well, and the perturbation (11.23) is also symmetric, so that there are non-zero first-order corrections to the energy eigenvalues. Note that the vanishing of $\hat{V}'_{kn}$ when $k + n$ is odd, (11.27), is due to the antisymmetry of the integrand concerned.

11.3 Second-order perturbations: anharmonic oscillations

As mentioned in Section 11.1, the perturbation expansion can only be used if the magnitude of the corrections decreases sufficiently rapidly as the order of the correction increases. So we expect the second-order corrections to the energy eigenvalues to be very much smaller than the first-order corrections, and correspondingly harder to confirm experimentally. However, in cases where the first-order correction is exactly zero the second-order corrections become important. For example, if a uniform electric field in the x direction is applied to a particle whose zeroth-order energy eigenfunctions are parity eigenstates then the first-order corrections to the energy levels vanish. In this case the **Stark effect** – the change in energy levels due to the applied electric field – is given to a first approximation by the second-order perturbation correction, which is proportional to the square of the perturbation, and thus to the square of the electric field. (See Problems 11.9 and 11.10.)

Second-order corrections are also important in determining the vibrational energy levels of a diatomic molecule: although the vibrations can be treated as simple harmonic oscillations when the amplitude is sufficiently small, anharmonic terms in the potential become increasingly important as the amplitude is increased. We mentioned this in Section 5.3, and Figure 5.5 shows how the potential in this case departs from the parabolic (x^2) form for large enough extensions beyond the equilibrium separation of the atoms. The potential can be expanded as a Taylor series about $x = 0$, and the anharmonic terms are those proportional to x^3 and higher powers of x. The potential energy operator for the oscillator becomes

$$\hat{V} = \tfrac{1}{2}m\omega^2 x^2 + \hat{V}' = \tfrac{1}{2}\hbar\omega\xi^2 + \hat{V}' \tag{11.29}$$

where ξ is defined by (5.2) and $\hat{V}'$ is a sum of terms in powers of x greater than the second. For an anharmonic correction proportional to x^3 we can write

$$\hat{V}' = C\left(\frac{m^3\omega^5}{\hbar}\right)^{1/2} x^3 = C\hbar\omega\xi^3 \tag{11.30}$$

where C is dimensionless. This potential can be treated as a perturbation provided C is small enough.

The zeroth-order problem is the simple harmonic oscillator problem, which we solved in Sections 5.1 and 5.2, so the zeroth-order eigenvalues $E_n^{(0)} = \epsilon_n^{(0)} \hbar \omega$ are given by (5.15), and the zeroth-order eigenfunctions $\psi_n^{(0)}$ by (5.26). The energy spectrum is again non-degenerate, and (according to (5.27) and (5.28)) the eigenfunctions form an orthonormal set:

$$\int_{-\infty}^{\infty} dx \, [\psi_k^{(0)}(\xi)]^* \psi_n^{(0)}(\xi) = \sqrt{\left(\frac{m\omega}{\hbar}\right)} \int_{-\infty}^{\infty} d\xi \, [\psi_k^{(0)}(\xi)]^* \psi_n^{(0)}(\xi) = \delta_{kn}$$

From (11.13) the first-order correction to the energy eigenvalue is

$$\epsilon_n^{(1)} = C\langle \xi^3 \rangle = C\sqrt{\left(\frac{m\omega}{\hbar}\right)} \int_{-\infty}^{\infty} d\xi \, [\psi_n^{(0)}(\xi)]^* \xi^3 \psi_n^{(0)}(\xi) \qquad \textbf{(11.31)}$$

where $\epsilon_n^{(1)} = E_n^{(1)}/\hbar\omega$ is the first-order correction in units of $\hbar\omega$. But this integral vanishes because the zeroth-order wave functions are parity eigenfunctions and ξ^3 is odd under parity inversion. Similarly, the expectation value of any odd power of ξ must be zero.

Although the first-order correction to the energy is zero, the first-order corrections to the eigenfunctions and the second-order corrections to the energy eigenvalues do not vanish, as we shall see. To calculate the coefficients $c_{nk}^{(1)}$, (11.14), and the second-order correction to the energy $E_n^{(2)}$, (11.20), we need all the off-diagonal elements $\hat{V}'_{kn}$, with $k \neq n$, of the matrix representing the perturbation. The easiest way to find these is by using the properties of the oscillator ladder operators given in (5.29)–(5.31). We used (5.4) to write $\hat{X}$ and $\hat{X}^2$ in terms of the ladder operators in (9.26) and (9.28), and similarly we can replace the algebraic variables ξ and ξ^3 by combinations of the ladder operators:

$$\xi = \sqrt{(\tfrac{1}{2})}(\hat{A} + \hat{A}^\dagger)$$

and

$$\begin{aligned} \xi^3 &= \tfrac{1}{2}\sqrt{(\tfrac{1}{2})}(\hat{A}^2 + \hat{A}\hat{A}^\dagger + \hat{A}^\dagger\hat{A} + \hat{A}^{\dagger 2})(\hat{A} + \hat{A}^\dagger) \\ &= \tfrac{1}{2}\sqrt{(\tfrac{1}{2})}(\hat{A}^2 + 2\hat{A}^\dagger\hat{A} + 1 + \hat{A}^{\dagger 2})(\hat{A} + \hat{A}^\dagger) \end{aligned}$$

To get the second line of this equation, we used the commutator of the ladder operators, (5.6), to write $\hat{A}\hat{A}^\dagger$ as $\hat{A}^\dagger\hat{A} + 1$. Expanding, with further use of the commutator, we obtain

$$\xi^3 = \tfrac{1}{2}\sqrt{(\tfrac{1}{2})}[\hat{A}^3 + 3(\hat{A} + \hat{A}^\dagger)\hat{A}^\dagger\hat{A} + 3\hat{A}^\dagger + \hat{A}^{\dagger 3}] \qquad \textbf{(11.32)}$$

From (5.31), $\hat{A}^\dagger\hat{A}\psi_n^{(0)} = n\psi_n^{(0)}$, so that

$$\xi^3 \psi_n^{(0)} = \tfrac{1}{2}\sqrt{(\tfrac{1}{2})}[\hat{A}^3 + 3n(\hat{A} + \hat{A}^\dagger) + 3\hat{A}^\dagger + \hat{A}^{\dagger 3}]\psi_n^{(0)} \qquad \textbf{(11.33)}$$

Repeated use of (5.29) and (5.30) then gives

$$\xi^3 \psi_n^{(0)} = \tfrac{1}{2} \sqrt{(\tfrac{1}{2})}\{[n(n-1)(n-2)]^{1/2} \psi_{n-3}^{(0)} + 3n^{3/2} \psi_{n-1}^{(0)} + 3(n+1)^{3/2} \psi_{n+1}^{(0)}$$
$$+ [(n+1)(n+2)(n+3)]^{1/2} \psi_{n+3}^{(0)}\} \qquad (11.34)$$

According to the definition (11.12), the matrix elements $\hat{V}'_{kn}$ are

$$\hat{V}'_{kn} = C\hbar\omega \sqrt{\left(\frac{m\omega}{h}\right)} \int_{-\infty}^{\infty} d\xi \, [\psi_k^{(0)}(\xi)]^* \xi^3 \, \psi_n^{(0)}(\xi)$$

If we substitute (11.34) in this integral and use the orthonormality of the zeroth-order eigenfunctions, we find that for a given n there are only four non-zero matrix elements:

$$\left.\begin{aligned}
\hat{V}'_{n-3,n} &= \tfrac{1}{2}C[\tfrac{1}{2}n(n-1)(n-2)]^{1/2}\hbar\omega \\
\hat{V}'_{n-1,n} &= \tfrac{3}{2}\sqrt{(\tfrac{1}{2})}Cn^{3/2}\hbar\omega \\
\hat{V}'_{n+1,n} &= \tfrac{3}{2}\sqrt{(\tfrac{1}{2})}C(n+1)^{3/2}\hbar\omega \\
\hat{V}'_{n+3,n} &= \tfrac{1}{2}C[\tfrac{1}{2}(n+1)(n+2)(n+3)]^{1/2}\hbar\omega
\end{aligned}\right\} \qquad (11.35)$$

Substituting these values in (11.20), with $E_n^{(0)} = (n + \tfrac{1}{2})\hbar\omega$, (5.15), we find that the second-order correction to the energy is

$$E_n^{(2)} = \sum_{k \neq n} \frac{|\hat{V}'_{kn}|^2}{E_n^{(0)} - E_k^{(0)}} = -\tfrac{15}{4}C^2(n^2 + n + \tfrac{11}{30})\hbar\omega \qquad (11.36)$$

Thus, to second order in perturbation theory, the energy of a simple harmonic oscillator perturbed by a small potential proportional to x^3 is

$$E_n \approx (n + \tfrac{1}{2})\hbar\omega - \tfrac{15}{4}C^2(n^2 + n + \tfrac{11}{30})\hbar\omega$$

These energy levels and those of the unperturbed oscillator are shown in Figure 11.2. The perturbation lowers the energy of the oscillator by an amount that increases with the level of excitation n, and the difference in energy between adjacent levels becomes

$$E_n - E_{n-1} = (1 - \tfrac{15}{2}nC^2)\hbar\omega \qquad (11.37)$$

The separation between a pair of adjacent levels is thus no longer the same for all such pairs, but decreases with increasing n. This decrease is evident in the vibrational spectra of diatomic molecules (see Section 5.3).

11.4 Degenerate perturbation theory: spin–orbit coupling

The energy eigenstates of a (spinless) particle moving in a one-dimensional potential are non-degenerate, and the perturbation procedure outlined in Section 11.1 can be used in a straightforward way, as we have

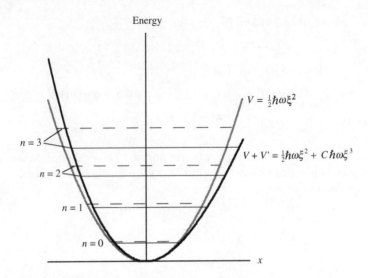

Figure 11.2 The lowest four energy levels of a simple harmonic oscillator subject to a cubic anharmonic perturbation. The unperturbed well is shown as a light curve, and the unperturbed levels are indicated by dashed lines.

seen in Sections 11.2 and 11.3. However, in most three-dimensional problems the symmetry properties of the unperturbed Hamiltonian lead to degeneracy in its energy spectrum, and as a result the expression (11.14) for the first-order expansion coefficients $c_{nj}^{(1)}$ may diverge for some values of n and j. To find out how to deal with this problem, we shall first examine the general case of a perturbation applied to a set of degenerate states, and then look at the particular example of spin–orbit coupling in hydrogen.

Consider a set of N states, described by the wave functions $\{\psi_{nk}^{(0)};$ $k = 1, \ldots, N\}$, which are eigenstates of the unperturbed Hamiltonian $\hat{H}_0$, all corresponding to the *same* energy $E_n^{(0)}$. These degenerate states are distinguished by N different eigenvalues of one or more operators (representing dynamical variables) that commute with the Hamiltonian $\hat{H}_0$. For example, they might be the $N = 2n^2$ degenerate states of a hydrogen atom in its nth energy level, distinguished by the angular momentum quantum numbers $\{l, m_l, m_s\}$. To keep the notation simple, we use the single label k to represent all quantum numbers that distinguish between the degenerate states. (In the case of hydrogen each k represents a different set of three numbers, l, m_l and m_s.) We wish to find the first-order corrections to the N degenerate eigenfunctions $\{\psi_{nk}^{(0)}\}$ due to a perturbation $\hat{V}'$. According to (11.8) and (11.14), the first-order correction to the wave function of the kth state is

$$\psi_{nk}^{(1)} = \sum_{n'j} c_{nkn'j}^{(1)} \psi_{n'j}^{(0)}$$

$$= \sum_{j \neq k} \frac{\hat{V}'_{njnk}}{E_n^{(0)} - E_n^{(0)}} \; \psi_{nj}^{(0)} + \text{ contributions from eigenfunctions of energy states other than the } n\text{th}$$

$$\tag{11.38}$$

where

$$\hat{V}'_{njnk} = \int_{\text{all space}} \mathrm{d}^3 r \; \psi_{nj}^{(0)*} \hat{V}' \psi_{nk}^{(0)}$$

is an element of the matrix representing $\hat{V}'$ evaluated between two of the degenerate states of the nth energy level. In (11.38) the contributions from the N eigenfunctions of the nth energy level clearly diverge, since the denominator vanishes. Equation (11.20) indicates that in this case the second-order correction to the energy will also be infinite! This means that the procedure outlined in Section 11.1 for obtaining corrections to the zeroth-order states has broken down.

From (11.38) it is evident that the divergence of $c_{nknj}^{(1)}$ can be avoided only if all the matrix element $\hat{V}'_{njnk}$ with $j \neq k$ are zero – that is, if the submatrix of $\hat{V}'$ that connects states within the same degenerate level is diagonal. Since $\hat{V}'$ must be Hermitian (in accordance with Postulate 2, Section 3.1) it can always be diagonalized (as mentioned in Section 10.4). This is equivalent to choosing a new basis in which all off-diagonal elements are zero, and the new basis states are independent linear combinations of the original ones [2]. Correspondingly, in terms of eigenfunctions, we define a new set of N mutually orthogonal functions $\{\phi_{nk'}^{(0)}; k' = 1, \dots, N\}$, which are linear combinations of the functions $\{\psi_{nk}^{(0)}; k = 1, \dots, N\}$ representing the states of the nth energy level:

$$\phi_{nk'}^{(0)} = \sum_{k=1}^{N} b_{k'k} \psi_{nk}^{(0)} \tag{11.39}$$

(Note that each of these wave functions corresponds to the same eigenvalue $E_n^{(0)}$ of $\hat{H}_0$ as the original degenerate set $\{\psi_{nk}^{(0)}\}$.) The coefficients $b_{k'k}$ are chosen so that $\hat{V}'$ is diagonal in the new basis:

$$\hat{V}'_{nj'nk'} = \int_{\text{all space}} \mathrm{d}^3 r \; \phi_{nj'}^{(0)*} \hat{V}' \phi_{nk'}^{(0)} = 0 \quad \text{for all } j' \neq k' \tag{11.40}$$

(We use primed labels to distinguish the new basis.)

The first-order corrections to the nth energy eigenvalue can be found from (11.11), expressed in terms of the new basis functions. The states labelled j and n in the original form of this equation are chosen to be the degenerate states represented by $\phi_{nj'}^{(0)}$ and $\phi_{nk'}^{(0)}$ respectively, and the equation can be written (for all possible j' and k') as

$$\int_{\text{all space}} \mathrm{d}^3 r \; \phi_{nj'}^{(0)*} \hat{V}' \phi_{nk'}^{(0)} - E_{nj'}^{(1)} \delta_{j'k'} = c_{nk'nj'}^{(1)} (E_n^{(0)} - E_n^{(0)})$$

The right-hand side of this equation vanishes (provided the coefficients $c^{(1)}_{nk'\,nj'}$ are finite), and the equation is consistent since, from (11.40), the off-diagonal matrix elements are zero. The first-order correction to the energy of the state represented by the eigenfunction $\phi^{(0)}_{nj'}$ is the diagonal matrix element

$$E^{(1)}_{nj'} = \int_{\text{all space}} \mathrm{d}^3 r \, \phi^{(0)*}_{nj'} \hat{V}' \phi^{(0)}_{nj'} = \hat{V}'_{nj'\,nj'} \tag{11.41}$$

In general, these corrections will depend on j', so the perturbation splits the degenerate level into a closely spaced multiplet of energy levels.

Spin–orbit coupling in hydrogen

The Hamiltonian for the hydrogen atom, including spin–orbit coupling, is given in (8.23), and is of the form $\hat{H} = \hat{H}_0 + \hat{V}'$, where $\hat{H}_0$ is the sum of the kinetic energy operator and the Coulomb potential and $\hat{V}'$ is the spin–orbit potential (8.22):

$$\hat{V}' = \frac{\hbar^2}{2m^2c^2} \frac{e^2}{4\pi\varepsilon_0} \frac{1}{r^3} \hat{\boldsymbol{L}} \cdot \hat{\boldsymbol{S}} \tag{11.42}$$

The order of magnitude of the energy due to this potential is $\alpha^2 E_R$ (see Problem 8.3a), where E_R is the Rydberg energy (7.11) and α is the fine structure constant which we introduced in Section 8.4. This estimate suggests that the contribution of the spin–orbit potential to the energy of hydrogen is small enough to be calculated, to a reasonably good approximation, using first-order perturbation theory. However, all the eigenstates of $\hat{H}_0$ are degenerate (since even the ground state has twofold degeneracy when we include spin), so we must use the technique for dealing with degenerate levels that we have just explained.

The zeroth-order problem was solved in Section 7.2, giving the energy spectrum (7.13) and the eigenfunctions (7.21). In order to find the first-order corrections to the nth level due to spin–orbit coupling, we define

$$E^{(0)}_n = -\frac{1}{n^2} E_R$$

$$\psi^{(0)}_{nk}(r, \theta, \phi) = \psi_{nlm_lm_s}(r, \theta, \phi) = R_{nl}(r) Y_{lm_l}(\theta, \phi) \chi_{m_s} \tag{11.43}$$

where n takes the fixed value appropriate to the degenerate level in which we are interested, and (l, m_l, m_s) is represented by k. We must diagonalize the $2n^2 \times 2n^2$ matrix that represents $\hat{V}'$ in the basis $\{\psi^{(0)}_{nk}\}$. The matrix elements are of the form

$$\hat{V}'_{k'k} = \frac{\hbar^2}{2m^2c^2} \frac{e^2}{4\pi\varepsilon_0} \int_{\text{all space}} \mathrm{d}^3 r \, \psi^{(0)\dagger}_{nk'}(r, \theta\, \phi) \frac{1}{r^3} \hat{\boldsymbol{L}} \cdot \hat{\boldsymbol{S}} \psi^{(0)}_{nk}(r, \theta, \phi)$$

$$\tag{11.44}$$

We can write $\hat{L} \cdot \hat{S}$ as

$$\hat{L} \cdot \hat{S} = \hat{L}_x \hat{S}_x + \hat{L}_y \hat{S}_y + \hat{L}_z \hat{S}_z = \tfrac{1}{2}(\hat{L}_+ \hat{S}_- + \hat{L}_- \hat{S}_+) + \hat{L}_z \hat{S}_z \quad (11.45)$$

where $\hat{L}_\pm$ and $\hat{S}_\pm$ are ladder operators for orbital angular momentum and spin respectively, defined in the same way as $\hat{J}_\pm$ in (6.6). According to Section 6.1, $\hat{J}_+$ or $\hat{J}_-$ acts on an eigenstate of $\hat{J}^2$ and $\hat{J}_z$ to produce a new eigenstate corresponding to the same eigenvalue of $\hat{J}^2$ but an eigenvalue of $\hat{J}_z$ differing by one unit from that in the original state. This applies equally in the cases of orbital angular momentum and spin. Therefore the operator $\hat{L} \cdot \hat{S}$ acting on $\psi_{nlm_l m_s}$ cannot alter the quantum number l (or s, which is equal to $\tfrac{1}{2}$ in hydrogen), so the values of l represented by k and k' in (11.44) are the same, and

$$\hat{V}'_{k'k} = \frac{\hbar^2}{2m^2c^2} \frac{e^2}{4\pi\varepsilon_0} \left\langle \frac{1}{r^3} \right\rangle_{nlnl} \langle \hat{L} \cdot \hat{S} \rangle_{m'_l m'_s m_l m_s} \quad (11.46)$$

where $\langle 1/r^3 \rangle_{nlnl}$ and $\langle \hat{L} \cdot \hat{S} \rangle_{m'_l m'_s m_l m_s}$ are given by

$$\left\langle \frac{1}{r^3} \right\rangle_{nlnl} = \int_0^\infty r^2 \, \mathrm{d}r \, R_{nl}^2 \frac{1}{r^3}$$

$$\langle \hat{L} \cdot \hat{S} \rangle_{m'_l m'_s m_l m_s} = \int_0^{2\pi} \mathrm{d}\phi \int_{-1}^{1} \mathrm{d}(\cos\theta) \, Y^*_{lm'_l}(\theta, \phi) \chi^\dagger_{m'_s} \hat{L} \cdot \hat{S} Y_{lm_l}(\theta, \phi) \chi_{m_s}$$

$$(11.47)$$

The problem of diagonalizing $\hat{V}'$ thus reduces to that of diagonalizing $\hat{L} \cdot \hat{S}$ for each possible value of l.

We have essentially solved the problem of diagonalizing $\hat{L} \cdot \hat{S}$ in Section 6.3, where we found that an eigenstate of $\hat{L} \cdot \hat{S}$ is simultaneously an eigenstate of $\hat{L}^2$, $\hat{S}^2$, $\hat{J}^2$ and $\hat{J}_z$, with $\hat{J} = \hat{L} + \hat{S}$. The eigenstates are of the form $\chi_{lsjj_z}(\theta, \phi)$ given in (6.42), and these are linear combinations of products of the form $Y_{lm_l}(\theta, \phi)\chi_{sm_s}$, with different values of m_l and m_s in each term. For hydrogen $s = \tfrac{1}{2}$, and, according to (6.33), there are two values of j for each value of l ($j = l + \tfrac{1}{2}$ and $l - \tfrac{1}{2}$) unless $l = 0$, when the only possibility is $j = \tfrac{1}{2}$. So the basis states that make the spin–orbit potential $\hat{V}'$ diagonal are of the form

$$\phi_{nk}^{(0)} = R_{nl}(r)\chi_{l,1/2,jj_z}(\theta, \phi) \quad (11.48)$$

where n is fixed, and k represents the three quantum numbers l, $j = l \pm \tfrac{1}{2}$ and j_z. Equation (6.35) allows us to write $\hat{L} \cdot \hat{S} = \tfrac{1}{2}(\hat{J}^2 - \hat{L}^2 - \hat{S}^2)$, which has eigenvalues

$$\langle \hat{L} \cdot \hat{S} \rangle_{kk} = \tfrac{1}{2}[j(j+1) - l(l+1) - s(s+1)] = \tfrac{1}{2}[j(j+1) - l(l+1) - \tfrac{3}{4}]$$

Using these in (11.46), we find that the eigenvalues of $\hat{V}'$, which give the first-order corrections to the energy of the nth level, are

$$E_{nlj}^{(1)} = \frac{\hbar^2}{2m^2c^2} \frac{e^2}{4\pi\varepsilon_0} \left\langle \frac{1}{r^3} \right\rangle_{nlnl} \tfrac{1}{2}[j(j+1) - l(l+1) - \tfrac{3}{4}] \quad (j = l \pm \tfrac{1}{2})$$

$$(11.49)$$

The expectation value of $1/r^3$ can be evaluated (using the mathematical properties of the Laguerre polynomials), and the result is

$$\left\langle \frac{1}{r^3} \right\rangle_{nlnl} = \frac{2}{r_B^3 n^3 l(l+1)(2l+1)} \tag{11.50}$$

Substituting this in (11.49), we get

$$E_{nlj}^{(0)} = \frac{\hbar^2}{2m^2c^2} \frac{e^2}{4\pi\varepsilon_0} \frac{1}{r_B^3} \frac{j(j+1) - l(l+1) - \frac{3}{4}}{n^3 l(l+1)(2l+1)}$$

$$= \alpha^2 E_R \frac{j(j+1) - l(l+1) - \frac{3}{4}}{n^3 l(l+1)(2l+1)} \tag{11.51}$$

Equation (11.51) indicates that the nth energy level of hydrogen is split into sublevels corresponding to different values of l and j. However, when the other relativistic corrections mentioned in Section 8.4 are included, the dependence on l disappears, and the correction to the nth level (to order α^2) is given by (8.25). This means that each level, except for the ground state, is split into two sublevels whose energy depends only on n and $j = l \pm \frac{1}{2}$. (For the ground state $l = 0$, and so j can only take the value $\frac{1}{2}$.) This is the fine structure of the levels, which gives rise to the fine structure of the optical spectrum of hydrogen. The fine structure of the $n = 2$ level is shown, above the label 'zero field', at the left of Figure 8.5. Spin–orbit coupling produces similar fine structure in the case of multi-electron atoms. In the case of the alkali metals, which have only one valence electron, each sublevel is split into two as for hydrogen, corresponding to the two possible values $j = l + \frac{1}{2}$ and $j = l - \frac{1}{2}$. (But sublevels with $l = 0$ are not split, because there the only possible value of j is $\frac{1}{2}$.) Transitions in sodium between the singlet 3s ground state ($n = 3, l = 0$) and the doublet 3p state ($n = 3, l = 1$) produce the D-line doublet in its optical spectrum, shown in Figure 8.3(a).

11.5 A variational method for finding the ground state of a bound particle

The expectation value of an operator in one of its eigenstates is the corresponding eigenvalue (Problem 9.9). So if the expectation value of the Hamiltonian for a bound particle is evaluated using the correct ground state eigenfunction in (9.13), the result is the ground state energy E_G. If we replace the ground state eigenfunction by any other wave function (satisfying the correct boundary conditions), the expectation value is greater than E_G, as we shall prove presently. This observation is the basis for variational methods for calculating the ground state eigenvalue and eigenfunction of a particle.

Remember that a wave function describing an arbitrary state of a particle can be expanded as a linear superposition of its energy eigenfunctions, as indicated by (9.1). So at $t = 0$ an arbitrary state of the particle can be represented by

$$\Psi(x, y, z, 0) = \sum_n c_n \psi_n(x, y, z) \qquad (11.52)$$

where $\psi_n(x, y, z)$ is the nth energy eigenfunction of the appropriate Hamiltonian $\hat{H}$. According to (9.21), the expectation value of the Hamiltonian in the state represented by $\Psi(x, y, z, 0)$ can be written in terms of the coefficients $\{c_n\}$ in the expansion (11.52):

$$\langle \hat{H} \rangle = \sum_n |c_n|^2 E_n \qquad (11.53)$$

Postulate 3 (Section 9.1) identified $|c_n|^2$ as the probability of obtaining the eigenvalue E_n when an energy measurement is made on the particle in a state described by $\Psi(x, y, z, 0)$ (provided all wave functions have been normalized). Since $|c_n|^2$ lies between 0 and 1 for all n, and $E_n \geqslant E_G$, (11.53) implies that

$$\langle \hat{H} \rangle \geqslant E_G \qquad (11.54)$$

For example, the expectation value of the energy in a state represented by the wave function (9.6) (which is not an energy eigenfunction) is greater than the ground state energy, as may be verified from (9.8) or (9.15). The equality can only hold if all the coefficients c_n are zero except for c_G, which refers to the ground state, in which case the wave function $\Psi(x, y, z, 0) = \psi_G(x, y, z)$ is identical to the ground state wave function.

Equation (11.54) implies that the expectation value of the Hamiltonian evaluated using an arbitrary wave function sets an upper bound on the ground state energy. The principle of the variational method is to search for an improved bound by varying the wave function, and the function that gives the lowest bound is the closest approximation to the ground state eigenfunction. In computer calculations the wave function can be varied numerically, but we shall describe a systematic analytical variational method.

Suppose that $\hat{H}$ is the Hamiltonian whose ground state we wish to find. We choose a trial wave function that depends on a continuous parameter λ (as well as on x, y and z), and calculate the expectation value $\langle \hat{H}(\lambda) \rangle$, which will be a function of λ. Then we minimize the expectation value with respect to λ by making

$$\frac{\partial \langle \hat{H}(\lambda) \rangle}{\partial \lambda} = 0 \qquad (11.55)$$

This determines a value $\lambda = \lambda_0$ of the parameter, and the minimum expectation value $\langle \hat{H}(\lambda_0) \rangle$ that our particular form of trial function can give.

The success of the variational method depends on choosing a good trial wave function. Of course it must remain finite everywhere, and should satisfy the appropriate boundary conditions, which in general require that it should tend to zero at large distances from the centre of the binding potential. In addition, we expect the ground state wave function to have no nodes in the classically allowed region, in accordance with the third general characteristic of bound states listed in Section 3.3. If the Hamiltonian is of the form $\hat{H} = \hat{H}_0 + \hat{V}'$, and we have been able to solve the eigenvalue equation for $\hat{H}_0$, then we can base our guess for a suitable trial wave function on the form of the eigenfunctions of $\hat{H}_0$.

To see how the method works, we shall calculate the change in the ground state energy of a one-dimensional simple harmonic oscillator with electric charge $-e$ when an electric field in the x direction is applied. The Hamiltonian is $\hat{H} = \hat{H}_0 + \hat{V}'$, where $\hat{H}_0$ is the simple harmonic oscillator Hamiltonian of (5.1), and the additional potential due to the electric field, $\hat{V}'$, is found from (8.3):

$$\left. \begin{aligned} \hat{H} &= \hat{H}_0 + \hat{V}' \\[2mm] \hat{H}_0 &= \frac{1}{2}\left(\xi^2 - \frac{d^2}{dx^2} \right)\hbar\omega \\[2mm] \hat{V}' &= e\mathcal{E}x = e\mathcal{E}\left(\frac{\hbar}{m\omega} \right)^{1/2}\xi = b\xi \end{aligned} \right\} \tag{11.56}$$

(We use the dimensionless variable ξ defined in (5.2), and for convenience introduce the single constant $b = e\mathcal{E}(\hbar/m\omega)^{1/2}$, which has the dimensions of energy.)

As a trial wave function we use

$$\psi(\xi, \lambda) = C(\lambda)e^{-\xi^2/2}e^{-\lambda\xi} = C(\lambda)e^{\lambda^2/2}e^{-(\xi+\lambda)^2/2} \tag{11.57}$$

This is the product of the simple harmonic oscillator ground state wave function given in (5.16) with an exponential factor. This factor does not introduce any nodes into the wave function, but effectively displaces the centre of the probability distribution by a distance proportional to $-\lambda$ in the direction of the electric field.

The normalization constant $C(\lambda)$ must be calculated, and depends on λ. We require

$$\int_{-\infty}^{\infty} dx\, \psi^*\psi = C^2(\lambda)e^{\lambda^2}\left(\frac{\hbar}{m\omega} \right)^{1/2}\int_{-\infty}^{\infty} d\xi\, e^{-(\xi+\lambda)^2} = 1 \tag{11.58}$$

where, as usual, we choose the normalization constant to be real. Since $d\xi = d(\xi + \lambda)$, the integral can be written as $\int_{-\infty}^{\infty} d(\xi + \lambda)\,e^{-(\xi+\lambda)^2}$, which has the value $\sqrt{\pi}$. So, from (11.58), the normalization constant is

$$C^2(\lambda) = \left(\frac{m\omega}{\pi\hbar} \right)^{1/2}e^{-\lambda^2} \tag{11.59}$$

The expectation value of $\hat{H}_0$ in the state with the wave function (11.57) is

$$\langle \hat{H}_0 \rangle = \tfrac{1}{2}\hbar\omega C^2(\lambda)e^{\lambda^2}\left(\frac{\hbar}{m\omega}\right)^{1/2}\int_{-\infty}^{\infty} d\xi\, e^{-(\xi+\lambda)^2/2}\left(\xi^2 - \frac{d^2}{d\xi^2}\right)e^{-(\xi+\lambda)^2/2}$$

$$= \frac{\hbar\omega}{2}\frac{1}{\sqrt{\pi}}\int_{-\infty}^{\infty} d\xi\, e^{-(\xi+\lambda)^2}[\xi^2 - (\xi+\lambda)^2 + 1]$$

$$= \frac{\hbar\omega}{2}\frac{1}{\sqrt{\pi}}\int_{-\infty}^{\infty} d(\xi+\lambda)\, e^{-(\xi+\lambda)^2}[-2\lambda(\xi+\lambda) + \lambda^2 + 1]$$

(We have used (11.59) for $C(\lambda)$.) Since the integral of any odd function of $\xi + \lambda$ between $-\infty$ and $+\infty$ vanishes, this expression becomes

$$\langle \hat{H}_0 \rangle = \frac{\hbar\omega}{2}\frac{1}{\sqrt{\pi}}(\lambda^2 + 1)\sqrt{\pi} = \tfrac{1}{2}\hbar\omega(\lambda^2 + 1) \qquad \textbf{(11.60)}$$

Similarly, the expectation value of $\hat{V}'$ is

$$\langle \hat{V}' \rangle = C^2(\lambda)e^{\lambda^2}\left(\frac{\hbar}{m\omega}\right)^{1/2}\int_{-\infty}^{\infty} d\xi\, e^{-(\xi+\lambda)^2}b\xi$$

$$= \frac{b}{\sqrt{\pi}}\int_{-\infty}^{\infty} d(\xi+\lambda)\, e^{-(\xi+\lambda)^2}[(\xi+\lambda) - \lambda]$$

$$= -b\lambda \qquad \textbf{(11.61)}$$

Combining (11.60) and (11.61), we obtain

$$\langle \hat{H}(\lambda) \rangle = \tfrac{1}{2}\hbar\omega(\lambda^2 + 1) - b\lambda \qquad \textbf{(11.62)}$$

Differentiating this expression with respect to λ, we find that

$$\frac{d\langle \hat{H}(\lambda) \rangle}{d\lambda} = 0 \quad \text{when } \lambda = \frac{b}{\hbar\omega} \qquad \textbf{(11.63)}$$

Substituting this value for λ in (11.62), we obtain the minimum expectation value for the Hamiltonian

$$\langle \hat{H} \rangle_{\min} = \tfrac{1}{2}\hbar\omega - \frac{b^2}{2\hbar\omega}$$

$$= \tfrac{1}{2}\hbar\omega - \frac{e^2\mathscr{E}^2}{2m\omega^2} \qquad \textbf{(11.64)}$$

(In the final expression we have rewritten b in terms of physical quantities, using the definition implicit in (11.56).) Note that there are no terms linear in the applied field. This is in agreement with perturbation theory, which gives a zero first-order correction to all energy eigenvalues. The second term in (11.64) is identical to the second-order correction to the ground state energy in perturbation theory (Problem 11.9a).

In fact, the solution that we have just found is exact. This can be confirmed by examining the Schrödinger equation in which the Hamiltonian is $\hat{H}$, (11.56). It can be rearranged so that the variable is $\xi - b/\hbar\omega$ rather than ξ, when it becomes an ordinary simple harmonic oscillator equation, but with total energy $E - b^2/2\hbar\omega$ instead of E.

The variational method can be used for excited states as well as for the ground state. Remember that eigenfunctions corresponding to different eigenvalues are orthogonal, and that the number of nodes tends to increase with the excitation of the state. So, assuming that we have found a good approximation to the ground state wave function, we choose a trial wave function for the first excited state that is orthogonal to this, and has at most one additional node in the classically allowed region. If a wave function is orthogonal to the ground state wave function ψ_G, its expansion, in the form (11.52), has zero contribution from this eigenfunction – that is, $c_G = 0$. Therefore the minimum possible value of $\langle \hat{H} \rangle$ evaluated using such a wave function is the energy eigenvalue for the first excited state. Consider, for example, the two wave functions $\Psi(z)$ and $\Phi(z)$ for a particle in an infinitely deep one-dimensional square well dealt with in Problems 2.1 and 2.2. They are orthogonal to one another, and $\Psi(z)$ has no nodes, whereas $\Phi(z)$ has one. Neither function is an energy eigenfunction, and we noted at the beginning of this section (and in Section 9.2) that the expectation value of the energy in the first state is close to but slightly greater than the exact ground state energy. If we had chosen $\Psi(z)$ as a good approximation to the ground state wave function then $\Phi(z)$ would be a sensible guess for an approximation to the wave function for the first excited state. The expectation value for the energy in this second state is slightly greater than the exact energy for the first excited state in the infinitely deep well (Problem 9.3b). Proceeding in a similar manner, it is possible to find approximations to the eigenvalues and eigenfunctions of successive excited states. The method can give very accurate results, and indeed it is used in standard calculations of the energy eigenstates and eigenfunctions of multi-electron atoms.

References

[1] For references to other perturbation and variational methods used in quantum mechanics see the bibliography at the end of Chapter 11 of Cohen-Tannoudji C., Diu B. and Laloë F. (1977). *Quantum Mechanics* Vol. 2. New York: Wiley.
[2] Kreyszig E. (1988). *Advanced Engineering Mathematics* 6th edn, Section 7.15. New York: Wiley.

Problems

11.1 (a) Use the perturbation expansion of the wave function, (11.4), to normalize ψ_n to first order, and verify (11.15). (Neglect all terms proportional to λ^2 and higher powers.)

(b) Now include terms proportional to λ^2 but neglect higher-order terms, and prove that if the wave function is normalized to second order, the coefficients $c_{nk}^{(2)}$ (in the expansion (11.18) of the second-order correction $\psi_n^{(2)}$) satisfy $c_{nn}^{(2)*} + c_{nn}^{(2)} = \sum_{k \neq n} |c_{nk}^{(1)}|^2$.

11.2 Perform the necessary integrations to verify (11.25) and (11.27).

11.3 Rewrite the derivation of the first-order perturbation correction to a non-degenerate energy level (Section 11.1) using Dirac notation (Section 10.6). (The expansion of a state $|\Psi\rangle$ in terms of a complete set of eigenstates $\{|n\rangle\}$ is $|\Psi\rangle = \sum_n c_n |n\rangle$. You might find it convenient to denote the kth-order correction to the nth eigenstate by $|n^{(k)}\rangle$.)

11.4 A particle in an infinitely deep one-dimensional square well extending from $z = 0$ to $z = L$ is perturbed by a potential that is zero except in the region $0 \leq z \leq \frac{1}{3}L$, where it takes the constant value C. Find the first-order correction to the energy of each of the first four levels, and show that the coefficients that determine the first-order corrections to the energy eigenfunctions are given by

$$c_{nk}^{(1)} = \frac{C}{E_1^{(0)}\pi} \frac{k \cos\frac{1}{3}k\pi \sin\frac{1}{3}n\pi - n \cos\frac{1}{3}n\pi \sin\frac{1}{3}k\pi}{(k^2 - n^2)^2}$$

$$E_1^{(0)} = \frac{h^2}{8mL^2}$$

For what values of C do you expect the first-order result to be a good approximation for levels up to $n = 10$?

11.5 A rigid rotator (with the Hamiltonian (4.47)) is perturbed by a potential $\hat{V}' = (C\hbar^2/2\mathcal{I})\cos^2\theta$, where $\mathcal{I}$ is the moment of inertia of the rotator and C is a small constant. Find the corrections to the eigenvalues and eigenfunctions to first order in perturbation theory. Estimate the maximum value of C for which this is a good approximation for the level characterized by the quantum number l. (You will find the properties of associated Legendre functions given in Table 4.1 useful.)

11.6 (a) Derive (11.32) from the expression $\xi = \sqrt{(\frac{1}{2})}(\hat{A} + \hat{A}^\dagger)$, using the commutation relation (5.6) for the ladder operators. Verify (11.33)–(11.35).

(b) Multiply (11.32) by $\xi = \sqrt{(\frac{1}{2})}(\hat{A} + \hat{A}^\dagger)$ and prove that

$$\xi^4 = \frac{1}{4}[\hat{A}^4 + 4(\hat{A}^2 + \hat{A}^{\dagger 2})\hat{A}^\dagger\hat{A} + 6(\hat{A}^\dagger\hat{A})^2 + 6\hat{A}^\dagger\hat{A} + 3 + \hat{A}^{\dagger 4} + 6\hat{A}^{\dagger 2} - 2\hat{A}^2]$$

(*Hint*: To simplify the algebra, recall the results of Problems 4.1(c) and 5.3(c).)

(c) Operate with ξ^4 on the unperturbed simple harmonic oscillator eigenfunction $\psi_n^{(0)}$, and prove that

$$\xi^4\psi_n^{(0)} = \tfrac{1}{4}[\hat{A}^4 + 2(2n-1)\hat{A}^2 + 2(2n+3)\hat{A}^{\dagger 2} + \hat{A}^{\dagger 4} + 6n^2 + 6n + 3]\psi_n^{(0)}$$

Use (5.29) and (5.30) to show that this gives

$$\xi^4\psi_n^{(0)} = \tfrac{1}{4}\{[n(n-1)(n-2)(n-3)]^{1/2}\psi_{n-4}^{(0)}$$
$$+ 2(2n-1)[n(n-1)]^{1/2}\psi_{n-2}^{(0)}$$
$$+ 2(2n+3)[(n+1)(n+2)]^{1/2}\psi_{n+2}^{(0)}$$
$$+ [(n+1)(n+2)(n+3)(n+4)]^{1/2}\psi_{n+4}^{(0)}$$
$$+ 6(n^2 + n + \tfrac{1}{2})\psi_n^{(0)}\}$$

11.7 The Hamiltonian of a one-dimensional anharmonic oscillator is $\hat{H} = \hat{H}_0 + \hat{V}'$, where $\hat{H}_0$ is the simple harmonic oscillator Hamiltonian and $\hat{V}' = Ch\omega\xi^4$, where C is a small dimensionless constant.

(a) Use first-order perturbation theory to calculate the correction to the energy and the eigenfunction of each of the first four levels of the oscillator. (Use the results of Problem 11.6(c).)

*(b) Use the computer program *Anharmonic oscillator* to find the first four eigenvalues (in units of $\hbar\omega$) of the complete Hamiltonian $\hat{H}$ for a range of values of the constant C: $C = 10^{-4}$, 10^{-3}, 10^{-3}, 10^{-1}, 1. Compare the results of your first-order perturbation calculation with the more accurate numerical results provided by the computer, and note how the accuracy of the perturbation calculation varies with the level n and the size of the constant C.

11.8 (a) If the eigenfunctions of the approximate Hamiltonian $\hat{H}_0$ are also parity eigenfunctions, show that the eigenfunctions of the Hamiltonian $\hat{H} = \hat{H}_0 + \hat{V}'$ that are correct to first order in perturbation theory will not be parity eigenfunctions unless $\hat{V}'$ commutes with the parity operator.

(b) Write down explicitly the integral expression for the first-order correction to the ground state energy of hydrogen due to a constant uniform electric field in the x direction (the **first-order Stark effect** in the ground state of hydrogen). Use a parity argument to show that this integral vanishes.

11.9 A weak electric field $\mathcal{E}$ in the x direction is applied to a one-dimensional simple harmonic oscillator with electric charge $-e$.

(a) Find the first- and second-order corrections to the ground state energy of the oscillator due to this perturbation, and compare your answers with the result of the variational calculation, (11.64). (*Hint*: You will need the result of Problem 5.4(c).)

(b) Find the energy eigenfunction correct to first order, and use it to calculate the expectation value of x, neglecting terms proportional to the second power of the weak field $\mathscr{E}$. (You will find $\langle x \rangle \propto \mathscr{E}$, showing that the electric field polarizes the oscillator.)

11.10 A weak electric field is applied to an electron in an infinitely deep one-dimensional square well extending from $z = 0$ to $z = \mathbb{L}$, producing a perturbation $\hat{V}' = e\mathscr{E}(z - \frac{1}{2}\mathbb{L})$. Find the first- and second-order corrections to the ground state energy of the electron.

11.11 Show that the energy levels of the rigid rotator in Problem 11.5 are split if the perturbation is changed to $\hat{V}' = (C\hbar^2/2\mathscr{I}) \cos^2 \theta \cos \phi$. In each of the cases $l = 0, 1, 2$ find the first-order correction to the energy of the rotator.

11.12 (a) Find the effect of spin–orbit coupling on the $n = 1$ and $n = 2$ levels of hydrogen by two methods:

   (i) using (11.51);
   (ii) diagonalizing $\hat{L} \cdot \hat{S}$ explicitly.
       (Use the results of Problem 10.6, and show that $\hat{L} \cdot \hat{S} Y_{00}(\theta, \phi)\chi_\pm = 0$ using the method of Problem 6.9.) (Equation (11.50) gives an expression for $\langle 1/r^3 \rangle$ for $l \neq 0$.)

   (b) The **Lyman-α line** (wavelength 121.6 nm) is due to electric dipole transitions between the $n = 2$ and $n = 1$ levels. Find the frequencies of the components of this spectral line that are predicted using

   (i) spin–orbit coupling alone;
   (ii) the complete relativistic correction (to order v/c) given by (8.25).

   Show that (i) and (ii) predict the same difference in the frequencies of the components of the line.

11.13 Calculate the Stark effect on the $n = 2$ level in hydrogen (ignoring spin–orbit coupling) due to an electric field in the z direction (which produces a potential $\hat{V}' = e\mathscr{E}r \cos \theta$. (*Hint*: Show that the matrix representing $\hat{V}'$ in the $n = 2$ level is that of Problem 10.5(b).)

11.14 A one-dimensional lattice contains N unit cells each of length $\mathbb{L}$. In the free-electron approximation $V = 0$, the energy eigenvalues and eigenfunctions are of the form

$$E_k = \frac{\hbar^2 k^2}{2m}, \qquad \psi_k(x) = \frac{1}{\sqrt{(N\mathbb{L})}} e^{ikx}$$

To improve on this approximation, a small potential V' with the periodicity of the lattice is added to the Hamiltonian, where $V' = V_0 \cos(2\pi x/\mathbb{L})$. Show that to first order in perturbation theory the correction to the energy is zero, unless $k = \pm \pi/\mathbb{L}$, when the problem of degeneracy arises (why?). For $k = \pm \pi/\mathbb{L}$ construct the appropriate 2×2 Hamiltonian matrix, and diagonalize it to find the first-order corrections to the degenerate level. Show that the corrected eigenfunctions are standing waves.

11.15   From (7.12), the hydrogen atom Hamiltonian can be written as

$$\hat{H} = - \frac{d^2}{d\rho^2} - \frac{2}{\rho}\frac{d}{d\rho} + \frac{l(l+1)}{\rho^2} - \frac{2}{\rho}$$

where ρ is the radial distance in units of the Bohr radius, and the energy is measured in units of the Rydberg energy.

(a) Put $l = 0$, and use the variational method with a trial radial wave function $R(\rho) = C(\lambda)e^{-\lambda\rho}$, where λ is the variational parameter, to estimate the ground state energy.

(b) Similarly estimate the lowest energy for a state with $l = 1$ using the trial wave function $R'(\rho) = C'(\lambda)\rho e^{-\lambda\rho}$. This function has one node, but need not be orthogonal to the function $R(\rho)$ of (a) – why not?

(The method gives the exact solution in both these cases.)

*11.16 Use the variational method to find an approximate value for the ground state energy of a particle in the triangular potential well

$$\hat{V} = \begin{cases} -V_0 x & (x \leqslant 0) \\ \infty & (x > 0) \end{cases}$$

Use the trial wave function $\psi(x) = C(\lambda)xe^{\lambda x}$, where λ is the variational parameter. Why must λ be positive? Compare your result with the value calculated numerically by the computer program *Triangular well*.

11.17   Two one-dimensional square wells A and B, of the same width and depth, are symmetrically situated on either side of the origin, so that the Hamiltonian for a particle moving in this structure is

$$\hat{H} = - \frac{\hbar^2}{2m}\frac{d^2}{dx^2} + \hat{V}_A + \hat{V}_B$$

where $\hat{V}_A$ is zero everywhere except for $-\mathbb{L} - b \leqslant x \leqslant -b$, where it is $-V_0$, and $\hat{V}_B$ is zero everywhere except for $b \leqslant x \leqslant b + \mathbb{L}$, where it is $-V_0$. To find an approximate ground state solution to the eigenvalue problem, use the variational method with trial wave function $\psi(x, \lambda) = \cos\lambda\,\psi_A(x) + \sin\lambda\,\psi_B(x)$, where ψ_A and ψ_B are the normalized ground state eigenfunctions in the case where only one square well is present:

$$\left(-\frac{\hbar^2}{2m}\frac{d^2}{dx^2} + \hat{V}_A\right)\psi_A = E\psi_A$$

$$\left(-\frac{\hbar^2}{2m}\frac{d^2}{dx^2} + \hat{V}_B\right)\psi_B = E\psi_B$$

Show that the expectation value of $\hat{H}$ is a minimum when $\lambda = \pm\frac{1}{4}\pi$, and the minimum value is

$$E + \langle A|\hat{V}_B|A\rangle \pm \tfrac{1}{2}\langle A|\hat{H}|B\rangle = E + \langle B|\hat{V}_A|B\rangle \pm \tfrac{1}{2}\langle B|\hat{H}|A\rangle$$

where we have used Dirac notation to simplify the expression. Which sign would you expect to give the lower value? Show that the corresponding ground state eigenfunction ψ_G is a parity eigenstate and determine its parity eigenvalue. The trial wave function is normalized only if $\langle A|B \rangle = 0$. This is not correct, but the magnitude is likely to be very small. Why? Show that neglecting $\langle A|B \rangle$ does not alter the parity of ψ_G.

CHAPTER 12

Time-Dependent Problems

12.1 The time-dependent Schrödinger equation

In Section 3.2 we introduced the time-dependent Schrödinger equation (3.13):

$$i\hbar \frac{\partial}{\partial t} \Psi(x, y, z, t) = \hat{H} \Psi(x, y, z, t) \qquad (12.1)$$

This was obtained by equating the effects on a wave function of two different operators representing the total energy of the particle: $\hat{E}_{\text{tot}}$, defined by (3.7), and the Hamiltonian $\hat{H}$, which is the sum of kinetic and potential energy terms, (3.12) (or, in the case of an electromagnetic interaction, (8.14)). In Section 3.2 we looked at the time dependence of energy eigenfunctions, which is of the exponential form given in (3.19). But (12.1) is assumed to be true in general, and is the equation of motion that determines how the wave function describing *any* state of a particle evolves with time. This is our fourth postulate.

Postulate 4

The way in which an arbitrary state of a non-relativistic quantum particle evolves with time is determined by its Hamiltonian, through the time-dependent Schrödinger equation.

The equation of motion for a classical particle moving in a potential is Newton's second law, which determines the rate of change of the momentum of the particle. At first sight the time-dependent Schrödinger equation bears no relation to this, but there is a correspondence between the quantum and classical equations of motion, which we can demonstrate by examining how the expectation value of an operator varies with time.

Suppose that we are interested in the measurement of a variable represented by the operator $\hat{A}$, on a particle in some state represented by the wave function $\Psi(x, y, z, t)$. The expectation value of the operator $\hat{A}$ in this state is defined by (9.13), and so, if the wave function is normalized, the rate of change of the expectation value is

$$\frac{d\langle \hat{A} \rangle}{dt} = \frac{d}{dt} \int_{\text{all space}} d^3r \, \Psi^*(x, y, z, t) \hat{A} \Psi(x, y, z, t)$$

$$= \int_{\text{all space}} d^3r \, \frac{\partial}{\partial t} [\Psi^*(x, y, z, t) \hat{A} \Psi(x, y, z, t)]$$

The definition of the operator $\hat{A}$ may contain time explicitly, in which case the partial derivative $\partial \hat{A}/\partial t$ is non-zero, and

$$\frac{d\langle \hat{A} \rangle}{dt} = \int_{\text{all space}} d^3r \left(\Psi^* \frac{\partial \hat{A}}{\partial t} \Psi + \frac{\partial \Psi^*}{\partial t} \hat{A} \Psi + \Psi^* \hat{A} \frac{\partial \Psi}{\partial t} \right)$$

$$(12.2)$$

Using (12.1), we can replace the last term in the integrand by $-(i/\hbar)\Psi^*\hat{A}\hat{H}\Psi$. To transform the second term in a similar way, we need to take the complex conjugate of (12.1):

$$-i\hbar \frac{\partial}{\partial t} \Psi^*(x, y, z, t) = [\hat{H} \Psi(x, y, z, t)]^* \qquad (12.3)$$

So the second term in the integrand in (12.2) is $(i/\hbar)(\hat{H}\Psi)^*\hat{A}\Psi$, and, since $\hat{H}$ is Hermitian, it is easy to demonstrate (Problem 10.7a) that

$$\int_{\text{all space}} d^3r \, (\hat{H}\Psi)^* \hat{A} \Psi = \int_{\text{all space}} d^3r \, \Psi^* \hat{H} \hat{A} \Psi$$

Then (12.2) can be written as

$$\frac{d\langle \hat{A} \rangle}{dt} = \int_{\text{all space}} d^3r \left[\Psi^* \frac{\partial \hat{A}}{\partial t} \Psi + \frac{i}{\hbar} (\hat{H}\Psi)^* \hat{A} \Psi - \frac{i}{\hbar} \Psi^* \hat{A} \hat{H} \Psi \right]$$

$$= \int_{\text{all space}} d^3r \left[\Psi^* \frac{\partial \hat{A}}{\partial t} \Psi + \frac{i}{\hbar} (\Psi^* \hat{H} \hat{A} \Psi - \Psi^* \hat{A} \hat{H} \Psi) \right]$$

This is the sum of the expectation value of $\partial \hat{A}/\partial t$ and $i/\hbar$ times the expectation value of the commutator of the Hamiltonian with $\hat{A}$:

$$\frac{d\langle \hat{A} \rangle}{dt} = \left\langle \frac{\partial \hat{A}}{\partial t} \right\rangle + \frac{i}{\hbar} \langle [\hat{H}, \hat{A}] \rangle \qquad (12.4)$$

For example, suppose that $\hat{A} = \hat{P}_x = -i\hbar \, \partial/\partial x$ represents the momentum of a particle moving in one dimension. In this case the operator is not explicitly dependent on time, and so $\partial \hat{P}_x/\partial t = 0$. Using the one-dimensional form of (3.12) for the Hamiltonian, we find that

$$[\hat{H}, \hat{P}_x] = -i\hbar \left[-\hbar^2 \frac{\partial^2}{\partial x^2} + \hat{V}(x), \frac{\partial}{\partial x} \right] = i\hbar \frac{d\hat{V}}{dx}$$

So, from (12.4),

$$\frac{\mathrm{d}\langle \hat{P}_x \rangle}{\mathrm{d}t} = -\left\langle \frac{\mathrm{d}\hat{V}}{\mathrm{d}x} \right\rangle \tag{12.5}$$

This is a quantum mechanical version of Newton's second law, equating the rate of change of the expectation value of the momentum to the expectation value of the operator $-\mathrm{d}\hat{V}/\mathrm{d}x$ that represents the force on the particle. The quantum mechanical equation of motion (12.1) thus implies a relationship between expectation values that corresponds to the classical equation of motion.

Equation (12.4) indicates that if an operator that does not depend explicitly on time commutes with the Hamiltonian then the expectation value of the corresponding variable is constant. This is the quantum mechanical equivalent of the statement that the variable is a constant of the motion. For example, any Hamiltonian (representing the total energy of a particle) commutes with itself, and so its expectation value is constant provided the potential energy does not depend explicitly on time. This applies whether or not the particle is in an energy eigenstate. Correspondingly, in classical mechanics the total energy is conserved if the potential energy does not depend explicitly on time. Similarly, if the potential is central, the orbital angular momentum is a constant of the classical motion, and in quantum mechanics both the square of the orbital angular momentum operator and its z component commute with $\hat{H}$, as we saw in Section 7.1. So in this case the expectation values of $\hat{\mathbf{L}}^2$ and $\hat{L}_z$ are constant, whether or not the particle is in an eigenstate of energy or of angular momentum.

For the remainder of this chapter we shall examine how a quantum particle behaves when at least one of the contributions to its potential energy varies explicitly with time, so that the total Hamiltonian is of the form

$$\hat{H} = \hat{H}_0 + \hat{V}(t) \tag{12.6}$$

(In general, the potential depends on the spatial variables as well as time, but it is the time dependence we wish to emphasize here, and so we show only the time dependence explicitly, and write $\hat{V}(t)$ rather than $\hat{V}(r, t)$). The total energy of a classical particle moving in a potential that is explicitly time-dependent ($\partial \hat{V}/\partial t \neq 0$) changes with time. Similarly, the total energy of a quantum particle with the Hamiltonian (12.6) varies with time.

In Section 8.6 we mentioned experiments in which the magnetic field acting on a sample is varied, though we did not treat the magnetic interaction term in the Hamiltonian as time-dependent. In such experiments the rate at which the field is changed should be very slow on a time scale $t = 2\pi/\omega_{jk} = h/(E_j - E_k)$ determined by the energy difference between eigenvalues of $\hat{H}_0$. Then it is a good approximation to assume that at each instant, an electron is in a particular energy eigenstate – the nth say. But, according to (8.31) and (8.32), the energy eigenvalues are proportional to the field strength, and so the energy of the electron in the nth eigenstate varies with time as the field is changed. Such very slow changes in the Hamiltonian are termed **adiabatic**. In an adiabatic approximation the time dependence of the Hamiltonian cannot produce transitions between different energy levels, but only results in a slow variation of each energy eigenvalue.

On the other hand, a non-adiabatic potential $\hat{V}(t)$ can cause transitions between different eigenstates of $\hat{H}_0$, the time-independent part of the Hamiltonian. For example, the transitions that produce optical spectra can be explained in terms of the rapidly varying potential experienced by an electron interacting with the oscillating electric field in a beam of light, as we shall discuss in Section 12.4. To see how such transitions can occur, we must solve the time-dependent Schrödinger equation (12.1). According to Postulate 3 (Section 9.1), the wave function representing an arbitrary state can always be expanded in terms of the complete set of eigenfunctions of $\hat{H}_0$, as in (9.1). In the most general case the expansion coefficients may vary with time (Problem 12.3), so we write

$$\Psi(x, y, z, t) = \sum_n c_n(t)\psi_n(x, y, z)e^{-iE_nt/\hbar} \qquad (12.7)$$

where

$$\hat{H}_0\psi_n(x, y, z) = E_n\psi_n(x, y, z) \qquad (12.8)$$

If we substitute (12.7) in the time-dependent Schrödinger equation, we find

$$i\hbar \frac{\partial}{\partial t} \sum_n c_n(t)\psi_n(x, y, z)e^{-iE_nt/\hbar} = [\hat{H}_0 + \hat{V}(t)]\sum_n c_n(t)\psi_n(x, y, z)e^{-iE_nt/\hbar}$$

and using (12.8) on the right-hand side this becomes

$$\sum_n \left[E_n c_n(t) + i\hbar \frac{dc_n}{dt} \right] \psi_n(x, y, z)e^{-iE_nt/\hbar}$$

$$= \sum_n c_n(t)[E_n + \hat{V}(t)]\psi_n(x, y, z)e^{-iE_nt/\hbar}$$

The first terms on each side of this equation cancel. Now we multiply on the left by ψ_k^*, integrate over all space, and use the orthonormality of the eigenfunctions of $\hat{H}_0$ to obtain

$$i\hbar \frac{dc_k}{dt} = \sum_n \hat{V}_{kn}(t)e^{i\omega_{kn}t}c_n(t) \quad (k = 1, 2, 3, \ldots) \qquad (12.9)$$

where $\hat{V}_{kn}(t)$ is an element of the matrix representing $\hat{V}(t)$ in the basis formed by the eigenstates of $\hat{H}_0$, and

$$\hbar\omega_{kn} = E_k - E_n \qquad (12.10)$$

Equation (12.9) is really nothing more than the time-dependent Schrödinger equation written in matrix notation: in the basis defined by the eigenfunctions of $\hat{H}_0$ at $t = 0$ the coefficients $c_n(t)e^{-iE_nt/\hbar}$ are the components of the vector representing the same state as the wave function $\Psi(x, y, z, t)$ of (12.7), and $\hat{V}_{kn}(t)$ is an element of the matrix representing the time-dependent potential.

In general, the operator $\hat{H}_0$ has an infinite number of eigenvalues, so (12.9) represents an infinite set of coupled first-order differential equations, and we must make some approximations in order to solve them. In principle, it is possible to solve a finite set of equations of this type, though the difficulty increases rapidly with the number of equations. It may be possible to reduce the set to a finite one by neglecting all but a small number of the coefficients c_n. This approach is discussed in Section 12.2. Alternatively, time-dependent perturbation theory provides expressions (which may involve infinite summations) for successive approximations to each coefficient. In both cases it is important that the time-dependent potential $\hat{V}(t)$ should be a small perturbation on the time-independent part of the Hamiltonian.

12.2 Resonant transitions between two energy levels

If the time-independent part of a Hamiltonian, $\hat{H}_0$, has only two eigenstates, the set of equations (12.9) reduces to two coupled equations for two coefficients, c_1 and c_2, which can be solved relatively easily. Clearly if the (Hermitian) time-dependent potential $\hat{V}$ has only two non-zero off-diagonal elements, $\hat{V}_{12}$ and $\hat{V}_{21} = \hat{V}_{12}^*$, the problem reduces to a two-level one, since the equations for all k other than $k = 1$ or 2 are uncoupled, and do not depend on the time-dependent part of the Hamiltonian. In this case only the two levels coupled by the potential need be considered when we discuss the effect of the time-dependent perturbation. In a real situation the off-diagonal matrix elements other than $\hat{V}_{12}$ and $\hat{V}_{21}$ will not all be identically zero, but, provided they are sufficiently small compared with $\hat{V}_{12}$ and $\hat{V}_{21}$, it is a good approximation to neglect them and treat the time-dependent problem as a two-level one.

Such a situation can occur when the time-dependent interaction varies with a frequency that is resonant with the angular frequency $\omega_{12} = (E_1 - E_2)/\hbar$ determined by the difference between two particular eigenvalues of $\hat{H}_0$. For example, the particle might be an atom interacting with a monochromatic beam of light whose frequency is resonant with the energy difference between two particular non-degenerate levels of the atom, or it could be a particle with total angular momentum $\frac{1}{2}$ in a magnetic resonance experiment (**electron spin resonance, ESR**, or **nuclear magnetic resonance, NMR**), when again just two energy levels are resonantly coupled, in this case by a time-dependent magnetic interaction.

In order to make our discussion concrete, we shall solve the two-level problem in the particular case of ESR, though the mathematical procedure is exactly the same for any two-level problem. In an ESR experiment the interaction of the intrinsic magnetic moment of the electron with a strong constant magnetic field $\mathcal{B}$ in the z direction is represented by the time-independent Hamiltonian $\hat{H}_0$, which has only two eigenvalues. Interaction with a weak magnetic field $\mathcal{B}'(t)$ that rotates in the (x, y) plane with angular frequency ω provides the time-dependent potential $\hat{V}(t)$. (In practice, the weak field is an oscillating field rather than a rotating one. However, such a field can be expressed as the sum of two components rotating in opposite directions, and only one component has an appreciable effect: from a classical point of view, the magnetic moment of the electron precesses about the constant magnetic field, and the component of the weak field that rotates in the opposite sense averages to zero in the rest frame of the electron, and can be neglected. This is the **rotating wave approximation**.)

The interaction of the electron's intrinsic magnetic moment with a magnetic field is given by (8.4) and (8.21), and the eigenvalue problem was discussed in Section 10.5. If we put the g factor of the electron equal to 2, the time-independent part of the Hamiltonian is

$$\hat{H}_0 = -\boldsymbol{\mu}_s \cdot \mathcal{B} = 2\mu_B \mathcal{B} \hat{S}_z = \hbar\omega_0 \hat{S}_z \qquad (12.11)$$

where we have used the definition (8.8) of the Bohr magneton, and ω_0 is twice the Larmor frequency ω_L defined by (8.5). The eigenvalues of $\hat{H}_0$ are $E_1 = \frac{1}{2}\hbar\omega_0$ and $E_2 = -\frac{1}{2}\hbar\omega_0$, and its eigenstates are the eigenstates of $\hat{S}_z$, represented by $\chi_{+1/2}$ and $\chi_{-1/2}$ respectively. It is convenient to use matrix notation, and in particular (10.26) for $\hat{S}_z$. Then $\hat{H}_0$ is represented as a diagonal matrix in the basis defined by its eigenstates:

$$\hat{H}_0 = \frac{1}{2}\hbar\omega_0 \begin{pmatrix} 1 & 0 \\ 0 & -1 \end{pmatrix} \qquad (12.12)$$

The weak field rotating anticlockwise in the (x, y) plane can be written in terms of unit vectors $\boldsymbol{x}$ and $\boldsymbol{y}$ along the coordinate axes as

$$\mathcal{B}'(t) = \mathcal{B}'(\cos \omega t \, \boldsymbol{x} + \sin \omega t \, \boldsymbol{y}) \qquad (12.13)$$

So the time-dependent potential is

$$\hat{V}(t) = -\hat{\boldsymbol{\mu}}_s \cdot \mathcal{B}'(t) = 2\mu_B \mathcal{B}'(\cos \omega t \, \hat{S}_x + \sin \omega t \, \hat{S}_y)$$
$$= \hbar\omega_V(\cos \omega t \, \hat{S}_x + \sin \omega t \, \hat{S}_y) \qquad (12.14)$$

where for convenience we have defined

$$\hbar\omega_V = 2\mu_B \mathcal{B}' \tag{12.15}$$

(Note that in this case the potential $\hat{V}(t)$ does not depend on the spatial variables, provided the magnetic field $\mathcal{B}'$ is uniform over the region occupied by the electron.) We can write $\hat{V}(t)$ as a matrix, using the matrices for $\hat{S}_x$ and $\hat{S}_y$ given in (10.36):

$$\hat{V}(t) = \tfrac{1}{2}\hbar\omega_V \begin{pmatrix} 0 & e^{-i\omega t} \\ e^{i\omega t} & 0 \end{pmatrix} \tag{12.16}$$

In the two-level case the set of equations (12.9) reduces to

$$\left.\begin{array}{l} i\hbar\dfrac{dc_1}{dt} = c_1 \hat{V}_{11} e^{i\omega_{11}t} + c_2 \hat{V}_{12} e^{i\omega_{12}t} \\[2mm] i\hbar\dfrac{dc_2}{dt} = c_1 \hat{V}_{21} e^{i\omega_{21}t} + c_2 \hat{V}_{22} e^{i\omega_{22}t} \end{array}\right\} \tag{12.17}$$

Inserting the eigenvalues of $\hat{H}_0$ in the definition (12.10) of ω_{kn} we find

$$\omega_{11} = \omega_{22} = 0, \qquad \omega_{12} = -\omega_{21} = \omega_0$$

The matrix elements of $\hat{V}$ can be read off from the matrix in (12.16), and so the pair of equations (12.17) becomes

$$\left.\begin{array}{l} \dfrac{dc_1}{dt} = -\tfrac{1}{2}i\omega_V e^{-i(\omega-\omega_0)t} c_2 \\[2mm] \dfrac{dc_2}{dt} = -\tfrac{1}{2}i\omega_V e^{i(\omega-\omega_0)t} c_1 \end{array}\right\} \tag{12.18}$$

To solve these equations, we must uncouple them: we differentiate the first, and then substitute from both equations to eliminate c_2. This gives

$$\frac{d^2 c_1}{dt^2} + i(\omega - \omega_0)\frac{dc_1}{dt} + (\tfrac{1}{2}\omega_V)^2 c_1 = 0 \tag{12.19}$$

which has the general solution

$$c_1 = C e^{-i(\omega-\omega_0)t/2} \sin(\tfrac{1}{2}\alpha t + \gamma), \qquad \alpha^2 = (\omega - \omega_0)^2 + \omega_V^2 \tag{12.20}$$

where C and γ are arbitrary constants. We can now calculate c_2 from the first of (12.18):

$$c_2 = \frac{C}{\omega_V} e^{i(\omega-\omega_0)t/2}[i\alpha\cos(\tfrac{1}{2}\alpha t + \gamma) + (\omega - \omega_0)\sin(\tfrac{1}{2}\alpha t + \gamma)] \tag{12.21}$$

An ESR experiment prepares a very large number of electrons in the same state $\Psi(t)$, which is represented (according to (12.7)) by the two-dimensional vector with components $c_1(t)e^{-iE_1 t/\hbar}$ and $c_2(t)e^{-iE_2 t/\hbar}$. The population of one of the energy levels at any instant is proportional to the

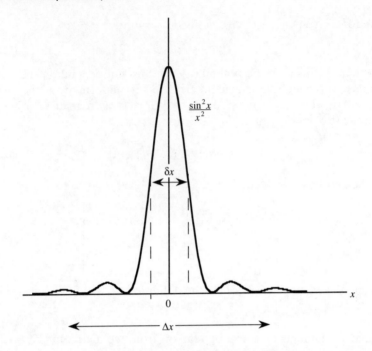

Figure 12.1 The form of the function $(\sin^2 x)/x^2$, indicating the width δx at half-height of the central maximum. The approximations made in the text in evaluating integrals involving this function are valid, provided the range of integration is at least the range Δx indicated on the diagram.

probability of finding an electron to have the energy appropriate to that particular level at that instant. So the populations of the two levels are determined by $|c_1|^2$ and $|c_2|^2$, and vary with time as indicated by (12.20) and (12.21). For example, if initially only the lower level $\chi_{-1/2}$ is populated then $c_1(0) = 0$ and $c_2(0) = 1$, and we can choose

$$\gamma = 0, \qquad C = -\mathrm{i}\,\frac{\omega_V}{\alpha} \qquad\qquad (12.22)$$

The populations of the two levels at any later time are determined by

$$|c_1(t)|^2 = \left(\frac{\omega_V}{\alpha}\right)^2 \sin^2 \tfrac{1}{2}\alpha t, \qquad |c_2(t)|^2 = 1 - |c_1(t)|^2 \qquad (12.23)$$

The expression for $|c_1(t)|^2$ can be written as the product of $\omega_V^2 t^2/4$ and a factor $(\sin^2 \tfrac{1}{2}\alpha t)/(\tfrac{1}{2}\alpha t)^2$, which is shown in Figure 12.1. There is a sharp maximum at $\tfrac{1}{2}\alpha t = 0$, and the subsidiary maxima are very small in comparison, so $|c_1(t)|^2$ is negligible unless αt is close to zero. From (12.20) we see that when $\omega = \omega_0$, α takes its minimum value ω_V, which is very much less than ω_0 provided the rotating field is very much weaker than

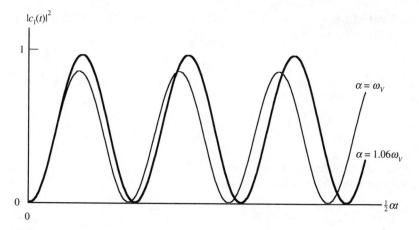

Figure 12.2 Oscillations in $|c_1(t)|^2$ (proportional to the population of the higher level) due to a harmonic perturbation on a two-level system, for $\alpha = \omega_V$ (resonant transition) and $\alpha = 1.06\omega_V$.

the constant field, $\mathcal{B}' \ll \mathcal{B}$ (see (12.11) and (12.15)). One of ω and ω_0 is kept constant, and the other varied until at resonance the two are equal, and there is a sharp increase in the average population of the upper level. From (12.23) this average is half the initial population of the lower level if the resonance condition is satisfied (since the average value of $\sin^2 \frac{1}{2}\alpha t$ is $\frac{1}{2}$). At resonance the probability of finding an electron in the upper level oscillates between 0 and 1, but off resonance the maximum probability will never reach 1, as shown in Figure 12.2

12.3 Time-dependent perturbation theory

Time-dependent perturbation theory provides the most generally useful method of solving the time-dependent Schrödinger equation. We follow a similar procedure to that introduced in Section 11.1 for time-independent perturbations, and rewrite the Hamiltonian (12.6) as

$$\hat{H} = \hat{H}_0 + \lambda \hat{V}(t) \quad (0 \leqslant \lambda \leqslant 1) \tag{12.24}$$

We then write a perturbation expansion for the solution to the time-dependent Schrödinger equation, but since we are solving the equation in the form of the set of equations (12.9), we expand the coefficients $c_n(t)$ rather than the wave function itself:

$$c_n(t) = c_n^{(0)} + \lambda c_n^{(1)} + \lambda^2 c_n^{(2)} + \ldots \qquad \textbf{(12.25)}$$

Substituting this expansion and the expression (12.24) for $\hat{H}$ in (12.9), we obtain (for each k)

$$i\hbar \frac{\mathrm{d}}{\mathrm{d}t} (c_k^{(0)} + \lambda c_k^{(1)} + \ldots) = \sum_n (c_n^{(0)} + \lambda c_n^{(1)} + \ldots)\lambda \hat{V}_{kn} \mathrm{e}^{\mathrm{i}\omega_{kn}t} \qquad \textbf{(12.26)}$$

As in the time-independent case, we now equate coefficients of each power of λ on either side of the equation:

$$i\hbar \frac{\mathrm{d}c_k^{(0)}}{\mathrm{d}t} = 0 \qquad \textbf{(12.27)}$$

$$i\hbar \frac{\mathrm{d}c_k^{(1)}}{\mathrm{d}t} = \sum_n \hat{V}_{kn}(t)\mathrm{e}^{\mathrm{i}\omega_{kn}t} c_n^{(0)}(t) \qquad \textbf{(12.28)}$$

$$i\hbar \frac{\mathrm{d}c_k^{(2)}}{\mathrm{d}t} = \sum_n \hat{V}_{kn}(t)\mathrm{e}^{\mathrm{i}\omega_{kn}t} c_n^{(1)}(t)$$

and in general the jth-order correction is determined by

$$i\hbar \frac{\mathrm{d}c_k^{(j)}}{\mathrm{d}t} = \sum_n \hat{V}_{kn}(t)\mathrm{e}^{\mathrm{i}\omega_{kn}t} c_n^{(j-1)}(t) \qquad \textbf{(12.29)}$$

Equation (12.27) shows that the zeroth-order coefficients are independent of time (as we should expect, since to this order we are neglecting the time-dependent perturbation $\hat{V}(t)$). If we assume that in the absence of the perturbation the particle would be in its ith energy eigenstate then

$$c_k^{(0)} = \delta_{ki} \quad (k = 1, 2, 3, \ldots) \qquad \textbf{(12.30)}$$

This means that (12.28) for the first-order correction reduces to

$$i\hbar \frac{\mathrm{d}c_k^{(1)}}{\mathrm{d}t} = \hat{V}_{ki}(t)\mathrm{e}^{\mathrm{i}\omega_{ki}t} \quad (k = 1, 2, 3, \ldots) \qquad \textbf{(12.31)}$$

This equation can now be integrated, once the zeroth-order energy levels and the form of the time-dependent perturbation are known. The second- and successive higher-order corrections can then be calculated using (12.29).

We are frequently interested in the effects produced by switching on a perturbation at some instant, which we shall call $t = 0$, and possibly switching it off again at some later time. The switching process itself makes the Hamiltonian time-dependent, but there may be further variations with time due to the time dependence of the perturbation $\hat{V}(t)$. We shall assume that we have Fourier-analysed the time-dependent potential, and consider only one Fourier component of the form

$$\hat{V}(t) = \begin{cases} 0 & (t < 0) \\ \hat{V}_0(r) \cos \omega t = \tfrac{1}{2}\hat{V}_0(r)(e^{i\omega t} + e^{-i\omega t}) & (t \geqslant 0) \end{cases} \quad \textbf{(12.32)}$$

(The spatial dependence of $\hat{V}(t)$ is determined by $\hat{V}_0$, as we have indicated explicitly by writing $\hat{V}_0(r)$.) The initial condition on the coefficients is that, long before the perturbation is switched on, the particle is in the ith eigenstate of $\hat{H}_0$, so that

$$c_k^{(1)}(t) \to 0 \quad \text{as } t \to -\infty \quad (k \neq i) \quad \textbf{(12.33)}$$

We can now integrate (12.31) from $t = -\infty$ to some time $t' > 0$:

$$c_k^{(1)}(t') = -\frac{i}{\hbar} \int_{-\infty}^{t'} dt\, \hat{V}_{ki}(t) e^{i\omega_{ki}t}$$

Using (12.32), we can write the matrix element of $\hat{V}(t)$ as

$$\hat{V}_{ki}(t) = \int_{\text{all space}} d^3r\, \psi_k^* \hat{V}_0(r) \cos \omega t\, \psi_i = \cos \omega t\, (\hat{V}_0)_{ki} \quad \text{(for } t \geqslant 0)$$

and so

$$c_k^{(1)}(t') = -\frac{i}{\hbar} \int_0^{t'} dt\, (\hat{V}_0)_{ki} \tfrac{1}{2}(e^{i(\omega_{ki}+\omega)t} + e^{i(\omega_{ki}-\omega)t})$$

Since $(\hat{V}_0)_{ki}$ is independent of t, this integration can be performed:

$$c_k^{(1)}(t') = -i\frac{(\hat{V}_0)_{ki}}{\hbar} \left[\frac{e^{i(\omega_{ki}+\omega)t'} - 1}{2i(\omega_{ki} + \omega)} + \frac{e^{i(\omega_{ki}-\omega)t'} - 1}{2i(\omega_{ki} - \omega)} \right]$$

$$= -i\frac{(\hat{V}_0)_{ki}}{\hbar} \left[e^{i(\omega_{ki}+\omega)t'/2} \frac{\sin\tfrac{1}{2}(\omega_{ki} + \omega)t'}{\omega_{ki} + \omega} \right.$$

$$\left. + e^{i(\omega_{ki}-\omega)t'/2} \frac{\sin\tfrac{1}{2}(\omega_{ki} - \omega)t'}{\omega_{ki} - \omega} \right] \quad \textbf{(12.34)}$$

Both terms in this equation contain factors of the form $(\sin x)/x$, with $x = \tfrac{1}{2}(\omega_{ki} \pm \omega)t'$, and like $(\sin^2 x)/x^2$ (Figure 12.1) this function is negligible except close to $x = 0$. If $\omega_{ki} + \omega \approx 0$, the term with $\omega_{ki} + \omega$ in the denominator takes its maximum value, and the other term is negligible in comparison. From (12.10) this resonance condition is $\hbar\omega \approx -\hbar\omega_{ki} = E_i - E_k$, so the energy of the initial state must be greater than that of the kth state, and the first term describes the emission of a unit of energy $\hbar\omega$. Similarly, if $\omega_{ki} - \omega \approx 0$, the first term in (12.34) is negligible compared with the second, and the second term describes the absorption of one unit of energy. The two resonance conditions are

$$\left. \begin{array}{ll} \hbar\omega \approx -\hbar\omega_{ki} = E_i - E_k & \text{(emission)} \\ \hbar\omega \approx \hbar\omega_{ki} = E_k - E_i & \text{(absorption)} \end{array} \right\} \quad \textbf{(12.35)}$$

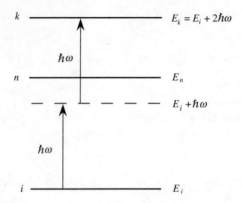

Figure 12.3 Schematic representation of a second-order transition $i \rightarrow k$ through an intermediate level n, produced by a harmonic perturbation with angular frequency ω. The arrows connect the initial or final level to a dashed line, rather than to the solid line indicating the intermediate level, in order to show that these transitions can be non-resonant (that is, the energy differences $E_n - E_i$ and $E_k - E_n$ need *not* be equal to $\hbar\omega$).

If we think of the time-dependent potential as transferring energy in units of $\hbar\omega$, these are just energy conservation requirements, equivalent to (1.4) which Bohr introduced as an assumption in his explanation of atomic spectra.

If either of the resonance conditions in (12.35) is satisfied, only one of the terms in (12.34) need be considered, and the probability of finding the particle in the kth eigenstate of $\hat{H}_0$ at time t' is

$$|c_k^{(1)}(t')|^2 = |(\hat{V}_0)_{ki}|^2 \frac{\sin^2 \frac{1}{2}(\omega - |\omega_{ki}|)t'}{\hbar^2 (\omega - |\omega_{ki}|)^2} \quad (k \neq i) \qquad \textbf{(12.36)}$$

This can be interpreted as a **transition probability**, and we think of a particle initially in the ith eigenstate of $\hat{H}_0$ making a transition to the kth through interacting with the time-dependent potential $\hat{V}(t)$. Since $\hat{V}(t)$ is supposed to be a weak perturbation, the probability of such a transition must be much less than one.

The set of coefficients $\{c_k^{(1)}\}$ gives the first-order correction to the state of the particle, and the matrix element $(\hat{V}_0)_{ki}$ provides a direct interaction between the ith and kth eigenstates of $\hat{H}_0$. The transition probability associated with the second-order correction involves products of matrix elements $(\hat{V}_0)_{kn}(\hat{V}_0)_{ni}$ (Problem 12.7), and can be visualized as arising from an indirect transition from the ith to the kth state through an intermediate state, as indicated schematically in Figure 12.3. All possible intermediate states must be taken into account, and in general each individual matrix element, $(\hat{V}_0)_{kn}$ or $(\hat{V}_0)_{ni}$, is not associated with a resonant transition – that is, energy need not be conserved in the interaction between the intermediate, nth, state and the ith or kth state, though it must be conserved overall in the transition from the ith state to the kth. The transition probability $|c_k^{(1)}(t')|^2$ is the probability for a **first-order transition**, and $|c_k^{(2)}(t')|^2$ is that for a **second-order transition**. An N**th-order transition** involves $N - 1$ intermediate states. Second- and higher-order transitions produce nonlinear optical effects [1], such as second-harmonic generation in which the frequency of light scattered by an atom is doubled.

12.4 Selection rules for electric dipole radiation spectra

As we saw in Section 12.3, the probability of a first-order transition from one state to another is very small unless one of the resonance conditions (12.35) is satisfied. However, even at resonance there is zero probability of the transition occurring if the relevant matrix element $(\hat{V}_0)_{ki}$ is zero. In practice, the symmetry properties of the perturbation make certain transitions impossible, and determine **selection rules** identifying which transitions can occur. We shall illustrate this by looking at the extremely important case of the electric dipole interaction of light with atoms, which gives rise to the most commonly observed atomic emission and absorption spectra. We have already used the electric dipole selection rules that we are about to derive in connection with molecular spectra (in Sections 4.7 and 5.3) and atomic spectra (in Section 8.2).

An electron in an atom interacts with the oscillating electric field $\mathcal{E}$ in a beam of light, and the potential representing the electric dipole part of the interaction is

$$\hat{V}(t) = e\mathbf{r} \cdot \mathcal{E}_0 \cos \omega t = \hat{V}_0 \cos \omega t \quad (t \geqslant 0) \tag{12.37}$$

$\mathcal{E}_0$ is independent of time, and we neglect its spatial dependence, which is reasonable if the wavelength of the light is very much greater than the dimensions of the atom in which the electron is confined. The potential (12.37) is identical to that of an electric dipole of moment $\mathcal{D} = -e\mathbf{r}$ in an electric field $\mathcal{E}(t) = \mathcal{E}_0 \cos \omega t$.

The polarization of the light determines the direction of the electric field, and we shall consider linear polarization in the z direction. The perturbation (12.37) is then of the form given in (12.32) with

$$\hat{V}_0 = e\mathcal{E}_0 z = e\mathcal{E}_0 r \cos \theta$$

This potential is not central, since it depends on the polar angle θ as well as on r, and so the spherical symmetry of the atom is broken, angular momentum is no longer conserved, and transitions can occur (as we shall show) between states with different orbital angular momentum quantum numbers. Consider an electron in a hydrogen atom initially in a state with orbital angular momentum quantum numbers (l, m_l). Its wave function is of the form $R_{nl}(r)Y_{lm_l}(\theta, \phi)$, and the matrix element of $\hat{V}_0$ between the initial state and the state with quantum numbers (n', l', m_l') is

$$(\hat{V}_0)_{n'l'm_l'nlm_l}$$

$$= \int_0^\infty r^2\,dr \int_0^{2\pi} d\phi \int_{-1}^1 d(\cos\theta)\, R_{n'l'}(r)Y^*_{l'm_l'}(\theta, \phi)e\mathcal{E}_0 r\cos\theta R_{nl}(r)Y_{lm_l}(\theta, \phi)$$

$$= e\mathcal{E}_0 \int_0^\infty dr\, R_{n'l'}(r)r^3 R_{nl}(r)\int_0^{2\pi} d\phi \int_{-1}^1 d(\cos\theta)\, Y^*_{l'm_l'}(\theta, \phi)\cos\theta\, Y_{lm_l}(\theta, \phi)$$

According to the expression for the spherical harmonics in (4.45), this matrix element is proportional to

$$\int_0^{2\pi} d\phi\, e^{-i(m_l'-m_l)\phi} \int_{-1}^1 d(\cos\theta)\, P_{l'}^{m_l'}(\cos\theta)\cos\theta\, P_l^{m_l}(\cos\theta)$$

The integral over ϕ is equal to $[-e^{-i(m_l'-m_l)\phi}/i(m_l'-m_l)]_0^{2\pi}$, which vanishes unless $m_l' = m_l$. The integration over $\cos\theta$ can be performed with the help of the first recurrence relation given in Table 4.1 for the associated Legendre functions. Putting $m_l' = m_l$, we obtain

$$\int_{-1}^1 d(\cos\theta)\, P_{l'}^{m_l}(\cos\theta)\cos\theta\, P_l^{m_l}(\cos\theta)$$

$$= \int_{-1}^1 d(\cos\theta)\, P_{l'}^{m_l}(\cos\theta)\left(\frac{l-|m_l|+1}{2l+1}\,P_{l+1}^{m_l} + \frac{l+|m_l|}{2l+1}\,P_{l-1}^{m_l}\right)$$

which is zero unless $l' = l \pm 1$ because the associated Legendre functions form an orthogonal set. Therefore the matrix element is zero unless

$$\Delta l = l' - l = \pm 1, \qquad \Delta m_l = m_l' - m_l = 0$$

These are the selection rules that tell us which transitions can occur when the atom interacts with light linearly polarized in the z direction through a dipole interaction of the form given in (12.37).

If we consider light that is linearly polarized in the x or y direction, the integral over ϕ vanishes unless $m_l' = m_l \pm 1$, and the integration over θ gives the same selection rule as before (Problem 12.8). We have neglected the spin of the electron, but in fact the spin state cannot be changed by a potential of the form (12.37). The complete set of selection rules for electric dipole transitions is therefore

$$\left.\begin{array}{l} \Delta l = l' - l = \pm 1 \\[4pt] \Delta m_l = m_l' - m_l = \pm 1, 0 \\[4pt] \Delta m_s = m_s' - m_s = 0 \end{array}\right\} \qquad \textbf{(12.38)}$$

Only the orbital angular momentum state of the electron was significant in deriving these selection rules, and so they apply to any particle described by a Hamiltonian $\hat{H}_0$ that is spherically symmetric (so that the eigenfunctions of $\hat{H}_0$ are simultaneously orbital angular momentum eigenfunctions). They apply, for example, to the rigid rotator described in Section 4.7, and we used the selection rule $\Delta l = \pm 1$ when discussing the rotational spectra of linear molecules in Sections 4.7 and 5.3.

12.5 Transition rates and Fermi's golden rule

Equation (12.36) for the probability of a first-order transition from the ith to the kth energy level was derived for a potential oscillating at a unique angular frequency ω, and refers to states with precisely defined energies E_i and E_k. In practice, the perturbation will contain a spread of frequencies – for example, a beam of light can never be completely monochromatic. In addition, there is a spread in the energy of any excited state, associated with the finite lifetime of that state, as we saw in Section 1.5. This produces a spread in the values of the energy difference $E_k - E_i$. Rather than attempt to deal with both these effects simultaneously, we shall consider the two extreme cases, in each of which one effect is negligible compared with the other.

First we shall assume that the uncertainties in the eigenvalues of $\hat{H}_0$ produce a spread in $\omega_{ki} = (E_k - E_i)/\hbar$ that is very much less than the spread in the frequency ω of the perturbation. This is often a good approximation in the case of stimulated electric dipole emission and absorption of light by atoms. We use (12.37) for the electric dipole interaction of an atomic electron with the electric field in the light, and find

$$|(\hat{V}_0)_{ki}|^2 = |e\mathscr{E}_0(\omega) \cdot \boldsymbol{r}_{ki}|^2 = |\mathscr{E}_0(\omega)|^2 |\hat{\mathscr{D}}_{kl}(\omega)|^2$$

Here $\hat{\mathscr{D}}_{ki}(\omega)$ is a matrix element of the component parallel to the electric field of the dipole moment operator for the atomic electron, $\hat{\mathscr{D}} = -e\boldsymbol{r}$. For monochromatic light of frequency ω the average energy density is [2]

$$\mathscr{E}(\omega) = \frac{1}{2\mu_0 c^2} |\mathscr{E}_0(\omega)|^2 = \tfrac{1}{2}\varepsilon_0 |\mathscr{E}_0(\omega)|^2$$

This can be used to eliminate the electric field $\mathscr{E}_0$ in our expression for $|(\hat{V}_0)_{ki}|^2$. Then (12.36) gives the first-order transition probability, for monochromatic light, as

$$|c_k^{(1)}(t)|^2 = \frac{2}{\varepsilon_0} \mathscr{E}(\omega)|\hat{\mathscr{D}}_{ki}(\omega)|^2 \frac{\sin^2 \tfrac{1}{2}(\omega - |\omega_{ki}|)t}{\hbar^2(\omega - |\omega_{ki}|)^2} \tag{12.39}$$

In the case where the incident light is not monochromatic the probability of a transition associated with light in the frequency range between ω and $\omega + d\omega$ is $|c_k^{(1)}(t)|^2 \, d\omega$, which can be obtained from (12.39) if $\mathscr{E}(\omega)$ is replaced by $\rho(\omega) \, d\omega$, the energy density in the frequency range between ω and $\omega + d\omega$. (The corresponding intensity is the energy in this frequency range crossing unit area perpendicular to the beam direction per unit time, or $c\rho(\omega) \, d\omega$, where c is the velocity of light.) The total probability for a first-order transition between the ith and kth states is obtained by integrating over $\Delta\omega$, the complete range of frequencies in the beam:

$$\int_{\Delta\omega} d\omega \, |c_k^{(1)}(t)|^2 = \int_{\Delta\omega} d\omega \, \frac{2}{\varepsilon_0} \, \rho(\omega) |\hat{\mathcal{D}}_{ki}(\omega)|^2 \, \frac{\sin^2 \frac{1}{2}(\omega - |\omega_{ki}|)t}{\hbar^2(\omega - |\omega_{ki}|)^2}$$

$$= \frac{t}{\varepsilon_0 \hbar^2} \int_{\Delta x} dx \, \rho(x) |\mathcal{D}_{ki}(x)|^2 \, \frac{\sin^2 x}{x^2}$$

where we have made the substitution

$$x = \tfrac{1}{2}t(\omega - |\omega_{ki}|), \qquad \Delta x = \tfrac{1}{2}t \, \Delta\omega \tag{12.40}$$

Because of the factor $(\sin^2 x)/x^2$, the integrand is negligibly small outside a narrow region of width δx centred on $x = 0$. We shall assume that the range Δx is much greater than δx, as indicated in Figure 12.1, so that it is a reasonable approximation to make the range of integration infinite. Although both the energy density of the light and the electric dipole matrix element are functions of x, they vary only slowly within the region of width δx, and can be approximated by their values exactly at resonance. We then obtain

$$\int_{\Delta\omega} d\omega \, |c_k^{(1)}(t)|^2 = t \, \frac{1}{\varepsilon_0 \hbar^2} \, \rho(\omega_{ki}) |\hat{\mathcal{D}}_{ki}(\omega_{ki})|^2 \int_{-\infty}^{\infty} dx \, \frac{\sin^2 x}{x^2}$$

$$= t \, \frac{\pi}{\varepsilon_0 \hbar^2} \, \rho(\omega_{ki}) |\hat{\mathcal{D}}_{ki}(\omega_{ki})|^2 \tag{12.41}$$

where we have used

$$\int_{-\infty}^{\infty} dx \, \frac{\sin^2 x}{x^2} = \pi \tag{12.42}$$

The total first-order transition probability in (12.41) is proportional to the time t for which the perturbation has been operating, so that the corresponding transition rate w_{ki} is constant:

$$w_{ki} = \frac{d}{dt} \int_{\Delta\omega} d\omega \, |c_k^{(1)}(t)|^2$$

$$= \frac{\pi}{\varepsilon_0 \hbar^2} \, \rho(\omega_{ki}) |\hat{\mathcal{D}}_{ki}(\omega_{ki})|^2 \tag{12.43}$$

Note that this transition rate is zero when the light intensity (proportional to $\rho(\omega_{ki})$) is zero. This is not surprising in the case of absorption, but in the case of emission the initial state must be an excited state, and we know that an atom in an excited state can decay in the absence of any external radiation, at a rate characterized by its lifetime. Such emission is **spontaneous emission**, while the emission described by (12.43) is **stimulated emission**. In non-relativistic quantum mechanics we treat electromagnetic radiation classically, but spontaneous emission can only be properly treated by a quantized theory of radiation.

Now we shall examine the opposite approximation, that the perturbation is monochromatic, but there is a spread $\Delta\omega_{ki}$ in the angular frequencies ω_{ki}, due to a spread in the energy of one or both of the levels concerned. For simplicity, let us assume that $\Delta\omega_{ki}$ is due to a spread in the energy E_k of the final state, so that it should be treated as a distribution of states with density $g(E_k)$. That is, the number of states with precise energy between E_k and $E_k + dE_k$ is $g(E_k)\,dE_k$. From (12.36), the total probability for a first-order transition from the initial state to one of this set of final states is therefore

$$\int_{\Delta E} dE_k\, g(E_k)|c_k^{(1)}(t)|^2 = \int_{\Delta\omega_{ki}} \hbar\,d\omega_{ki}\, g(\omega_{ki})|(\hat{V}_0)_{ki}|^2 \frac{\sin^2\frac{1}{2}(\omega - |\omega_{ki}|)t}{\hbar^2(\omega - |\omega_{ki}|)^2}$$

The variable can be changed to $x = \frac{1}{2}(\omega - |\omega_{ki}|)t$ as before, and once again the integrand is essentially zero everywhere except in a narrow region of width δx centred on $x = 0$. The density of states and the matrix element can be taken outside the integral and given their values at $\omega_{ki} = \omega$, and, provided $\delta x \ll \Delta x = \frac{1}{2}t\,\Delta\omega_{ki}$, the range of integration can be extended to $(-\infty, \infty)$. With the value for the integral given in (12.42) we again find that the total transition probability increases linearly with time, and the transition rate is

$$w_{ki} = \frac{\pi}{2\hbar}\, g(E_k)|(\hat{V}_0)_{ki}|^2 = \frac{2\pi}{\hbar}\, g(E_k)|\tfrac{1}{2}(\hat{V}_0)_{ki}|^2 \qquad \textbf{(12.44)}$$

where the matrix element and the density of states are to be evaluated at $E_k = E_i \pm \hbar\omega$ (depending on whether we have absorption or emission). The second form on the right-hand side of this equation reminds us that the transition rate has been calculated using only one of the terms in the potential (12.32): either $\frac{1}{2}\hat{V}_0 e^{i\omega t}$, corresponding to energy emission, or $\frac{1}{2}\hat{V}_0 e^{-i\omega t}$, corresponding to absorption. Equation (12.44) is known as **Fermi's golden rule**.

In our derivations of both (12.44) and (12.43) we made the assumption that $\Delta x \gg \delta x$, or in other words that the range of integration was much greater than the width of the central peak of the function $(\sin^2 x)/x^2$. This is not true for very short times, since $\Delta x = \frac{1}{2}t\,\Delta\omega_{ki}$, while δx is fixed, and for times short enough that $\sin x \approx x$ the transition rate increases linearly with time instead of being constant.

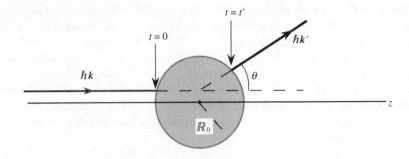

Figure 12.4 Scattering of a particle of momentum $\hbar k$ by a potential of finite range $\mathbb{R}_0$. The particle enters the interaction region at time $t = 0$ and leaves it at time t', with momentum $\hbar k'$. The angle between the final and initial directions of the momentum is the scattering angle θ.

12.6 High-energy elastic scattering by a finite-range potential

Consider a particle approaching a target with which it interacts through some force that depends on the separation r of the particle from the centre of the target, and suppose that the particle does not feel the effect of the target until it comes within a distance $\mathbb{R}_0$ of its centre, as shown in Figure 12.4. That is, the potential in which the incident particle moves due to its interaction with the target is zero outside a sphere of radius $\mathbb{R}_0$ centred on the origin. The potential is said to be of **finite range** provided $\mathbb{R}_0$ is finite, and in this case the particle can be treated as free when it is outside the interaction region. (This is not true for the Coulomb interaction, for which $\mathbb{R}_0$ is infinite, but it is a good approximation in the case of the nuclear interaction, with $\mathbb{R}_0 \approx 10^{-15}$ m). From the point of view of the incident particle, there is an instant, which we label $t = 0$, at which it enters the region where it interacts with the target and the potential is effectively switched on. It is switched off again when the particle emerges from the interaction region, at $t = t'$ say. So the interaction potential is time-dependent, and, provided it is sufficiently small, the effect of the target on the particle can be found using first-order time-dependent perturbation theory. In this context 'sufficiently small' means that the potential energy is very much less than the zeroth-order energy of the particle, which is its initial kinetic energy. So the perturbation treatment will be a good approximation for scattering at sufficiently high incident energies.

The situation we have just described is a scattering event, such as we discussed in Section 2.5 and illustrated in Figure 2.2. The particle is one of an incident beam of free particles, all prepared in the same eigenstate of energy and momentum, and, assuming the beam direction is along the z axis, the wave function describing this initial state is of the form (2.9):

$$\Psi_i(z, t) = \sqrt{(N)}\,e^{i(kz-\omega t)} = \sqrt{(N)}\,e^{i(k \cdot r-\omega t)} \tag{12.45}$$

where the momentum and energy eigenvalues are

$$p_i = \hbar k, \qquad E_i = \hbar\omega = \frac{\hbar^2 k^2}{2m} \tag{12.46}$$

(The normalization constant was identified in (2.12) by integrating the probability density over a finite volume V containing, on average, one particle, and if the incident beam has density N then $V = 1/N$.) The incident beam is scattered by the target, and the total and differential cross-sections, defined by (2.14) and (2.16), are measured experimentally. The differential cross-section is determined by the probability per unit time that a particle is scattered in a particular direction (θ, ϕ). This means that we are interested in transitions from the initial state to a final state in which the particle is once again free (beyond the range of the interaction with the target) and has momentum $p_f = \hbar k'$ in the direction (θ, ϕ). This final state is an energy and momentum eigenstate described by a wave function of similar form to the initial state wave function (12.45):

$$\Psi_f(x, y, z, t) = \sqrt{(N)}\,e^{i(k' \cdot r-\omega' t)} \tag{12.47}$$

with

$$E_f = \hbar\omega' = \frac{\hbar^2 k'^2}{2m} \tag{12.48}$$

Since the final state wave function is assumed to refer to the same single particle as the incident wave function, the normalization constant must be the same. The energy and momentum of a free particle both have continuous spectra, so each final state is one of a continuous set of states, and the transition rate in which we are interested can be found using Fermi's golden rule (12.44).

We shall assume the simplest possible case, in which the only time dependence of the potential is that produced by its being switched on at time $t = 0$ and off at time $t = t'$:

$$\hat{V}(t) = \begin{cases} 0 & (t < 0, t > t') \\ \hat{V}_0(r, \theta, \phi) & (0 \leqslant t \leqslant t') \end{cases} \tag{12.49}$$

This is a special case of the harmonically varying potential (12.32), with angular frequency $\omega = 0$. So the first-order amplitude for a transition from a unique initial state Ψ_i to a particular final state Ψ_f can be found from (12.34) by putting $\omega = 0$:

$$c_f^{(1)}(t') = -2i \frac{(\hat{V}_0)_{fi}}{\hbar} e^{i\omega_{fi}t'/2} \frac{\sin\frac{1}{2}\omega_{fi}t'}{\omega_{fi}} \qquad (12.50)$$

(For times $t > t'$ there can be no further transitions, and $c_f^{(1)}(t) = c_f^{(1)}(t')$.) The two terms in (12.34) are equal when $\omega = 0$, so that (12.50) contains only one term and an additional factor 2. Because of the factor of the form $(\sin x)/x$, the transition amplitude is negligible unless

$$\hbar\omega_{fi} = E_f - E_i \approx 0 \qquad (12.51)$$

This means that there is no net exchange of energy with the target, so that the scattering caused by a perturbation with the time dependence (12.49) is elastic: $E_f \approx E_i$, and $k' = k$.

The transition probability $|c_f^{(1)}(t)|^2$ calculated from (12.50) contains a factor of the form $(\sin^2 x)/x^2$ with $x = \frac{1}{2}\omega_{fi}t'$, similar to that given by (12.36) in the case of a harmonically varying perturbation. So if there is a distribution of final states with density $g(E_f)$, we can calculate the total probability for a transition to one of them, and find a constant transition rate, just as we did in deriving Fermi's golden rule (12.44). We obtain

$$w_{fi} = \frac{2\pi}{\hbar} g(E_f) |(\hat{V}_0)_{fi}|^2 \qquad (12.52)$$

where the density of states and the matrix element of the perturbation are evaluated at $E_f = E_i$. The density of states for a single free particle can be found by the method that we used to determine the density of states in a solid in Section 7.6. We have to treat the particle as if it moved in a very large but finite volume $\mathbb{V}$ and apply periodic boundary conditions similar to (7.39). In fact, we have already assumed that the particle exists in a finite volume $\mathbb{V} = 1/N$ in order to normalize the free particle wave function. The number of different momentum states in the interval d^3k is $dn(k)$, given by (7.44), and for a spinless particle the density of states in volume $\mathbb{V} = 1/N$ is

$$g(E_f) = \frac{dn(k)}{dE} = \frac{\mathbb{V}}{(2\pi)^3} k^2 d\Omega \frac{dk}{dE_f} = \frac{1}{N(2\pi)^3} k^2 d\Omega \frac{dk}{dE_f}$$

(The density of states (7.45) is defined in a unit volume, $\mathbb{V} = 1$, and contains an extra factor 2 to take account of the electron spin.) The derivative dk/dE_f can be found from (12.48) with $k' = k$, and the density of states becomes

$$g(E_f) = \frac{1}{N(2\pi)^3} \frac{m}{\hbar^2} k \, d\Omega \qquad (12.53)$$

Using (12.45) and (12.47) for the initial and final state wave functions, we find that the matrix element of the interaction potential is

$$(\hat{V}_0)_{\text{fi}} = \int_{\text{all space}} d^3r \, N e^{i(k'-k)\cdot r} \hat{V}_0(r, \theta, \phi) \tag{12.54}$$

So the probability per unit time that a particle makes a transition from the initial (incident) state to a final state in which the momentum direction lies within the element of solid angle $d\Omega$ is, from (12.52), (12.53) and (12.54),

$$w_{\text{fi}} = \frac{1}{(2\pi)^2} \frac{m}{\hbar^3} k \, d\Omega \, N \left| \int_{\text{all space}} d^3r \, e^{i(k'-k)\cdot r} \hat{V}_0(r, \theta, \phi) \right|^2$$

According to the definition (2.16), the differential cross-section is

$$\frac{d\sigma}{d\Omega} = \frac{1}{\text{incident flux}} \frac{w_{\text{fi}}}{d\Omega}$$

and the incident flux is given by (2.13). The final expression for the differential cross-section for the elastic scattering of a spinless particle, at incident energies high enough for first-order perturbation theory to be valid, is therefore

$$\frac{d\sigma}{d\Omega} = \frac{m^2}{(2\pi)^2\hbar^4} \left| \int_{\text{all space}} d^3r \, e^{i(k'-k)\cdot r} \hat{V}_0(r, \theta, \phi) \right|^2 \tag{12.55}$$

The integral in equation (12.55) is the Fourier transform of the perturbing potential. Since the scattering is elastic, the magnitude of the momentum is unchanged, $\hbar|k| = \hbar|k'| = \hbar k$, but its direction has changed, and $\hbar q = \hbar(k' - k)$ is the momentum transferred to the particle by the target. If the angle through which the particle has been scattered is θ then

$$q^2 = (k' - k)^2 = |k'|^2 + |k|^2 - 2k' \cdot k = 2k^2(1 - \cos\theta)$$
$$= 4k^2 \sin^2\tfrac{1}{2}\theta \tag{12.56}$$

In the case of a spherically symmetric potential the integration over the angular variables can be performed, giving

$$\int_{\text{all space}} d^3r \, e^{i(k'-k)\cdot r} \hat{V}_0(r) = \int_0^\infty r^2 \, dr \, \hat{V}_0(r) 2\pi \int_{-1}^{1} d(\cos\theta') \, e^{iqr\cos\theta'}$$
$$= 4\pi \int_0^\infty r^2 \, dr \, \hat{V}_0(r) \frac{\sin qr}{qr} \tag{12.57}$$

A typical differential cross section for scattering at high incident energies is shown in Figure 12.5. The factor $(\sin qr)/qr$ in the integral in (12.57) is responsible for the large peak at $\theta = 0$ and the oscillations in $d\sigma/d\Omega$ as the scattering angle θ varies. These oscillations are diffraction effects, due to the wave-like nature of the scattered particles, and this type of scattering is known as **diffraction scattering**.

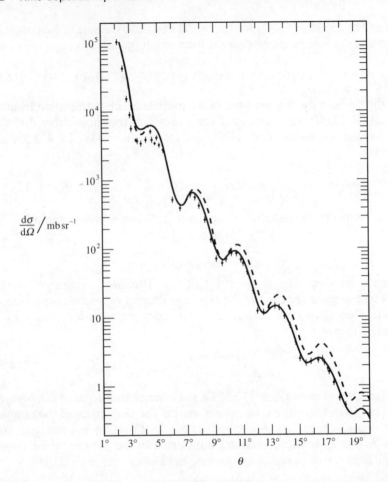

$$\frac{\mathrm{d}\sigma}{\mathrm{d}\Omega} \Big/ \mathrm{mb\,sr}^{-1}$$

θ

Figure 12.5 The differential cross-section for the elastic scattering of protons with incident energy 1050 MeV by a lead nucleus [3]. The scattering angle θ refers to the centre-of-mass frame, in which the centre of mass of the proton–nucleus system is at rest.

References

[1] For a popular review see Giordmaine J.A. (1964). Nonlinear optics. *Scientific American* **210** (4) 38. For further information see Baldwin G.C. (1969). *An Introduction to Nonlinear Optics*. New York: Plenum.
[2] Lorrain P., Corson D.P. and Lorrain F. (1988). *Electromagnetic Fields and Waves* 3rd edn, Section 28.3. San Francisco: W.H. Freeman.
[3] Igo G. (1975) In *High Energy and Nuclear Structure* (Nagle D. *et al.*, eds). New York: American Institute of Physics.

Problems

12.1 (a) For a particle whose motion is described by a Hamiltonian of the form

$$\hat{H} = \frac{1}{2m}\, \hat{P}_x^2 + \hat{V}(x)$$

prove that

$$\frac{d}{dt}\langle \hat{X} \rangle = \frac{1}{m}\langle \hat{P}_x \rangle$$

(b) A rigid rotator (with Hamiltonian (4.47)) is subjected to a constant perturbation $\hat{V} = V_0 \cos\theta$. Show that the expectation value of the z component of orbital angular momentum is constant, and find the rate of change of the expectation value of $\hat{L}^2$.

(c) If the same rotator is perturbed by the explicitly time-dependent potential $\hat{V} = V_0 \cos\theta \cos\omega t$, how are the results of (b) changed?

12.2 A very slowly increasing magnetic field in the z direction is applied to an atom in an energy eigenstate that is also an eigenstate of orbital angular momentum and of spin, with quantum numbers $L = 1$, $S = 0$. Sketch the behaviour of the energy eigenvalues with time. (*Hint*: Refer back to Section 8.2.)

12.3 Show that, if the expansion coefficients c_n in (12.7) do *not* depend on time, the expectation value of the energy in the state $\Psi(x, y, z, t)$ (which is not, in general, an energy eigenstate) is constant in time.

12.4 (a) Work through the derivation of the form (12.9) of the time-dependent Schrödinger equation, starting from the form (12.1).

(b) Express (12.9) in Dirac notation.

(c) Write out the set of equations represented by (12.9) in full for a two-level system, and show that it is the set given in (12.17). Similarly, write out in full the set of equations for a three-level and for a four-level system.

12.5 Derive (12.19) from (12.18), and show that the form of c_1 given in (12.20) is a solution. In a similar way, derive a second-order differential equation for c_2 from (12.18), and show that (12.21) is a solution to it.

12.6 The energies of two non-degenerate levels of an atom are E_1 and E_2. If these levels are coupled by an interaction $\hat{V}_0 \cos\omega t$, and ω is close to the resonant frequency, find the populations of the two levels at time t, assuming that they are equal initially. Show that the state of an atom oscillates between the two levels (these are **Rabi oscillations**), and calculate the frequency of this oscillation at resonance (the **Rabi frequency**).

12.7 A particle initially in the ith eigenstate of the unperturbed Hamiltonian $\hat{H}_0$ is subject to a time-dependent perturbation $\hat{V}(t) = \frac{1}{2}\hat{V}_0 e^{i\omega t}$ that is switched on at time $t = 0$. Show that the second order amplitude for the transition $i \rightarrow k$ is

$$c_k^{(2)}(t) = \left(\frac{1}{\hbar}\right)^2 \sum_n \frac{(\hat{V}_0)_{kn}(\hat{V}_0)_{ni}}{4(\omega + \omega_{ni})} \left(\frac{e^{i(2\omega + \omega_{kn} + \omega_{ni})t} - 1}{2\omega + \omega_{kn} + \omega_{ni}} - \frac{e^{i(\omega + \omega_{kn})t} - 1}{\omega + \omega_{kn}}\right)$$

Assuming that $(\hat{V}_0)_{ki} = 0$, and that $\omega + \omega_{kn} \neq 0$, show that the probability for the second-order transition is

$$|c_k^{(2)}(t)|^2 = \left|\sum_n \frac{(\hat{V}_0)_{kn}(\hat{V}_0)_{ni}}{2\hbar(\omega + \omega_{ni})}\right|^2 \frac{\sin^2\frac{1}{2}(2\omega + \omega_{ki})t}{\hbar^2(2\omega + \omega_{ki})^2}$$

and that this is negligible unless $2\omega + \omega_{ki} \approx 0$.

12.8 Prove the selection rules $\Delta l = \pm 1$, $\Delta m_l = \pm 1$ for an electric dipole interaction with light linearly polarized in the x or y direction. What are the corresponding selection rules for circularly polarized light?

12.9 A time-dependent perturbation $\hat{V}(t) = V_0 e^{i\omega t}$, where V_0 is a constant, acts on an electron in a Bloch state in a one-dimensional periodic lattice of lattice constant L. Prove the selection rule that the crystal momentum can only change by a reciprocal lattice vector (7.43). That is, prove that $\Delta k = k' - k = 0$ or an integer multiple of $2\pi/L$. (*Hint*: Since the periodic part of the Bloch function, $u_k(x)$, has the periodicity of the lattice, it can be expanded in a complex Fourier series $u_k(x) = L^{-1/2}\sum_{n=-N}^{N} c_n e^{2\pi i n x/L}$.)

12.10 The **Einstein A and B coefficients** are defined by

$$w_{ik} = B_{ik}\rho(\nu)N_i$$

$$w_{ki} = [A_{ki} + B_{ki}\rho(\nu)]N_k$$

where w_{ik} is the rate at which atoms are excited from the ith to the kth level, w_{ki} is the rate at which they decay back to the ith level, $\rho(\nu)\,d\nu$ is the energy density of radiation present in the frequency range $(\nu, \nu + d\nu)$, and N_i and N_k are the numbers of atoms in the two levels. The B coefficients refer to stimulated emission and absorption, and the A coefficient to spontaneous emission. For electric dipole transitions prove that

$$B_{ik} = \frac{\pi}{3\varepsilon_0\hbar^2} |\hat{\mathcal{D}}_{ki}(\omega_{ki})|^2 = B_{ki}$$

and find an expression for the rate of spontaneous emission by the atom. (Note that the A and B coefficients are defined in the case where the direction of the field in the radiation is randomly oriented.)

12.11 Show that, at very short times t after a time-dependent perturbation is switched on, the transition probability given by (12.36) is proportional to t^2, and hence the transition rate increases linearly with time (*Hint*: Expand the sine function in a power series.)

12.12 A constant perturbation is switched on at time $t = 0$:

$$\hat{V}(t) = \begin{cases} \hat{V}_0(r) & (t \geqslant 0) \\ 0 & (t < 0) \end{cases}$$

Substitute this potential in (12.31), and integrate the equation, applying the initial conditions (12.33), to find the probability of a first-order transition between the ith and kth eigenstates of the unperturbed Hamiltonian $\hat{H}_0$.

12.13 Calculate the differential cross-section (as a function of the scattering angle θ) for the scattering of a particle by the potential

$$\hat{V}(r) = V_0 \frac{e^{-r/r_0}}{r}$$

at high incident energies, where V_0 is a constant. (This is the **Yukawa potential** representing the nuclear interaction between two nucleons, which has range r_0.) Integrate your result over all directions to obtain the total cross-section for the scattering.

The Yukawa potential becomes the Coulomb potential in the limit as $r_0 \to \infty$. Show that in this limit your expression for the differential cross-section becomes that for Rutherford scattering, which is inversely proportional to $\sin^4 \frac{1}{2}\theta$.

CHAPTER 13

Many-Particle Systems

13.1 The wave function for a system of non-interacting particles

When we introduced the wave function in Chapter 2 we stressed that it described the state of a single particle. However, most of the real physical systems in which we are interested, such as atoms and molecules, solids, and nuclei, contain many particles. We tried to understand a little about the electronic structure of atoms and solids (in Sections 7.3 and 7.6 respectively) through independent-particle models in which each electron is treated as if it moves independently in a static potential representing the average effect of all the other particles in the system. But we must also be able to describe the state of the system as a whole – we need a wave function that determines the probability of simultaneously finding particle 1 at position (x_1, y_1, z_1), particle 2 at (x_2, y_2, z_2), particle 3 at (x_3, y_3, z_3), and so on for all the particles in the system. That is, we need a wave function whose modulus squared gives a **joint probability distribution** for all the particles in the system. Such a wave function must clearly be a function not just of one set of position coordinates, but of position coordinates for each particle, so for an N-particle system it is a function of $3N$ coordinates: $\Psi(x_1, y_1, z_1; x_2, y_2, z_2; \ldots; x_N, y_N, z_N)$.

In this chapter we are concerned with the energy eigenvalues and eigenfunctions of a system of particles, which can be found, in principle, by solving a Schrödinger equation of the form

$$\hat{H}\,\Psi(\mathbf{r}_1, \mathbf{r}_2, \ldots, \mathbf{r}_N) = E\,\Psi(\mathbf{r}_1, \mathbf{r}_2, \ldots, \mathbf{r}_N) \qquad (13.1)$$

(In order to make the notation less cumbersome, we have used $\mathbf{r}_i$ instead of x_i, y_i, z_i to indicate the dependence of the wave function on the position of the ith particle.) The Hamiltonian $\hat{H}$ must include a kinetic energy operator for each particle, terms accounting for the effects of any external forces on the system, and terms representing the interactions of all the particles with one another. Clearly it is not possible to solve such a

Schrödinger equation, in general, without making some approximations. However, if the particles in the system do not interact with one another, the many-particle wave function $\Psi(r_1, r_2, \ldots, r_N)$ can be expressed as a product of single-particle wave functions, and the total energy of the system is the sum of the energies of the individual particles. We shall demonstrate this in the case of a system of two non-interacting particles.

Consider two particles that do not interact with one another, but move in potentials $\hat{V}_1$ and $\hat{V}_2$ respectively, representing their interactions with a fixed particle, or system of particles, external to the two-body system. For non-interacting particles the potential $\hat{V}_1$ in which particle particle 1 moves depends only on the position r_1 of particle 1, and is totally independent of the position r_2 of particle 2. Similarly, $\hat{V}_2$ depends only on r_2. So the Hamiltonian for the two-particle system is of the form

$$\hat{H} = -\frac{\hbar^2}{2m_1} \nabla_1^2 + \hat{V}_1(r) - \frac{\hbar^2}{2m_2} \nabla_2^2 + \hat{V}_2(r_2) \qquad (13.2)$$

where m_i is the mass and $-i\hbar\nabla_i$ the momentum operator for particle $i = 1$ or 2. For example, the two particles could be the electrons in a helium atom in an approximation where we neglect the electron–electron repulsion. Each electron is attracted to the nucleus (which carries charge $+2e$) through a potential of the form

$$\hat{V}_i = -\frac{2e^2}{4\pi\varepsilon_0} \frac{1}{r_i} \qquad (i = 1, 2)$$

where r_i is the distance of electron i from the nucleus. The Hamiltonian (13.2) is the sum of two terms, each of which is the Hamiltonian of a single particle:

$$\left.\begin{aligned} \hat{H} &= \hat{H}_1 + \hat{H}_2 \\ \hat{H}_i &= -\frac{\hbar^2}{2m_i} \nabla_i^2 + \hat{V}_i(r_i) \quad (i = 1, 2) \end{aligned}\right\} \qquad (13.3)$$

A function of x_2, y_2 and z_2 behaves like a constant under the action of the operator ∇_1 with components $(\partial/\partial x_1, \partial/\partial y_1, \partial/\partial z_1)$, and similarly a function of x_1, y_1 and z_1 behaves like a constant under the action of the operator ∇_2. Therefore the single-particle Hamiltonians (13.3) commute with one another and with the complete Hamiltonian $\hat{H}$:

$$[\hat{H}_1, \hat{H}_2] = 0, \qquad [\hat{H}, \hat{H}_i] = 0 \quad (i = 1, 2)$$

This means that an eigenstate of $\hat{H}$ can be simultaneously an eigenstate of both $\hat{H}_1$ and $\hat{H}_2$. In the same way that we expressed the simultaneous eigenfunctions of energy and orbital angular momentum in product form, (7.4), we can write a simultaneous eigenfunction of $\hat{H}_1$ and $\hat{H}_2$ as the product

$$\Psi(r_1, r_2) - \psi^{(1)}(r_1)\psi^{(2)}(r_2) \qquad (13.4)$$

where, for $i = 1, 2$, $\psi^{(i)}(r_i)$ is an eigenfunction of $\hat{H}_i$ corresponding to the eigenvalue $E^{(i)}$:

$$\hat{H}_i \psi^{(i)}(r_i) = E^{(i)} \psi^{(i)}(r_i) \quad (i = 1, 2) \tag{13.5}$$

The eigenfunction $\psi^{(1)}(r_1)$ behaves like a constant under the action of the operator $\hat{H}_2$, and similarly $\psi^{(2)}(r_2)$ behaves like a constant when acted on by $\hat{H}_1$. So, using the complete Hamiltonian $\hat{H} = \hat{H}_1 + \hat{H}_2$ on the product wave function (13.4), we obtain (with the help of (13.5))

$$\hat{H}\Psi(r_1, r_2) = (\hat{H}_1 + \hat{H}_2)\psi^{(1)}(r_1)\psi^{(2)}(r_2)$$
$$= [E^{(1)}\psi^{(1)}(r_1)]\psi^{(2)}(r_2) + \psi^{(1)}(r_1)[E^{(2)}\psi^{(2)}(r_2)]$$
$$= (E^{(1)} + E^{(2)})\psi^{(1)}(r_1)\psi^{(2)}(r_2)$$

That is,

$$\left.\begin{array}{r}\hat{H}\Psi(r_1, r_2) = E\Psi(r_1, r_2)\\E = E^{(1)} + E^{(2)}\end{array}\right\} \tag{13.6}$$

which shows that the product wave function (13.4) is an eigenfunction of the complete Hamiltonian $\hat{H}$ of (13.3), corresponding to an eigenvalue E that is just the sum of the energy eigenvalues of the two separate particles.

In general, if a system contains any number, N say, of non-interacting particles, its Hamiltonian is the sum of N single-particle Hamiltonians, similar to the form (13.3). Each single-particle Hamiltonian commutes with every other, and thus with the complete Hamiltonian for the system. Then, as we stated at the beginning of this section, an energy eigenfunction for the system as a whole can be written as a product of single-particle energy eigenfunctions:

$$\Psi(r_1, r_2, \ldots r_N) = \psi^{(1)}(r_1)\psi^{(2)}(r_2) \cdots \psi^{(N)}(r_N) \tag{13.7}$$

The eigenvalues of each single-particle Hamiltonian determine the possible energies of the corresponding particle, and the total energy of the system is just the sum of the energies of the individual particles:

$$E = \sum_{k=1}^{N} E^{(k)} \tag{13.8}$$

In the case of particles that interact with one another the result we have just obtained would not be correct. For example, in the case of two particles interacting through a potential $\hat{V}(r_1 - r_2)$ that depends on their separation $r_1 - r_2$ the complete Hamiltonian $\hat{H}$ of (13.3) would become $\hat{H} = \hat{H}_1 + \hat{H}_2 + \hat{V}(r_1 - r_2)$. The single-particle Hamiltonians $\hat{H}_1$ and $\hat{H}_2$ do not commute with $\hat{V}(r_1 - r_2)$, and so they no longer commute with $\hat{H}$. Thus eigenstates of the two-particle system would no longer be simultaneously eigenstates of the single-particle Hamiltonians $\hat{H}_1$ and

$\hat{H}_2$, and an energy eigenfunction of the system as a whole could no longer be written as a product of the form (13.4).

However, even when the particles in a system interact with one another, it may be possible to treat them as non-interacting to a first approximation, and then to include the effects of the interactions using perturbation theory or the variational method. We shall treat the two electrons in a helium atom in this way in Section 13.4. In more complex systems, such as many-electron atoms, an independent-particle model may include some effects of interparticle interactions, through defining an effective potential for each particle that depends only on the position of that particular particle. In this type of approximation all the effectively single-particle Hamiltonians commute with one another and with the complete Hamiltonian of the system. So the energy eigenfunctions and eigenvalues for the many-particle system are determined by equations like (13.7) and (13.8). A suitable approximation method can then be used to calculate corrections due to parts of the interaction that have been neglected, or averaged to zero, in constructing the effective potentials.

13.2 The Born–Oppenheimer approximation

An important variation on independent-particle models of the type we have discussed is the **Born–Oppenheimer approximation**, which is the basis for calculations of molecular structure, and is also widely used in the theory of solids. In this approximation the wave function describing the molecule is written as a product of two wave functions, one describing the state of the electrons and the other the state of the nuclei. However, in this case it is clear that the electronic wave function cannot be completely independent of the positions of the nuclei, so that the factorization is not automatic. It is justified by a further physical approximation, based on the great difference between the time scales on which the motions of the nuclei and those of the electrons change.

The Hamiltonian for a molecule containing N nuclei and n electrons is of the general form

$$\hat{H} = \hat{H}_{\text{nucl}} + \hat{H}_{\text{el}} + \hat{H}_{\text{e-n}} \qquad (13.9)$$

Here $\hat{H}_{\text{nucl}}$ is the Hamiltonian representing the kinetic energy of the nuclei and the interaction between them:

$$\hat{H}_{\text{nucl}} = -\sum_{i=1}^{N} \frac{\hbar^2}{2m_i} \nabla_i^2 + \sum_{j<i} \sum_{i=1}^{N} \hat{V}_{\text{n-n}}(\boldsymbol{R}_i - \boldsymbol{R}_j) \qquad (13.10)$$

where $\boldsymbol{R}_i$ is the position of the ith nucleus, m_i is its mass and $\boldsymbol{R}_i - \boldsymbol{R}_j$ is the vector from the jth nucleus to the ith. The second term in $\hat{H}_{\text{nucl}}$ is the sum of the interactions between all possible pairs of nuclei, and depends on the separation of the nuclei in each pair. (The summation is over all i, and $j < i$ so as to avoid counting the same pair twice.) Similarly, the Hamiltonian representing the kinetic energy of all the electrons and their mutual repulsion can be written as

$$\hat{H}_{\text{el}} = -\sum_{i=1}^{n} \frac{\hbar^2}{2m_{\text{e}}} \nabla_i^2 + \sum_{j<i} \sum_{i=1}^{n} \hat{V}_{\text{e-e}}(r_i - r_j) \tag{13.11}$$

where m_{e} is the electron mass and r_i is the position of the ith electron. The final contribution to the Hamiltonian $\hat{H}$ in (13.9) is the term representing the interaction between the electrons and nuclei:

$$\hat{H}_{\text{e-n}} = \sum_{j=1}^{n} \sum_{i=1}^{N} \hat{V}_{\text{e-n}}(r_j - R_i) \tag{13.12}$$

The interaction $\hat{H}_{\text{e-n}}$ is clearly of paramount importance in determining the electronic part of the wave function, since the state of an electron must depend critically on its Coulomb attraction to each nucleus in the molecule. The inertia of the electrons is much less than that of the nuclei – remember that each proton or neutron is nearly 2000 times more massive than an electron. So the electrons respond very rapidly to changes in the nuclear positions, while the motion of the nuclei is on a much longer time scale. This means that changes in $\hat{H}_{\text{e-n}}$ due to the motion of the nuclei can be treated as adiabatic (Section 12.1). To a good approximation, the electronic state at each instant is an eigenstate of the Hamiltonian $\hat{H}' = \hat{H}_{\text{el}} + \hat{H}_{\text{e-n}}$ for the corresponding instantaneous arrangement of the nuclei.

The influence of the electrons on the state of the nuclei is approximated by a potential that, for each arrangement of the nuclei, is the electron–nucleus interaction $\hat{H}_{\text{e-n}}$ averaged over the positions of the electrons. This effective potential therefore depends on the positions of the nuclei but not on the positions of individual electrons, and correspondingly the nuclear part of the wave function is a function only of $\{R_i \ldots R_N\}$. In principle, the solution to the energy eigenvalue problem for the nuclei is obtained by replacing the rapidly varying potential describing the attraction of a nucleus to the rapidly moving electrons by a static potential $\hat{V}_{\text{eff}}(R_1, R_2, \ldots, R_N)$, representing its interaction with an average negative charge distribution. Since it does not depend on the positions of individual electrons, this effective potential commutes with the electronic part of the Hamiltonian, $\hat{H}' = \hat{H}_{\text{el}} + \hat{H}_{\text{e-n}}$, and so in this approximation we can express the energy eigenfunctions of the molecule as simultaneous eigenfunctions of $\hat{H}'$ and $\hat{H}_{\text{nucl}} + \hat{V}_{\text{eff}}$. (However, the form of $\hat{V}_{\text{eff}}$ depends on the average distribution of the electrons, and so the eigenvalue equation for $\hat{H}'$ must be solved first.)

Thus the essence of the Born–Oppenheimer approximation is to treat the dynamical state of the nuclei and that of the electrons separately, and the molecular wave function is approximated by a product of electronic and nuclear wave functions:

$$\Psi(r_1, \ldots, r_n; R_1, \ldots, R_N)$$
$$= \Psi_{el}(r_1, \ldots, r_n; R_1, \ldots, R_N)\Psi_{nucl}(R_1, \ldots, R_N) \quad \text{(13.13)}$$

For each fixed arrangement $\{R_1, \ldots, R_N\}$ of the nuclei the electronic wave function $\Psi_{el}(r_1, \ldots, r_n; R_1, \ldots, R_N)$ satisfies the Schrödinger equation for the electrons:

$$(\hat{H}_{el} + \hat{H}_{e-n})\Psi_{el}(r_1, \ldots, r_n; R_1, \ldots, R_N)$$
$$= E_{el}\Psi_{el}(r_1, \ldots, r_n; R_1, \ldots, R_N) \quad \text{(13.14)}$$

The nuclear positions $\{R_1, \ldots, R_N\}$ are treated as fixed parameters as far as the electronic wave function $\Psi_{el}(r_1, \ldots, r_n; R_1, \ldots, R_N)$ is concerned. Equation (13.12) shows that $\hat{H}_{e-n}$ is different for each set of values $\{R_1, \ldots, R_N\}$, so (13.14) is solved for a range of different (fixed) nuclear arrangements, and the energy of each electronic level, $E_{el}(R_1, \ldots, R_N)$, becomes a function of the N variables representing the nuclear positions.

Using the complete Hamiltonian (13.9), with the product form (13.13) for the wave function, we can now write the Schrödinger equation for the molecule as a whole, and use (13.14) to replace $\hat{H}_{el} + \hat{H}_{e-n}$ by the electronic energy $E_{el}(R_1, \ldots, R_N)$:

$$(\hat{H}_{nucl} + \hat{H}_{el} + \hat{H}_{e-n})\Psi_{el}(r_1, \ldots, r_n; R_1, \ldots, R_N)\Psi_{nucl}(R_1, \ldots, R_N)$$
$$= [\hat{H}_{nucl} + E_{el}(R_1, R_N)]\Psi_{el}(r_1, \ldots, r_n; R_1, \ldots, R_N)\Psi_{nucl}(R_1, \ldots, R_N)$$
$$= E\Psi_{el}(r_1, \ldots, r_n; R_1, \ldots, R_N)\Psi_{nucl}(R_1, \ldots, R_N)$$

Here E is the energy eigenvalue of the molecule as a whole. Since we are treating the nuclear positions as fixed parameters as far as the electronic part of the wave function is concerned, $\Psi_{el}(r_1, \ldots, r_n; R_1, \ldots, R_N)$ behaves like a constant under the action of $\hat{H}_{nucl}$, so

$$[\hat{H}_{nucl} + E_{el}(R_1, \ldots, R_N)]\Psi_{el}(r_1, \ldots, r_n; R_1, \ldots, R_N)\Psi_{nucl}(R_1, \ldots, R_N)$$
$$= \Psi_{el}(r_1, \ldots, r_n; R_1, \ldots, R_N)[\hat{H}_{nucl} + E_{el}(R_1, \ldots, R_N)]\Psi_{nucl}(R_1, \ldots, R_N)$$

The Schrödinger equation may then be divided throughout by the electronic wave function to obtain the equation

$$[\hat{H}_{nucl} + E_{el}(R_1, \ldots, R_N)]\Psi_{nucl}(R_1, \ldots, R_N) = E\Psi_{nucl}(R_1, \ldots, R_N)$$

$$\text{(13.15)}$$

This is just a Schrödinger equation for the nuclei, with the energy $E_{el}(R_1, \ldots, R_N)$ now playing the role of an additional potential. (Remember that $\hat{H}_{nucl}$ contains the potential representing the interaction

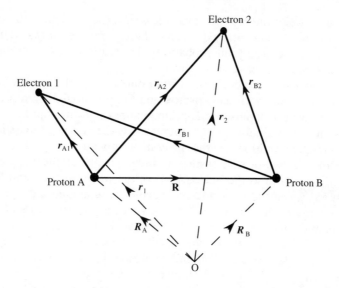

Figure 13.1 Position vectors for the two electrons and two protons in a hydrogen molecule.

between the nuclei.) So $E_{el}(\boldsymbol{R}_1, \ldots, \boldsymbol{R}_N)$, which varies with the arrangement of the nuclei, acts as an effective potential representing the interaction of the nuclei with the electrons.

The simplest application of the Born–Oppenheimer approximation is to the hydrogen molecule (containing two protons and two electrons and shown schematically in Figure 13.1). First the Schrödinger equation (13.14) for the electrons must be solved. Because of the presence of the term V_{e-e} (representing the electron–electron repulsion) in the Hamiltonian H_{el} of (13.11), this equation cannot be solved exactly. The variational method discussed in Section 11.5 can be used to calculate approximate expressions for the electronic wave functions and energy eigenvalues. The calculation must be repeated for different values of R, the separation between the protons, to obtain the function $E_{el}(R)$, which acts as an effective potential in the Schrödinger equation, of the form (13.15), for the two protons.

The approximate solution of the electron Schrödinger equation (13.14) shows that in the ground state of the hydrogen molecule the two-electron wave function is symmetric about a point midway between the two protons, where the joint probability distribution has a maximum. Clearly when the electron distribution is concentrated between the protons in this way it has a shielding effect, partially hiding the positive charge on one proton from the other. It is intuitively reasonable for a distribution of this type to be associated with the lowest-energy state, since the effective repulsion between the two protons is less in this case than if the electron distribution were concentrated outside the central region (as would happen if the two-electron joint probability distribution were antisymmetric, with a zero midway between the protons).

For the hydrogen molecule the Hamiltonian $\hat{H}_{nucl}$ of (13.10) consists of a kinetic energy term for each proton and the potential $\hat{V}_{n-n}$ representing the proton–proton repulsion. The kinetic energy terms can be rewritten as the sum of the kinetic energy operator for the centre of mass and that for the relative motion of the protons (Appendix A1). The internal motion of the two-proton system is independent of the motion of its centre of mass, and so the nuclear part of the wave function and the total internal energy of the molecule are calculated retaining only the relative kinetic energy term in (13.15). This becomes a one-dimensional Schrödinger equation with total potential energy $\hat{V}(R) = \hat{V}_{n-n}(R) + E_{el}(R)$. If the form of $E_{el}(R)$ appropriate to the ground state of the hydrogen molecule is added to the Coulomb repulsion $\hat{V}_{n-n}(R)$ between the protons, the shape of $\hat{V}(R)$ is similar to that shown in Figure 5.5. It approaches a constant value for large separations between the protons, when the system behaves approximately like two independent hydrogen atoms, and reaches a minimum at a separation $R = r_0$, of the order of a Bohr radius. This is the equilibrium separation of the two protons. At smaller separations the curve rises very steeply because the proton–proton repulsion $\hat{V}_{n-n}(R)$ becomes dominant. The potential $\hat{V}_{n-n}(R) + E_{el}(R)$ in the ground states of other diatomic molecules is of a similar shape. The motions of the nuclei are generally of small amplitude compared with those of the electrons, and a small departure of the nuclear separation R from its equilibrium value r_0 can be treated approximately as a simple harmonic oscillation, giving rise to the vibrational spectrum discussed in Section 5.3.

13.3 Identical particles and the Pauli exclusion principle

All electrons have exactly the same electric charge, intrinsic magnetic moment and rest mass, and there is no feature that distinguishes one electron from another. Similarly, all protons are identical to one another, as are all nuclei of a particular isotope of the same element. Although the concept of identical particles also occurs in classical mechanics, it has much deeper implications in quantum theory.

Think first of two identical particles from a classical point of view. For example, two white billiard balls – provided they have been made to standard specifications – are identical. They should have the same size and mass, and even a close inspection of the colour and smoothness of their surfaces may not provide any distinguishing features. Suppose two such balls are placed in a box, one by person A and the other by person B, the lid is closed, and the box shaken around. If the box is then opened and one of the balls allowed to roll out, we cannot tell whether it is A's ball or B's ball – the two balls are indistinguishable. But according to a classical way of thinking, this does not imply that the question of whether the ball is A's or B's is meaningless: each ball is assumed to follow its own unique trajectory inside the box, whether or not we observe it. So the two balls retain their distinct identities, and there is a unique correct answer (even if we don't known what it is) to the question of which ball it is that emerges. Of course this point of view is reasonable for billiard balls because their motion is on a scale for which classical physics is a good approximation, as discussed in Section 1.6.

However, for particles that must be treated quantum mechanically the situation is completely different. In this case the particle cannot be assumed to follow a unique continuous trajectory through space, as emphasized in Chapters 1 and 2. If two identical quantum particles are in states described by wave functions Ψ_1 and Ψ_2 that overlap in some region of space then either particle might be found within this region. Suppose that a measurement detects one particle at a point in the region. Is this particle the one in the state described by Ψ_1 or is it the other? There is no possible way to tell. The probability that it is the first particle is determined by $|\Psi_1|^2$ (evaluated at the point concerned), and similarly $|\Psi_2|^2$ gives the probability that it is the second. Since neither particle can be assumed to follow a unique well-determined trajectory, it is not possible to assume that there *is* a unique answer to the question, and the two particles cannot be treated as though they retain completely distinct identities: two identical quantum particles are indistinguishable not only in practice but also *in principle*.

To see what this fundamental indistinguishability of quantum particles implies, consider two particles bound by the same potential (for example two electrons in the same atom, or two protons or neutrons in the same nucleus). In an independent-particle approximation the particles are treated as non-interacting. As we saw in Section 13.1, in this approximation the energy eigenfunction for the two-particle state can be written as a product of energy eigenfunctions for the individual particles, and the total energy is the sum of the individual energies. Suppose that the probability distributions of particles 1 and 2 are functions of r_1 and r_2 respectively, and that particle 1 is in a state labelled by a set of quantum numbers $\{n\}$, while particle 2 is in a state labelled by $\{n'\}$. A product wave function for the two-particle state can be written as

$$\Psi^{(1,2)}(r_1, r_2) = \psi_{\{n\}}(r_1)\psi_{\{n'\}}(r_2) \qquad (13.16)$$

The total energy of the pair is then the sum of the energies of the two single particles, $E_{\{n\}} + E_{\{n'\}}$, and similarly the other quantum numbers describing the state (such as the orbital angular momentum or the spin quantum numbers) are determined by adding the quantum numbers for the single-particle states in the appropriate way.

But now consider the wave function

$$\Psi^{(2,1)}(r_1, r_2) = \psi_{\{n\}}(r_2)\psi_{\{n'\}}(r_1) \qquad (13.17)$$

This product has been obtained from that in (13.16) by exchanging the two particles (as indicated in the superscript): it is now particle 2 that is in the state labelled $\{n\}$, and particle 1 that is in the state labelled $\{n'\}$. The two wave functions $\Psi^{(1,2)}(r_1, r_2)$ and $\Psi^{(2,1)}(r_1, r_2)$, are *different* functions of r_1 and r_2, unless the primed and unprimed sets of quantum numbers happen to be identical. However, they represent two-particle states with the *same* energy $E_{\{n\}} + E_{\{n'\}}$ and the same values of all other quantum numbers that are relevant for labelling the states. So we have two different wave functions representing exactly the same physical state. Because the particles are identical, there is no answer to the question of which of them is in the particular state labelled $\{n\}$, and which in the other state.

It should be emphasized that the wave functions (13.16) and (13.17) do not represent two degenerate two-particle states, because there is no dynamical variable whose value can distinguish one state from the other. What we have are two different wave functions to represent a single state, which is unsatisfactory: the two wave functions determine two different joint probability distributions, but a unique state should be represented by a unique joint probability distribution. So we must pick one function or the other, or some linear superposition of them, that correctly describes the physically observed joint probability distribution. Since there is, as we have said, no way to associate a particular particle with a particular set of quantum numbers, *the joint probability distribution should be unchanged when we exchange the two particles*. This is not the case for either of the states in (13.16) and (13.17) alone, so let us take some linear superposition of them:

$$\Psi(r_1, r_2) = c_1\Psi^{(1,2)}(r_1, r_2) + c_2\Psi^{(2,1)}(r_1, r_2)$$

(Assuming that $\Psi^{(1,2)}$ and $\Psi^{(2,1)}$ are normalized, this combination will also be normalized if $|c_1|^2 + |c_2|^2 = 1$.) Under exchange of the two particles,

$$\Psi(r_1, r_2) \rightarrow c_1\Psi^{(2,1)}(r_1, r_2) + c_2\Psi^{(1,2)}(r_1, r_2)$$

and for the joint probability distribution to remain unchanged we require

$$|c_1\Psi^{(1,2)}(r_1, r_2) + c_2\Psi^{(2,1)}(r_1, r_2)|^2 = |c_1\Psi^{(2,1)}(r_1, r_2) + c_2\Psi^{(1,2)}(r_1, r_2)|^2$$

or

$$c_1\Psi^{(1,2)}(r_1, r_2) + c_2\Psi^{(2,1)}(r_1, r_2) = \pm[c_1\Psi^{(2,1)}(r_1, r_2) + c_2\Psi^{(1,2)}(r_1, r_2)]$$

This is satisfied by

$$c_2 = \pm c_1$$

so that the possible, normalized, forms for the two-particle wave function are

$$\Psi^{(\pm)}(r_1, r_2) = \sqrt{(\tfrac{1}{2})}[\Psi^{(1,2)}(r_1, r_2) \pm \Psi^{(2,1)}(r_1, r_2)] \qquad (13.18)$$

If we choose the plus sign, the two-particle wave function is invariant under exchange of the two particles, whereas if we choose the minus sign the two-particle wave function changes sign: the two-particle wave function is either symmetric or antisymmetric under exchange of the particles.

Although both the wave functions (13.18) satisfy the requirement that the probability distribution be invariant under exchange of the two particles, we are still faced with two different wave functions for a single two-particle state. It is not possible to identify which is the correct one to use on the basis of non-relativistic quantum theory alone. Quantum field theory, however, imposes the following symmetry restriction on the wave function describing any system of quantum particles:

Symmetrization requirement

The complete wave function describing a many-particle state is antisymmetric under the exchange of any pair of identical fermions, and symmetric under the exchange of any pair of identical bosons.

As we explained at the end of Section 6.2, fermions and bosons are particles with half-integer and integer spins respectively.

Since electrons are fermions, the symmetrization requirement means that the complete wave function for a system containing more than one electron must be antisymmetric under the exchange of any pair of electrons. This immediately leads to the Pauli exclusion principle: suppose that two electrons are in states described by identical sets of quantum numbers $\{n\} \equiv \{n'\}$. In this case (13.16) and (13.17) show that $\Psi^{(1,2)}(r_1, r_2) \equiv \Psi^{(2,1)}(r_1, r_2)$, so that from (13.18) the antisymmetric wave function $\Psi^{(-)}(r_1, r_2)$ vanishes identically. Therefore two electrons within the same many-particle system cannot exist in identical states, which means they cannot have identical sets of quantum numbers. This is the **Pauli exclusion principle**, which applies not only to electrons in an atom but to any pair of identical fermions in the same many-particle system.

To write down a totally antisymmetric wave function for a system of N identical fermions, we make use of the properties of determinants [1]: a determinant changes sign under the exchange of any pair of rows (or any pair of columns). The expansion of a determinant contains $N!$ terms, formed by taking all possible combinations of N elements, one from each row and each column. So, for example, if we have three identical fermions labelled 1, 2 and 3, in three states labelled by $\{n\}$, $\{n'\}$ and $\{n''\}$, we construct the completely antisymmetric three-particle state from the determinant

$$\Psi^{(-)}(r_1, r_1, r_3) = \frac{1}{\sqrt{6}} \det \begin{pmatrix} \psi_{\{n\}}(r_1) & \psi_{\{n'\}}(r_1) & \psi_{\{n''\}}(r_1) \\ \psi_{\{n\}}(r_2) & \psi_{\{n'\}}(r_2) & \psi_{\{n''\}}(r_2) \\ \psi_{\{n\}}(r_3) & \psi_{\{n'\}}(r_3) & \psi_{\{n''\}}(r_3) \end{pmatrix}$$

$$= \sqrt{(\tfrac{1}{6})}[(\psi_{\{n\}}(r_1)\psi_{\{n'\}}(r_2)\psi_{\{n''\}}(r_3) - \psi_{\{n\}}(r_1)\psi_{\{n''\}}(r_2)\psi_{\{n'\}}(r_3)$$

$$+ \psi_{\{n'\}}(r_1)\psi_{\{n''\}}(r_2)\psi_{\{n\}}(r_3) - \psi_{\{n'\}}(r_1)\psi_{\{n\}}(r_2)\psi_{\{n''\}}(r_3)$$

$$+ \psi_{\{n''\}}(r_1)\psi_{\{n\}}(r_2)\psi_{\{n'\}}(r_3) - \psi_{\{n''\}}(r_1)\psi_{\{n'\}}(r_2)\psi_{\{n\}}(r_3)]$$

Note that each of the $N! = 3! = 6$ terms is the product of three single-particle wave functions, one for each particle, and each corresponding to a different set of quantum numbers. It is easy to check that interchanging any pair of particle labels multiplies $\Psi^{(-)}(r_1, r_2, r_3)$ by -1. The factor $\sqrt{\tfrac{1}{6}} = 1/\sqrt{(3!)}$ is included to normalize the expansion. Similarly, the normalized totally antisymmetric wave function for N identical fermions, in the N states labelled by quantum numbers $\{n_1\}$, ..., $\{n_N\}$, can be written as

$$\Psi^{(-)}(r_1, \ldots, r_2) = \frac{1}{\sqrt{(N!)}} \det \begin{pmatrix} \psi_{\{n_1\}}(r_1) & \psi_{\{n_2\}}(r_1) & \cdots & \psi_{\{n_N\}}(r_1) \\ \psi_{\{n_1\}}(r_2) & \psi_{\{n_2\}}(r_2) & \cdots & \psi_{\{n_N\}}(r_2) \\ \vdots & \vdots & & \vdots \\ \psi_{\{n_1\}}(r_N) & \psi_{\{n_2\}}(r_N) & \cdots & \psi_{\{n_N\}}(r_N) \end{pmatrix}$$

$$(13.19)$$

13.4 Systems containing two identical particles

The symmetrization requirement introduced in Section 13.3 applies to the complete wave function describing a system containing identical particles. In many cases of interest it is possible to represent the many-particle state by the product of a space-dependent wave function and a spinor describing the spin state of the system. For example, we might represent the state of two identical particles by

$$\Psi^{(\pm)}(r_1, r_2) = \Psi_{\text{space}}(r_1, r_2)\chi_{\text{spin}} \qquad (13.20)$$

where for $\Psi^{(+)}(r_1, r_2)$ the spatial and spin parts must be either both symmetric or both antisymmetric under exchange of the particles, while for $\Psi^{(-)}(r_1, r_2)$ the spatial and spin parts must have opposite symmetry properties. Symmetric and antisymmetric forms of the space-dependent wave function are of the form (13.18):

$$\Psi_{\text{space}}^{(\pm)}(r_1, r_2) = C[\psi_{\{n\}}(r_1)\psi_{\{n'\}}(r_2) \pm \psi_{\{n\}}(r_2)\psi_{\{n'\}}(r_1)] \quad \textbf{(13.21)}$$

The wave function is normalized if $C = \sqrt{\frac{1}{2}}$, unless the two sets of quantum numbers are identical. If this is the case, $\Psi_{\text{space}}^{(-)}$ vanishes identically, and the two terms in $\Psi_{\text{space}}^{(+)}$ are identical, so that for normalization $C = \frac{1}{2}$. As in Section 13.3, $\psi_{\{n\}}(r_i)$ represents the wave function for particle $i = 1$ or 2 in the single-particle state labelled by the set of quantum numbers $\{n\}$, but now this set does not include spin quantum numbers. For an electron in an atom, for example, $\{n\}$ means the set $\{n, l, m_l\}$. Similarly, the normalized symmetric and antisymmetric spin states are

$$\chi_{\text{spin}}^{(\pm)} = \sqrt{(\tfrac{1}{2})}(\chi_{sm_s}^{(1)}\chi_{sm_s'}^{(2)} \pm \chi_{sm_s}^{(2)}\chi_{sm_s'}^{(1)}) \quad \textbf{(13.22)}$$

(though, as in the case of $\Psi_{\text{space}}^{(+)}$, the normalization constant for $\chi_{\text{spin}}^{(+)}$ should be $\frac{1}{2}$ if $m_s = m_s'$). In (13.22) $\chi_{sm_s}^{(i)}$ is the eigenspinor for particle i in the state with spin quantum numbers $\{s, m_s\}$. These eigenspinors may represent the spin state of either a boson or a fermion.

For a system containing two identical particles orbiting one another in a state of relative orbital angular momentum characterized by the quantum numbers $\{l, m_l\}$ the symmetry of the space-dependent part of the wave function is determined by the quantum number l. As in the classical case discussed in Appendix A2, the Hamiltonian for the system can be separated into terms $\hat{H}_{\text{cm}}$ and $\hat{H}_{\text{rel}}$ representing the centre-of-mass energy and the internal energy respectively. Since $\hat{H}_{\text{cm}}$ commutes with $\hat{H}_{\text{rel}}$, the two-particle wave function can be rewritten as the product

$$\Psi_{\text{space}}^{(+)}(r, R) = \psi_{\text{cm}}(R)\psi_{nlm_l}(r) \quad \textbf{(13.23)}$$

where the coordinate R refers to the centre of mass, $\psi_{\text{cm}}(R)$ describes the state of motion of the centre of mass, and $r = r_1 - r_2$ is the relative position vector. The internal state of the two-particle system is represented by $\psi_{nlm_l}(r)$, which has an angular dependence given by the spherical harmonic $Y_{lm_l}(\theta, \phi)$, (4.45), and needs the third quantum number n to label the energy level. The centre of mass of a system of two identical particles is at $R = \frac{1}{2}(r_1 + r_2)$, midway between them, and since the expression for R is symmetric under exchange of r_1 and r_2, the wave function $\psi_{\text{cm}}(R)$ must also be symmetric. Therefore it is the wave function $\psi_{nlm_l}(r, \theta, \phi)$, describing the relative motion, that determines whether the two-particle wave function (13.23) is symmetric or antisymmetric. From Figure 13.2 we can see that exchanging the position vectors r_1 and r_2 is equivalent to inverting the direction of the relative position vector $r = r_1 - r_2$, and this has exactly the same effect on the relative coordinates (r, θ, ϕ) as parity inversion, (4.30). In Section 4.5 we noted that the spherical harmonic $Y_{lm_l}(\theta, \phi)$ is an eigenstate of parity corresponding to the eigenvalue $(-1)^l$, so that $\psi_{nlm_l}(r, \theta, \phi)$ is symmetric

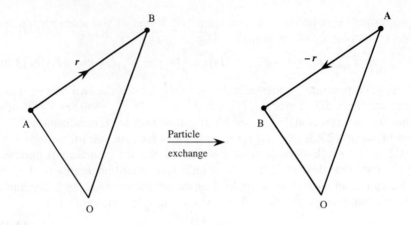

Figure 13.2 The transformation of the vector joining particle A to particle B under exchange of the particles. This is equivalent to the change $r \rightarrow -r$ produced by parity inversion.

under exchange of the two particles if l is even or zero, and antisymmetric if l is odd.

In the case of two identical spin-zero particles the total spin of the system is also zero, and the state of the pair of particles is completely described by the wave function (13.23). Since a spinless particle is a boson, the symmetrization requirement states that the wave function must be symmetric under exchange of the particles, which means that their relative orbital angular momentum quantum number must be even or zero. This has experimentally detectable consequences. For example, it can restrict the way in which a particle decays. One of the subatomic particles produced in high-energy scattering experiments is ρ, the rho meson [2], which is rather like the pion that we met in Section 4.6, except that it has spin 1 whereas the pion has spin 0. Like the pion, the rho occurs with electric charge +1, 0, and −1 (in units of the electron charge e). The charged particles, ρ^+ and ρ^-, decay rapidly (with a lifetime of the order of 10^{-23} s) into one charged and one neutral pion:

$$\rho^+ \rightarrow \pi^+ + \pi^0, \qquad \rho^- \rightarrow \pi^- + \pi^0$$

The neutral rho also decays into two pions, conserving electric charge, so that one might expect to see two decay channels:

$$\rho^0 \rightarrow \pi^+ + \pi^-, \qquad \rho^0 \rightarrow \pi^0 + \pi^0$$

However, the decay into two neutral pions is never observed. This is because the relative orbital angular momentum of the two spinless pions must be $l = 1$, to conserve total angular momentum in the decay of the rho with spin 1. (Spin is transformed into orbital angular momentum in the decay process.) But this means that the final two-pion state is anti-symmetric under exchange of the pions, and this is forbidden for identical bosons, so the decay $\rho^0 \to \pi^0 + \pi^0$ cannot occur. However, π^+ and π^- have different electric charge, so they are not identical particles. In this case, therefore, the symmetry requirement does not apply, and ρ^0 can decay into $\pi^+ + \pi^-$.

If a system consists of two identical spin-$\frac{1}{2}$ particles, the rules for addition of angular momenta that we discussed in Section 6.3 indicate that the total spin state of the system has quantum number $S = 1$ or 0. (See (6.33) and Problem 6.5(a).) The eigenspinors for the two-particle system, χ_{SM_S}, are given in terms of the eigenspinors $\chi_{m_s}^{(i)}$ ($i = 1, 2$) for spin-$\frac{1}{2}$ particles by

$$
\left.
\begin{aligned}
\chi_{1,+1} &= \chi_{+1/2}^{(1)} \chi_{+1/2}^{(2)} \\
\chi_{10} &= \sqrt{(\tfrac{1}{2})}(\chi_{+1/2}^{(1)} \chi_{-1/2}^{(2)} + \chi_{+1/2}^{(2)} \chi_{-1/2}^{(1)}) \\
\chi_{1,-1} &= \chi_{-1/2}^{(1)} \chi_{-1/2}^{(2)}
\end{aligned}
\right\}
\qquad (13.24)
$$

for the $S = 1$ states, and

$$
\chi_{00} = \sqrt{(\tfrac{1}{2})}(\chi_{+1/2}^{(1)} \chi_{-1/2}^{(2)} - \chi_{+1/2}^{(2)} \chi_{-1/2}^{(1)}) \qquad (13.25)
$$

for $S = 0$. Note that the three $S = 1$ spinors are all symmetric under exchange of the particle labels 1 and 2, while the $S = 0$ state is antisymmetric.

The helium atom

The helium atom is a three-particle system containing a nucleus and two identical electrons, but we reduce it to a two-particle problem by treating the nucleus as at rest at the origin. If we neglect the repulsion between the two electrons, the Hamiltonian for the helium atom is of the form (13.2), with each electron moving in a potential

$$
\hat{V}_i = -\frac{2e^2}{4\pi\varepsilon_0} \frac{1}{r_i} \qquad (i = 1, 2)
$$

The spatial dependence of the energy eigenfunction for each electron should therefore be of the same general form (7.16) as for hydrogen:

$$
\psi_{\{n\}}(r_i) \equiv \psi_{nlm_l}(r_i, \theta_i, \phi_i) = R_{nl}(r_i) Y_{lm_l}(\theta_i, \phi_i) \qquad (i = 1, 2) \quad (13.26)
$$

But since the charge on the helium nucleus is twice that for hydrogen, the explicit forms for the single-electron eigenfunctions in helium are obtained from those for hydrogen by replacing the Bohr radius r_B (defined in (7.11)) by $\frac{1}{2}r_B$. Similarly, the Rydberg energy E_R must be replaced by $4E_R$ in (7.13) to obtain the single-electron energy levels in helium:

$$
E_n^{(i)} = -\frac{4}{n^2} E_R \qquad (n = 1, 2, \ldots) \qquad (13.27)
$$

The total wave function for the helium atom (neglecting the motion of the nucleus) can be factorized into a spatial and a spin part, as in (13.20). Substituting (13.26) in (13.21), we find that when one electron is in the state with quantum numbers $\{n, l, m_l\}$ and the other is in the state labelled $\{n', l', m_l'\}$ the symmetric and antisymmetric space-dependent wave functions for the two-electron system are

$$\Psi_{\text{space}}^{(\pm)}(\mathbf{r}_1, \mathbf{r}_2) = C[R_{nl}(r_1)Y_{lm_l}(\theta_1, \phi_1)R_{n'l'}(r_2)Y_{l'm_l'}(\theta_2, \phi_2)$$
$$\pm R_{nl}(r_2)Y_{lm_l}(\theta_2, \phi_2)R_{n'l'}(r_1)Y_{l'm_l'}(\theta_1, \phi_1)] \quad \textbf{(13.28)}$$

The spin part of the two-electron wave function is either the symmetric product corresponding to total spin quantum number $S = 1$, (13.24), or the antisymmetric product corresponding to $S = 0$, (13.25). Since electrons are fermions, the wave function for the helium atom must be antisymmetric under exchange of the two electrons, and so we must have either a symmetric spatial part with total spin $S = 0$, or an antisymmetric spatial part with $S = 1$. Equations (13.6) and (13.27) give the energy spectrum of helium, in the approximation that the electrons are non-interacting:

$$E = -4E_R\left(\frac{1}{n^2} + \frac{1}{n'^2}\right) \quad \textbf{(13.29)}$$

For the ground state of helium we expect both electrons to be in the lowest level: $n = n' = 1$, $l = l' = 0$ and $m_l = m_l' = 0$. According to (13.29), the ground state energy of the atom is

$$E_G = -8E_R \quad \textbf{(13.30)}$$

Since the quantum numbers $\{n, l, m_l\}$ are the same for the two electrons, the antisymmetric wave function of the form (13.28) vanishes, and the spatial part of the ground state wave function is the symmetric combination

$$\Psi_{\text{space}}^{(+)}(\mathbf{r}_1, \mathbf{r}_1) = \psi_{100}(r_1, \theta_1, \phi_1)\psi_{100}(r_2, \theta_2, \phi_2)$$
$$= \left(\frac{8}{\pi r_B^3}\right)^2 \exp\left(-2\frac{r_1 + r_2}{r_B}\right) \quad \textbf{(13.31)}$$

(We have used the form for the hydrogen ground state wave function given in Table 7.1, with r_B replaced by $\frac{1}{2}r_B$.) Because the spatial part of the wave function is symmetric, the ground state of helium must have spin 0. (This is in accord with the Pauli principle, since in the spin-0 state, (13.25), the electrons have opposite values of m_s.)

So far we have neglected the Coulomb repulsion between the electrons,

$$\hat{V}_{\text{int}} = \frac{e^2}{4\pi\varepsilon_0}\frac{1}{|\mathbf{r}_1 - \mathbf{r}_2|}$$

and we now calculate the effect of this interaction using first-order perturbation theory. Equation (11.13) identifies the first-order correction to the energy spectrum (13.29) as the expectation value of $\hat{V}_{int}$ in the unperturbed (zeroth-order) state. For the ground state (with zeroth-order wave function given by (13.31)) this is

$$E_G^{(1)} = \frac{e^2}{4\pi\varepsilon_0} \int_{\text{all space}} d^3r_1 \int_{\text{all space}} d^3r_2 \left(\frac{8}{\pi r_B^3}\right)^4 \exp\left(-4\,\frac{r_1 + r_2}{r_B}\right) \frac{1}{|r_1 - r_2|}$$

$$\approx 2.5 E_R$$

(The integral is not trivial to perform, but may be evaluated using properties of Legendre polynomials [3].) The corrected ground state energy is therefore

$$E_G + E_G^{(1)} \approx -5.5 E_R \approx -74.8 \, \text{eV}$$

This should be compared with the experimentally measured ground state energy of helium, which is approximately $-78.98 \, \text{eV}$. Better agreement can be obtained using the variational method rather than first-order perturbation theory.

Although the spin of the ground state of helium must be zero, excited states may have spin 1 or 0. If the interaction between the electrons is neglected, the energy levels (13.29) depend only on the pair of principal quantum numbers that specify the single-electron states, and so states characterized by the same pair of numbers $\{n, n'\}$ but with different total spins are degenerate. When we take account of the interaction between the electrons this degeneracy is lifted.

According to perturbation theory, the first-order correction to the energy of an excited state of helium is

$$E^{(1)} = \frac{e^2}{4\pi\varepsilon_0} \int_{\text{all space}} d^3r_1 \int_{\text{all space}} d^3r_2$$

$$\times \, [\Psi_{\text{space}}^{(\pm)}(r_1, r_2)]^* \frac{1}{|r_1 - r_2|} \, \Psi_{\text{space}}^{(\pm)}(r_1, r_2)$$

$$= \frac{e^2}{4\pi\varepsilon_0} \int_{\text{all space}} d^3r_1 \int_{\text{all space}} d^3r_2$$

$$\times \, \sqrt{(\tfrac{1}{2})}[\psi_{\{n\}}(r_1)\psi_{\{n'\}}(r_2) \pm \psi_{\{n\}}(r_2)\psi_{\{n'\}}(r_1)]^*$$

$$\times \, \frac{1}{|r_1 - r_2|} \sqrt{(\tfrac{1}{2})}[\psi_{\{n\}}(r_1)\psi_{\{n'\}}(r_2) \pm \psi_{\{n\}}(r_2)\psi_{\{n'\}}(r_1)]$$

where the $+$ signs apply if the spatial wave function is symmetric and $S = 0$, while the $-$ signs apply if the spatial wave function is antisymmetric and $S = 1$. (We have normalized the states by putting the constant C equal to $\sqrt{\tfrac{1}{2}}$.) This correction can be written in the form

$$E^{(1)} = \tfrac{1}{2}(\Delta E_D \pm \Delta E_E) \tag{13.32}$$

where

$$\tfrac{1}{2}\Delta E_D = \frac{e^2}{4\pi\varepsilon_0} \int_{\text{all space}} d^3r_1 \int_{\text{all space}} d^3r_2$$

$$\times \; \psi^*_{\{n\}}(r_1)\psi^*_{\{n'\}}(r_2) \frac{1}{|r_1 - r_2|} \psi_{\{n\}}(r_1)\psi_{\{n'\}}(r_2)$$

$$= \frac{e^2}{4\pi\varepsilon_0} \int_{\text{all space}} d^3r_1 \int_{\text{all space}} d^3r_2$$

$$\times \; \psi^*_{\{n\}}(r_2)\psi^*_{\{n'\}}(r_1) \frac{1}{|r_1 - r_2|} \psi_{\{n\}}(r_2)\psi_{\{n'\}}(r_1)$$

$$\tfrac{1}{2}\Delta E_E = \frac{e^2}{4\pi\varepsilon_0} \int_{\text{all space}} d^3r_1 \int_{\text{all space}} d^3r_2$$

$$\times \; \psi^*_{\{n\}}(r_1)\psi^*_{\{n'\}}(r_2) \frac{1}{|r_1 - r_2|} \psi_{\{n\}}(r_2)\psi_{\{n'\}}(r_1)$$

$$= \frac{e^2}{4\pi\varepsilon_0} \int_{\text{all space}} d^3r_1 \int_{\text{all space}} d^3r_2$$

$$\times \; \psi^*_{\{n\}}(r_2)\psi^*_{\{n'\}}(r_1) \frac{1}{|r_1 - r_2|} \psi_{\{n\}}(r_1)\psi_{\{n'\}}(r_2)$$

Clearly each zeroth-order energy level is split into two, separated by the **exchange energy** ΔE_E and distinguished by the value of the total spin quantum number, even though the perturbing potential is just the Coulomb interaction between electrons, which does not depend on the spin. The exchange energy is so named because it takes the form of a matrix element of the perturbation between a product state $\psi_{\{n\}}(r_2)\psi_{\{n'\}}(r_1)$ and the state $\psi_{\{n\}}(r_1)\psi_{\{n'\}}(r_2)$ obtained from it by exchanging the electrons. On the other hand, the matrix element that determines ΔE_D is the normal, direct, expectation value of the perturbation in a state represented by an unsymmetrized product wave function (of the form (13.16) or (13.17)).

There are thus two sets of energy levels in helium, one with $S = 0$ and the other with $S = 1$, represented in Figure 13.3 by solid and dashed lines respectively. The n and l values indicated in the figure refer to the state of one electron while the other remains in the $n = 1$ state, and the continuum in the energy spectrum corresponds to states where one electron is free, and the helium ion He$^+$ is in its ground state. The optical atomic spectrum of helium shows two sets of lines, which can be explained by the fact that spin is not changed in an electric dipole transition, in accordance with the selection rules (12.38). One set of lines is produced by transitions between levels with $S = 0$ and the other by transitions between levels with $S = 1$. In Figure 13.3 three transitions between $S = 0$ levels and two between $S = 1$ levels are indicated.

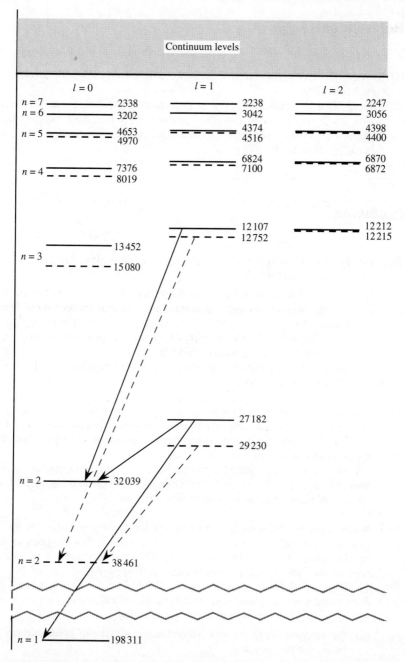

Figure 13.3 The helium spectrum, with singlet (spin-0) levels shown as solid lines, and triplet (spin-1) levels shown as dashed lines. The number beside each level indicates its depth below the continuum in cm^{-1}. All possible electric dipole transitions from the $n = 3$, $l = 1$ to the $n = 2$, $l = 0$ levels, and from the $n = 2$, $l = 1$ to the $n = 2$, $l = 0$ and $n = 1$ levels are shown.

References

[1] Kreyszig E. (1988). *Advanced Engineering Mathematics* 6th edn, Section 7.10. New York: Wiley.

[2] Eisberg R. and Resnick R. (1985). *Quantum Physics of Atoms, Molecules, Solids, Nuclei and Particles* 2nd edn, Section 17.7, p. 652. New York: Wiley.

[3] Jackson J.D. (1975). *Classical Electrodynamics* 2nd edn, Section 3.6, Equation (3.70). New York: Wiley.

Problems

13.1 (a) For the Hamiltonians defined in (13.3), prove that $[\hat{H}_1, \hat{H}_2] = 0$ and $[\hat{H}, \hat{H}_i] = 0$ $(i = 1, 2)$.

   (b) Two particles with independent Hamiltonians $\hat{H}_1$ and H_2 (of the form (13.3)) interact through a potential $\hat{V}_{int}$, so that the complete Hamiltonian for the two-particle system is $\hat{H} = \hat{H}_1 + \hat{H}_2 + \hat{V}_{int}$. If $\hat{V}_{int} = V_0/|r_1 - r_2|$, where r_i $(i = 1, 2)$ are the position vectors associated with the two particles, find $[\hat{H}_i, \hat{V}_{int}]$ and $[\hat{H}, \hat{V}_{int}]$, and explain the physical significance of your results. (Remember that $|r_1 - r_2| = \sqrt{(r_1^2 + r_2^2 - 2r_1 \cdot r_2)}$.).

13.2 Two spinless non-interacting particles, with equal masses m, move in the same one-dimensional simple harmonic oscillator potential, of the form $\hat{V} = \frac{1}{2}m\omega^2 x^2$. Find the ground state energy eigenvalue and eigenfunction of the two-particle system.

   How is your result altered if the particles have different masses? Repeat your calculation of the ground state energy eigenvalue and eigenfunction in the case of three spinless non-interacting particles.

13.3 Write down the Schrödinger equation for the hydrogen molecular ion H_2^+ (one electron and two protons), in the approximation that the protons are a fixed distance R apart. Find the electronic wave function for the ground state of the system, using a variational method similar to that of Problem 11.17: use a trial wave function $\psi = C(\lambda)(\sin \lambda \, \psi_1 + \cos \lambda \, \psi_2)$, where ψ_i is the ground state wave function for a hydrogen atom formed by the electron and proton $i = 1, 2$. Show that in the ground state

   (a) the electron wave function is symmetric about the point midway between the two protons;

   (b) the energy of the molecule when the separation between the protons is R can be expressed in the form

$$\langle E \rangle = -E_R + \frac{2r_B}{R} E_R - \frac{\langle 1|V_2|1 \rangle + \langle 1|V_1|2 \rangle}{1 + \langle 1|2 \rangle}$$

where

$$V_i = \frac{e^2}{4\pi\varepsilon_0} \frac{1}{r_i}$$

and r_i is the distance from proton $i = 1, 2$ to the electron.

(Note that, as in Problem 11.17, we have used Dirac notation in the expression for $\langle E \rangle$.)

13.4 Use the results of Problem 13.3 to find an expression for the ground state energy of the *neutral* hydrogen molecule in an approximation where the interaction between the two electrons is neglected. Write down the corresponding two-electron ground state wave function, and give the total spin of the two electrons. How (qualitatively) would you expect the energy eigenvalue and eigenfunction to be modified when electron–electron repulsion is included?

13.5 Let $\hat{X}_{AB}$ be an operator that exchanges the position coordinates of particles A and B. Show that the symmetrization principle requires the state of a multiparticle system to be an eigenstate of $\hat{X}_{AB}$ for each pair of identical particles A and B. What is the eigenvalue of $\hat{X}_{AB}$ if A and B are (a) fermions and (b) bosons?

13.6 Two identical non-interacting particles of mass m are trapped in the same infinitely deep one-dimensional square well of width $\mathbb{L}$. If the total energy of the two-particle state is $5\hbar^2/8m\mathbb{L}^2$, write down the two-particle wave function in each of the following cases:

(a) the particles are spinless;
(b) the particles are electrons and the spin of the state is $S = 0$;
(c) the particles are electrons and the spin of the state is $S = 1$.

When the Coulomb repulsion between the electrons is taken into account in (b) and (c), which spin state will have the lower energy?

13.7 From Figure 13.3, find the energy necessary to ionize helium from

(a) the ground state;
(b) the lowest $S = 1$ level.

What further energy must be added to doubly ionize the atom? Calculate the wavelengths of the electric dipole transitions between levels with $n = 2$ and $2 < n \leqslant 7$, and compare the series you obtain with the Balmer series in hydrogen.

13.8 In the hydrogen molecule ^{1}H–^{1}H, where the ground state wave function of the electrons is symmetric about the centre of mass of the hydrogen nuclei, only rotational levels with $l = 0, 2, 4, \ldots$ occur. However, rotational levels corresponding to all integer values of l occur in the ^{1}H–^{2}H molecule (where ^{2}H is deuterium). Explain these observations. In the diatomic oxygen molecule ^{16}O–^{16}O the rotational levels that occur are those with $l = 1, 3, 5, \ldots$. What can you deduce about the electronic state of the molecule?

CHAPTER 14

Coherence in Quantum Mechanics

14.1 Coherence in a system containing many identical particles

In the two-slit experiment, which we discussed at length in Chapters 1 and 2, light from a single source is split into two beams by the slit screen (Figure 1.4), and an interference pattern is observed on the detector screen in the region where the beams overlap. The two slits act as two sources of light, but if the single source and the slit screen were replaced by two separate sources then no interference fringes would be seen. There is clearly some particular characteristic of the two beams arising from a single source that is not shared by beams from two separate sources. This characteristic is **coherence**.

The classical idea of coherence is illustrated in Figure 14.1. If two beams are coherent, the phase difference between them, at any point where they overlap, remains constant in time. Clearly, unless the condition is satisfied throughout some sufficiently long time interval, the positions of the interference fringes will shift too rapidly for the pattern to be visible. It is also necessary for the phase differences between the beams at different points in space to be correlated, so that a regular rather than a random distribution of maxima and minima appears in the region where the beams overlap. Two purely monochromatic beams of the same frequency can be represented as infinitely long wave trains with fixed frequency and phase, and such waves always satisfy the criteria for coherence, as shown in Figure 14.1(a). But in practice no source is completely monochromatic, and the spread in frequency of the light means that its phase is not precisely defined. Furthermore, light is emitted from most sources in discrete bursts, with the frequency content and phase of the wave changing randomly from one burst to the next (Figure 14.1b). Two such beams can be coherent only if each random

(a)

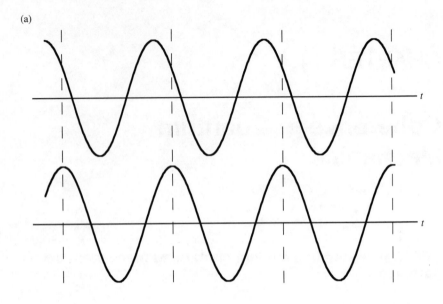

(b)

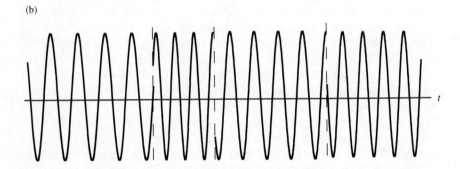

Figure 14.1 (a) Sections of two coherent waves, showing their amplitudes at the same point in space as a function of time. The dashed lines mark the times at which maxima of the lower wave occur, and the upper wave is at the same phase of its motion at each of these instants. (b) A wave emitted by a real source, which changes frequency and phase abruptly at the random instants indicated by the dashed lines.

phase change in one is matched by a corresponding change in the other, and so to observe interference the beams must be derived from the same source by splitting the light with a **beam splitter** – a slit screen or diffraction grating, or a half-silvered mirror. The average time between random phase changes in light from a source is called the **coherence time**.

For a sodium lamp, for example, this is of the order of 10^{-10} s, which means that there is a correlation between the phases at different points in the beam over a length of about 3 cm (the coherence time multiplied by the velocity of light). This is the **coherence length**. The closest approach in practice to a monochromatic beam of light is that emitted by a laser operating continuously on a single mode, and light from a He–Ne laser can have a coherence length as long as 600 km!

Because quantum particles can show wave-like characteristics, the concept of coherence is relevant in quantum mechanics. In the discussion of two-slit diffraction of quantum particles in Section 2.2 we emphasized that interference occurs between two parts of the wave function describing the state of a *single* particle, so that the probability distribution for each individual particle is the two-slit diffraction pattern. So the two parts of the wave function that interfere always arise from the same source. But the pattern is visible only when a sufficiently large number of particles have been detected on the final screen, and unless the positions of minima in all the single-particle probability distributions coincide closely enough the pattern will not be evident. So the observation of the diffraction pattern not only depends on the coherence of the two components of each single-particle probability wave but also requires all the particles to be in approximately the same state, so that the probability distributions associated with different particles are, nearly enough, the same. This demonstrates the importance of two types of coherence in quantum phenomena: the coherence of different components of a single wave function, and the coherence associated with a large number of particles in states described by identical wave functions.

The symmetrization principle of Section 13.3 (essentially the Pauli exclusion principle) implies that it is not possible for two identical fermions whose probability distributions overlap to have identical wave functions. This means that not more than two electrons (in different spin states) can have identical momentum and energy, and it is not possible to produce a completely parallel beam of electrons all with the same energy. But since the momentum spectrum of a free electron is continuous, there are a very large number of different momentum states within a very small range of magnitudes and directions. So the difference between the momenta of different electrons in the beam can be made sufficiently small for the minima in the single-electron probability distributions to coincide closely enough, over a large enough area, for the diffraction pattern to be observed. In such circumstances it is a reasonably good approximation to represent the wave function of each particle by the same plane monochromatic wave, of the form given in (2.9).

In the case of bosons the symmetrization principle does not forbid two particles to be in the same state. Therefore in a system consisting of a very large number of identical bosons it is possible for all the particles to be in identical states. Quantum statistical mechanics predicts that in such a system, at sufficiently low temperatures, a large proportion of the bosons may condense into their ground state. This is known as **Bose–Einstein condensation** [1], and it can lead to very dramatic coherence effects on a macroscopic scale. The escape of superfluid helium from its container by flowing up the sides is one example, and the way a small piece of superconductor can float above the pole of a magnet (due to the Meissner effect [2]) is another.

In the case of helium the isotope ^{4}He is a boson. At a temperature of 2.18 K (about 2 K below the boiling point at normal pressure) a component of liquid helium becomes superfluid: that is, the viscosity associated with linear flow vanishes, and its thermal conductivity becomes very large. The superfluid contains only ^{4}He, though natural helium also contains a small proportion of the isotope ^{3}He. But below 3×10^{-3} K there is another phase transition, and ^{3}He also becomes superfluid. Although the ^{3}He atom is a fermion, at these very low temperatures the atoms tend to form correlated pairs, ^{3}He–^{3}He, which are bosons. Similarly, correlated pairs of electrons form the bosons responsible for low-temperature superconductivity. In this case the correlation is due to an effective attraction between a pair of electrons through the interaction of each of them with the crystal lattice. These correlated pairs of electrons are called **Cooper pairs** [3], and the condensation into their ground state of large numbers of Cooper pairs is responsible for the dramatic disappearance of electrical resistance that characterizes a superconductor. Quantum field theory is necessary for a proper discussion of superfluidity and superconductivity, and so we shall not discuss them further.

14.2 Successive Stern–Gerlach experiments

We now turn to effects associated with the coherence of different components of the same single-particle wave function. For example, the production of a two-slit diffraction pattern by a beam of quantum particles is due ultimately to the interference between different components of the single-particle wave function of each individual particle, as discussed in Section 1.4. In this case each of the components corresponds to a trajectory through a particular slit. In this section we shall examine a different example, in which two components of the state of a spin-$\frac{1}{2}$ particle interfere, since this illustrates the quantum mechanical basis of such effects particularly clearly.

Consider an atom with total angular momentum quantum number $j = \frac{1}{2}$ passing between the pole pieces of successive Stern–Gerlach (SG) magnets. Remember that in an SG experiment (Section 8.3) a beam of atoms is passed through a region in which the magnetic field changes rapidly in the direction parallel to the field. The resulting force on an atom is proportional to the component of its magnetic moment in the field direction, and in general the beam is split into $2j + 1$ components, where j is the total angular momentum quantum number. So when $j = \frac{1}{2}$ the beam is split into two by the magnetic field. In the first SG apparatus the magnetic field $\mathcal{B}_1$ is in the z direction, and the magnetic moments of the atoms in the beam entering the apparatus are randomly oriented. So the incident beam is split into two, each associated with a particular eigenvalue of $\hat{J}_z$. We can prepare a beam of atoms in a state with a particular z component of angular momentum by allowing one beam to emerge from the apparatus and blocking off the other.

Now suppose that the beam of atoms in the state with $j_z = +\frac{1}{2}$ is selected from the first SG apparatus, and passed between the pole pieces of a second SG magnet (as in Problem 8.2). If the field in the second magnet is parallel to that in the first (Figure 14.2a) then the selected beam will not be split again, since all the atoms are in the same eigenstate of $\hat{J}_z$. (We are assuming that the beam is completely isolated from external interactions as it passes from the first SG apparatus to the second, so that the state of the atoms is not altered.) But what happens if the field of the second magnet, $\mathcal{B}_2$, is aligned in the x direction, as shown in Figure 14.2(b)? Since the operator $\hat{J}_x$ does not commute with $\hat{J}_z$, the value of $\hat{J}_x$ is indeterminate for an atom in an eigenstate of $\hat{J}_z$, in the sense discussed in Section 1.5. So we cannot say definitely whether an atom entering the second SG apparatus will emerge from it in the $j_x = +\frac{1}{2}$ or the $j_x = -\frac{1}{2}$ beam. The result is that the beam is split by the second magnet into components corresponding to the two different eigenvalues of $\hat{J}_x$. The relative intensities of the two components are determined by the probabilities that an atom will be found in one eigenstate of $\hat{J}_x$ or the other.

The state of an atom entering the second SG apparatus, at time $t = 0$ say, can be represented by the eigenvector $\boldsymbol{b}_+ - \begin{pmatrix} 1 \\ 0 \end{pmatrix}$, corresponding to the eigenvalue $+\frac{1}{2}$ of $\hat{J}_z$. (For the sake of clarity, we shall use the matrix and Dirac notations of Chapter 10 throughout this chapter.) To find the probability that the atom will emerge from the second apparatus in a particular beam, the initial state must be expanded in terms of the eigenvectors of $\hat{J}_x$. In the basis formed by the eigenvectors of $\hat{J}_z$, the eigenvectors of $\hat{\boldsymbol{J}} \cdot \boldsymbol{n}$, where $\boldsymbol{n}$ is a unit vector in the direction (θ, ϕ), are (Problem 10.10)

$$\left. \begin{aligned} \boldsymbol{c}_+ &= \cos\tfrac{1}{2}\theta\, \boldsymbol{b}_+ + \sin\tfrac{1}{2}\theta\, e^{i\phi} \boldsymbol{b}_- \\ \boldsymbol{c}_- &= -\sin\tfrac{1}{2}\theta\, e^{-i\phi} \boldsymbol{b}_+ + \cos\tfrac{1}{2}\theta\, \boldsymbol{b}_- \end{aligned} \right\} \tag{14.1}$$

These are eigenvectors of $\hat{J}_x$ when $\theta = \frac{1}{2}\pi$, $\phi = 0$. The eigenvector $\boldsymbol{b}_+$ can be found in terms of $\boldsymbol{c}_+$ and $\boldsymbol{c}_-$ by inverting this pair of equations.

(a)

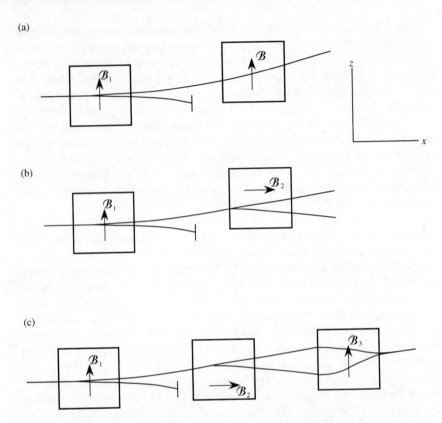

(b)

(c)

Figure 14.2 Schematic representation of an atomic beam passing through success-ive Stern–Gerlach magnets. In each case the lower beam emerging from the first magnet is stopped, and only one beam enters the second magnet. (a) The fields $\mathcal{B}_1$ and $\mathcal{B}$ in the first and second magnets are both in the z direction. (b) The fields $\mathcal{B}_1$ and $\mathcal{B}_2$ are in the z and x directions respectively. (c) The same as (b), but with a third magnet with field $\mathcal{B}_3$ in the z direction. The case where the two beams from the second magnet regenerate a single beam is illustrated.

To keep track of the component of $\hat{\boldsymbol{J}}$ to which the eigenvectors refer, it is more convenient to use Dirac notation. Let $|+\frac{1}{2}; \boldsymbol{n}\rangle$ denote the eigenst-ate of $\hat{\boldsymbol{J}} \cdot \boldsymbol{n}$ corresponding to the eigenvalue $+\frac{1}{2}$, $|+\frac{1}{2}; z\rangle$ the eigenstate of $\hat{J}_z$ corresponding to the eigenvalue $+\frac{1}{2}$, and so on. In this notation (14.1) becomes

$$\left.\begin{array}{l} |+\tfrac{1}{2};\, \boldsymbol{n}\rangle = \cos\tfrac{1}{2}\theta\, |+\tfrac{1}{2};\, z\rangle + \sin\tfrac{1}{2}\theta\, e^{i\phi}|-\tfrac{1}{2};\, z\rangle \\[2mm] |-\tfrac{1}{2};\, \boldsymbol{n}\rangle = -\sin\tfrac{1}{2}\theta\, e^{-i\phi}|+\tfrac{1}{2};\, z\rangle + \cos\tfrac{1}{2}\theta\, |-\tfrac{1}{2};\, z\rangle \end{array}\right\} \qquad \textbf{(14.2)}$$

Choosing n to be along the x axis, and inverting the equations, we find that the state of an atom entering the second SG apparatus is

$$|+\tfrac{1}{2}; z\rangle = \sqrt{(\tfrac{1}{2})} \, (|+\tfrac{1}{2}; x\rangle - |-\tfrac{1}{2}; x\rangle) \qquad (14.3)$$

So the probability that the atom is in the $j_x = +\tfrac{1}{2}$ beam is $|\sqrt{\tfrac{1}{2}}|^2 = \tfrac{1}{2}$, equal to the probability that it is in the $j_x = -\tfrac{1}{2}$ beam. This means that the intensities of the two beams should be equal.

We have not taken into account the time dependence of the state of the atom, but in fact our conclusion is not affected by this. The evolution of the state with time is determined by the energy eigenstates, which are the eigenstates of $\hat{J}_x$ when the atom is in the field $\mathcal{B}_2$. They vary with time as $e^{-i\omega t}$ for the state with $j_x = +\tfrac{1}{2}$ and $e^{+i\omega t}$ for that with $j_x = -\tfrac{1}{2}$, where ω is proportional to $\mathcal{B}_2$. The state of an atom entering the second SG apparatus at $t = 0$ is $|\Psi(0)\rangle = |+\tfrac{1}{2}; z\rangle$, and its expansion in terms of the energy eigenstates at time $t = 0$ is given by (14.3). So at any later time t the state of the atom will be

$$|\Psi(t)\rangle = \sqrt{(\tfrac{1}{2})}(e^{-i\omega t}|+\tfrac{1}{2}; x\rangle - e^{i\omega t}|-\tfrac{1}{2}; x\rangle) \qquad (14.4)$$

(This is in accord with our discussion of the energy eigenstates of a spin-$\tfrac{1}{2}$ particle in a magnetic field in an arbitrary direction in Section 10.5, and can be obtained by inverting the pair of equations (10.39) and inserting the appropriate values for θ and ϕ.) According to (14.4), the probability that an atom is in the $j_x = +\tfrac{1}{2}$ beam at time t is $|\sqrt{(\tfrac{1}{2})}e^{-i\omega t}|^2$, which is equal to $\tfrac{1}{2}$, independently of the time. So we expect the intensities of the two beams emerging from the second SG apparatus to be equal, no matter how long the atoms take to pass through the magnetic field.

The linear superposition (14.4) is a coherent combination of states corresponding to different eigenvalues of $\hat{J}_x$, and this coherence can be demonstrated by passing the two beams emerging from the second SG apparatus through a third, in which the magnetic field $\mathcal{B}_3$ is in the z direction, parallel to $\mathcal{B}_1$ (Figure 14.2c). Again we assume that the two beams are completely free as they travel from one SG apparatus to the next, so that the state of the atoms is not disturbed. Suppose that an atom leaves the region where the field is $\mathcal{B}_2$ at time t' and enters the third field at t''. The phase of each component of the state vector $|\Psi(t')\rangle$ depends on $\gamma = \omega t'$, which is determined by the magnitude of the field $\mathcal{B}_2$ and the length of time t' that the atom has spent in it. In the region of zero field between the two magnets the energies of the two beams are degenerate, and equal to zero on the energy scale we are using, so that the factor $e^{-iEt/\hbar}$ is equal to 1 for both beams in the time interval $t' \leqslant t \leqslant t''$. Then the state of an atom at $t = t''$ entering the third SG apparatus is the same as it was on leaving the second, $|\Psi(t'')\rangle = |\Psi(t')\rangle$, and, according to (14.4), this is

$$|\Psi(t'')\rangle = \sqrt{(\tfrac{1}{2})}(e^{-i\gamma}|+\tfrac{1}{2}; x\rangle - e^{i\gamma}|-\tfrac{1}{2}; x\rangle) \qquad (14.5)$$

This can be expanded in terms of the eigenstates of $\hat{J}_z$ (the component parallel to the field $\mathcal{B}_3$ in the third SG apparatus), and, from (14.2),

$$|\Psi(t'')\rangle = \cos\gamma\,|+\tfrac{1}{2};\,z\rangle + i\sin\gamma\,|-\tfrac{1}{2};\,z\rangle \qquad (14.6)$$

If γ is an integer multiple of π, the state of the atom is once again an eigenstate of $\hat{J}_z$ corresponding to the eigenvalue $+\tfrac{1}{2}$. The state in which it left the first apparatus has been regenerated by recombining coherently the two beams from the second SG apparatus. In this case the beam is not split again when it passes through the third SG apparatus. Similarly, if γ is an odd integer multiple of $\tfrac{1}{2}\pi$, the beam is not split, though in this case the eigenvalue of $\hat{J}_z$ is $-\tfrac{1}{2}$. For other values of γ the beam is split into two, and the relative intensities of the beams are determined by γ. The intensity of each beam oscillates between 0 and a maximum as γ is varied (by varying the magnitude of the field $\mathcal{B}_2$ for example), and these oscillations are out of phase by π in the two beams, so that when one is a maximum the other is zero, and vice versa. Clearly these intensity oscillations are due to interference between the $j_x = +\tfrac{1}{2}$ and $j_x = -\tfrac{1}{2}$ components of the state vector $|\Psi(t'')\rangle$, and such interference can only occur when the $j_x = +\tfrac{1}{2}$ and $j_x = -\tfrac{1}{2}$ beams are coherent.

The interference we have just described occurs between two components of the same single-particle state, so that the interference pattern is present in the probability distribution of each atom. This is similar to the two-slit diffraction of quantum particles. In Sections 1.5 and 1.6 we associated the observation of the diffraction pattern with complete uncertainty as to which slit a given particle had passed through. Similarly, in the present case, the fact that the state of each atom can be expressed, through (14.5), as a superposition of components with $j_x = +\tfrac{1}{2}$ and $j_x = -\tfrac{1}{2}$ implies that the atom cannot be assigned definitely to one beam emerging from the second SG apparatus or to the other. But suppose that we perform a measurement between the second and third SG magnets, to determine to which beam each atom belongs. This corresponds to identifying the slit through which each particle passed in the two-slit diffraction case, and the interference effects are destroyed.

So if we can identify the beam to which each atom entering the third SG apparatus belongs, the two beams lose their coherence, though the physical mechanism through which this loss occurs may not always be clear. Before the measurement each atom is in the coherent superposition of states $|\Psi(t')\rangle = \sqrt{(\tfrac{1}{2})}(e^{-i\gamma}|+\tfrac{1}{2};\,x\rangle - e^{i\gamma}|-\tfrac{1}{2};\,x\rangle)$, and its value of $\hat{J}_x$ is indeterminate. The result of the measurement on a particular atom shows that it is *either* in the $j_x = +\tfrac{1}{2}$ beam *or* the $j_x = -\tfrac{1}{2}$ one, so that its state is either $|\Psi_+\rangle = |+\tfrac{1}{2};\,x\rangle$ or $|\Psi_-\rangle = |-\tfrac{1}{2};\,x\rangle$. The effect of the measurement can be represented as

$$|\Psi(t')\rangle \to |\Psi_+\rangle \qquad \text{or} \qquad |\Psi(t')\rangle \to |\Psi_-\rangle \qquad (14.7)$$

(This is the **collapse of the wave packet**.) After the measurement there is no longer any indeterminacy in the value of $\hat{J}_x$ (though now the value of $\hat{J}_z$ is indeterminate), and there is no correlation between the phases of the states $|\Psi_+\rangle$ and $|\Psi_-\rangle$. The states of the two beams can be expanded in terms of eigenstates of $\hat{J}_z$ through (14.2):

$$\left. \begin{aligned} |\Psi_+\rangle &= \sqrt{(\tfrac{1}{2})}(|+\tfrac{1}{2};\, z\rangle + |-\tfrac{1}{2};\, z\rangle) \\ |\Psi_-\rangle &= \sqrt{(\tfrac{1}{2})}(-|+\tfrac{1}{2};\, z\rangle + |-\tfrac{1}{2};\, z\rangle) \end{aligned} \right\} \qquad \textbf{(14.8)}$$

So in the field $\mathcal{B}_3$ each beam independently splits into two beams of equal intensity. There can be no interference between $|\Psi_+\rangle$ and $|\Psi_-\rangle$, and so the variation in intensity of the beams with γ implied by (14.6) does not occur – in this case two beams of equal intensity always leave the third SG apparatus whatever the value of γ.

We have seen that the interference between different components of the state $|\Psi(t')\rangle$ is associated with indeterminacy in the value of $\hat{J}_x$. This means that the position of an atom emerging from the second SG apparatus is also indeterminate, because the two beams are separated spatially, and the atom is not definitely in one beam or the other. In other words, the atom is **non-localized**. Does this mean that the atom is 'smeared out' in the x direction over a distance at least as great as the separation between the beams? The question must be phrased in such a way that it is possible to design a physical measurement to answer it (assuming that we are looking for a physical rather than a purely philosophical answer!). We have already mentioned such a measurement – the measurement that determines the x component of the position of an atom between the second and third SG apparatus. Essentially this measures the value of $\hat{J}_x$, and the result is always either $+\tfrac{1}{2}$ or $-\tfrac{1}{2}$, but never an intermediate value. That is, an atom will always be found in one beam or the other, but never smeared out between the two. But of course this measurement simultaneously destroys the interference effects, and the non-locality that gave rise to the question. Single-particle interference experiments cannot provide any further information about the meaning of this quantum non-locality. However, interference in correlated two-particle systems does provide further insight into the concept, and so in Section 14.3 we turn to an example of such two-particle correlations.

14.3 Two-particle correlation experiments

If a pair of particles is produced in a single process, such as the decay of a rho into two pions (Section 13.4) or the decay of an excited state of an atom in which two photons are emitted, the states of the two particles are correlated, and the pair may be described by a two-particle wave function. As an example, consider a pair of spin-half particles produced in a state with total spin zero [4]. Suppose that the particles move off in opposite directions, and do not interact with one another once they have left the source O. (For example, at low energy and for large scattering angles proton–proton scattering satisfies this criterion.) Then the spin state of the pair is given by (13.25), which can be expressed in Dirac notation as

$$|AB, 0\rangle = \sqrt{(\tfrac{1}{2})}(|A+;\, n\rangle|B-;\, n\rangle - |A-;\, n\rangle|B+;\, n\rangle) \qquad (14.9)$$

Here $|A\pm;\, n\rangle$ and $|B\pm;\, n\rangle$ are eigenstates corresponding to the eigenvalue $\pm\tfrac{1}{2}$ of the spin component $\hat{s} \cdot n$, for particles A and B respectively. (Lower-case $\hat{s}$ refers to the spin of a single particle, while capital $\hat{S}$ refers to the spin of the pair.) The direction of n does not matter, but it is the same in each of the four single-particle states that appear on the right-hand side of (14.9).

The spin-0 pairs are produced from their source O, and the two particles move off towards detectors D_A and D_B, as shown in Figure 14.3. (Of course not all particle pairs produced by O are necessarily emitted towards these two detectors, but we are only concerned with those that are.) In front of each detector is a filter, F_A or F_B, which allows particles with one eigenvalue of a chosen spin component to pass, but stops those with the opposite eigenvalue. For example, let filter F_A stop particles with spin components $-\tfrac{1}{2}$ in the direction defined by unit vector a (just as, in Section 14.2 we envisaged selecting atoms with $j_z = +\tfrac{1}{2}$ by blocking off the $j_z = -\tfrac{1}{2}$ beam emerging from the first SG apparatus). Then only the arrival of particles whose $\hat{s} \cdot a$ eigenvalue is $+\tfrac{1}{2}$ will be recorded by detector D_A. Similarly, we arrange that only particles whose $\hat{s} \cdot b$ eigenvalue is $+\tfrac{1}{2}$ will pass filter F_B and be detected by D_B. The settings a and b of the two filters can be varied independently.

From (14.9), the probability that particle A has a value of the spin component $\hat{s} \cdot n$ equal to $+\tfrac{1}{2}$ is $\tfrac{1}{2}$, for any direction n. So in 50% of the events in which one particle is emitted towards detector D_A the particle will pass filter F_A and be recorded by D_A, whatever the direction a. We call such an event a 'hit', while a 'miss' is an event in which the particle is stopped by the filter. (We assume that an observer monitoring the detector can also monitor the source, and knows each time a pair is emitted with one particle travelling towards D_A.) The observer at D_A will see a random pattern of hits and misses. Similarly, 50% of the particles emitted towards detector D_B will pass filter F_B, whatever the direction b, and again the pattern of hits and misses is random. However, interesting interference effects become apparent when correlations between the responses of the two detectors are examined.

First, suppose that the filters are set so that $b = a$. We label any particle A if it enters filter F_A. The arrival of a particle in the state $|A+;\, a\rangle$ is

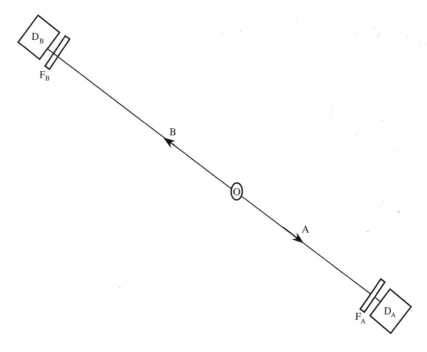

Figure 14.3 Schematic representation of the arrangement for a two-particle correlation experiment. O is the source of pairs, A and B mark the paths of two particles belonging to the same pair, F_A and F_B are the filters, and D_A and D_B are the detectors.

registered as a hit, while a particle in the state $|A-; \, a\rangle$ registers a miss. Putting $n = a$ in (14.9), we see that if particle A passes filter F_A then the other member of the pair (the corresponding B particle) must be in the state $|B-; \, a\rangle$, and will fail to pass the filter F_B. So a hit at D_A is always correlated with a miss at D_B, and vice versa. This means that, although the observers at D_A and D_B each see a random pattern of hits and misses, when they compare results they will find a 100% anticorrelation between the two random patterns: a hit in one of the patterns always coincides with a miss in the other. If the setting at F_B is changed to $b = -a$ then only particles in the state $|B+; \, -a\rangle$ register a hit at D_B, and this state is identical to $|B-; \, a\rangle$. So in this case each pair of particles would produce either a pair of hits or a pair of misses, and there is complete correlation between the two sets of results.

The degree of correlation between the two random patterns of hits and misses observed at D_A and D_B can be represented by a **correlation function** $\mathbb{E}$ defined by

$$\mathbb{E}(a, b) = \mathbb{P}(++) + \mathbb{P}(--) - \mathbb{P}(+-) - \mathbb{P}(-+) \qquad \textbf{(14.10)}$$

Here $\mathbb{P}(++)$ is the probability that hits will be recorded at both D_A and D_B when the filter settings are a and b respectively. Correspondingly $\mathbb{P}(--)$ is the probability of a pair of misses, $\mathbb{P}(+-)$ is the probability of a hit at D_A and a miss at D_B, and $\mathbb{P}(-+)$ is the probability of a miss at D_A and a hit at D_B. For example, there is a probability equal to $\frac{1}{2}$ that a hit will be recorded at D_A, and in the previous paragraph we found that when $b = a$ a hit at D_A is always accompanied by a miss at D_B. So in this case $\mathbb{P}(++) = 0$ and $\mathbb{P}(+-) = \frac{1}{2}$. The complete set of probabilities in the two cases we have discussed is

$$\left. \begin{array}{ll} \mathbb{P}(++) = \mathbb{P}(--) = 0, & \mathbb{P}(+-) = \mathbb{P}(-+) = \frac{1}{2} \quad (b = a) \\ \mathbb{P}(++) = \mathbb{P}(--) = \frac{1}{2}, & \mathbb{P}(+-) = \mathbb{P}(-+) = 0 \quad (b = -a) \end{array} \right\} \quad \textbf{(14.11)}$$

Substituting each of these sets in (14.10), we obtain

$$\left. \begin{array}{l} E(a, a) = 0 + 0 - \frac{1}{2} - \frac{1}{2} = -1 \\ E(a, -a) = \frac{1}{2} + \frac{1}{2} - 0 - 0 = +1 \end{array} \right\} \quad \textbf{(14.12)}$$

so that the correlation function takes the value $+1$ for complete correlation and -1 for complete anticorrelation. If there were no correlation at all between the results at D_A and those at D_B, the correlation function would be zero, since in that case one particular combination of results would be as likely as any other, and so, for any pair of directions (a, b), we should find

$$\mathbb{P}(++) = \mathbb{P}(--) = \mathbb{P}(+-) = \mathbb{P}(-+) = \frac{1}{4}, \qquad E(a, b) = 0 \quad \textbf{(14.13)}$$

What happens if the second filter setting is not parallel to the first? Suppose that A registers a hit at D_A, so that its spin state is $|A+; a\rangle$. Then the spin state of B must be $|B-; a\rangle$. If the setting of F_B is $b \neq a$, we must expand $|B-; a\rangle$ in terms of the eigenstates of $\hat{s} \cdot b$ to predict the result at D_B. If the z axis is defined to be the direction of a and the direction of b is determined by the polar angles (θ, ϕ) then (14.2) gives the spin states $|B+; b\rangle$ and $|B-; b\rangle$ in terms of the eigenstates $|B+; a\rangle$ and $|B-; a\rangle$. Inverting the equations, we find

$$\left. \begin{array}{l} |B+; a\rangle = \cos\frac{1}{2}\theta \, |B+; b\rangle - \sin\frac{1}{2}\theta \, e^{i\phi} |B-; b\rangle \\ |B-; a\rangle = \sin\frac{1}{2}\theta \, e^{-i\phi} |B+; b\rangle + \cos\frac{1}{2}\theta \, |B-; b\rangle \end{array} \right\} \quad \textbf{(14.14)}$$

So if a hit is registered at D_A, the probability of a miss at D_B (from the expression for $|B-; a\rangle$) is $\cos^2\frac{1}{2}\theta$, while the probability of a hit is $\sin^2\frac{1}{2}\theta$. Similarly, if the result at D_A is a miss, the state of B must be $|B+; a\rangle$, so that the probability of a hit at D_B is $\cos^2\frac{1}{2}\theta$, and the probability of a miss is $\sin^2\frac{1}{2}\theta$. The four joint probabilities in this case are

$$\mathbb{P}(++) = \mathbb{P}(--) = \frac{1}{2}\sin^2\frac{1}{2}\theta, \qquad \mathbb{P}(+-) = \mathbb{P}(-+) = \frac{1}{2}\cos^2\frac{1}{2}\theta \quad \textbf{(14.15)}$$

where θ is the angle between a and b. The corresponding value of the correlation function $\mathbb{E}(a, b)$ is

$$\mathbb{E}(a, b) = \sin^2\tfrac{1}{2}\theta - \cos^2\tfrac{1}{2}\theta = -\cos\theta = -a \cdot b \qquad (14.16)$$

(This reproduces the values given in (14.12) for $b = a$ and $b = -a$, and gives zero correlation when the two directions are perpendicular.)

The correlation function $\mathbb{E}(a, b)$ is a quantity that can be measured experimentally, and there is no difficulty in using quantum mechanics to predict the result: (14.16) predicts a unique value for $\mathbb{E}$ for any possible choice of the directions a and b. But, just as in the case of single-particle interference effects, there is an intuitive problem if we attempt to visualize the experiment in terms of the behaviour of individual particles.

Suppose that the filter and detector F_A and D_A are closer to the source O than F_B and D_B. Then particle A reaches F_A before B reaches F_B, and the result obtained for particle A can be used to predict the result to be expected for B. According to quantum mechanics, the individual spin state of each particle is indeterminate when it leaves the source (though the total spin state, $S = 0$, is fixed). When A enters the filter F_A its spin state is changed to one of the eigenstates of $\hat{s} \cdot a$, in a process similar to that represented in (14.7). That is, for whatever setting a has been chosen for the filter, either a hit or a miss is registered. But this means that the two-particle state (14.9) must also collapse to one of its components:

$$|AB, 0\rangle \rightarrow |A+; a\rangle\, |B-; a\rangle \quad \text{or} \quad |AB, 0\rangle \rightarrow |A-; a\rangle\, |B+; a\rangle$$

$$(14.17)$$

If a hit is observed, the first of these must have occurred, and we can predict with absolute certainty that the result at D_B, if b is chosen parallel to a, is a miss. On the other hand, if A records a miss, the alternative collapse has occurred, and the state of B must be $|B-; a\rangle$. The spin state of B was indeterminate before A reached filter F_A, and the hit or miss event at D_A simultaneously changes the states of both A and B. The two particles are non-interacting, and can be as far apart as we like when the measurement on A is made, yet it appears that as soon as a measurement of $\hat{s} \cdot a$ is made on particle A, the spin component parallel to a of the other particle in the pair, B, becomes well defined! Furthermore, the spin state of B immediately after the measurement on A depends on the choice of the setting a of filter F_A: if the setting had been chosen to be a' instead of a, the state of B would become either $|B+; a'\rangle$ or $|B-; a'\rangle$, so that the value of $\hat{s} \cdot a'$ rather than $\hat{s} \cdot a$ is well defined. This means that the probability of a hit at D_B (for a fixed choice of b) changes if the direction of the setting at the distant filter F_A is changed.

Our intuitive feeling is that if the spin state of particle B is changed by a measurement on A, there must be some kind of interaction between the pair. But the quantum mechanical prediction is completely independent of the separation between the events at D_A and D_B. The separation in time and space between the arrival of A at F_A and the arrival of B at F_B could be made so great that there is no time for a light signal to propagate from the hit (or miss) event at D_A to the particle B before it arrives at F_B. Indeed, it could be arranged that B reaches filter F_B almost immediately after the event at D_A, and, since the prediction for the result at D_B applies as soon as the state of A is determined, it seems that any signal connecting one event with the other would have to be almost instantaneous. So an interaction connecting the two events would have to be some kind of 'action at a distance' – that is, a non-local interaction. Do such non-local effects really exist, or is there some way of interpreting the events in terms of purely local interactions? This question was made susceptible to experimental investigation by the work of John Bell, as we shall discuss in Section 14.4.

14.4 Determinism, locality and Bell's inequality

The intuitive difficulties in understanding quantum mechanics are associated with the indeterminacy and non-locality that we discussed in Sections 14.2 and 14.3. These characteristics of the theory have led to suggestions that a quantum mechanical description of physical phenomena is not complete, and that there is an underlying theory (not yet discovered) that reproduces the correct predictions of quantum mechanics but that is deterministic and local [5]. We shall re-examine the single-particle and two-particle correlation experiments from a classical point of view, to try to understand what such a theory would entail, and finally look at what we can learn from experiment.

First of all let us look again at the beam of atoms passing through consecutive SG magnets. An atom emerging from the first SG apparatus in Figure 14.2 is in an eigenstate of $\hat{J}_z$ corresponding to the eigenvalue $+\frac{1}{2}$, and from the classical point of view this means that the atom emerges with its magnetic moment aligned parallel to the z direction. When the beam passes through the second SG apparatus we know (from experiment) that it splits in two, and quantum mechanics successfully predicts this, and distinguishes the states of atoms in the two beams by the eigenvalue of $\hat{J}_x$. Classically this means that the atoms have their magnetic moments either parallel or antiparallel to the field $\mathcal{B}_2$. But it is impossible for a classical magnetic moment aligned with the z axis to be simultaneously either parallel or antiparallel to the x axis, so the direction of the magnetic moment, and thus the angular momentum, of each atom must change when it enters the second magnetic field. In some cases it flips to the x direction, and in others to the $-x$ direction. (This behaviour is *not* predicted by classical electromagnetic theory.)

Quantum mechanics cannot predict in which eigenstate of $\hat{J}_x$ a particular atom will be found – the x component of angular momentum is indeterminate as it enters the second SG apparatus. But a deterministic theory would identify dynamical variables whose values determine precisely how the atom behaves. Since no measurements can determine in advance which way the angular momentum of a particular atom will flip, the variables that determine the behaviour uniquely must be ones that are not accessible to our measurements, and are termed **hidden variables**. According to this view, the 'true' state of a particle is only completely specified if the values of all the hidden variables are known, in addition to the information contained in the quantum mechanical wave function: the quantum mechanical description of the state of a particle is incomplete.

A deterministic theory makes the fundamental assumption that an atom in the second SG apparatus is 'really' in one of the two beams, even if we cannot know which – in complete contrast to the quantum point of view explained in Section 1.5. But in practice a physicist who believes in determinism cannot predict in which beam a given atom will be found with any greater certainty than a quantum theorist. The former prefers to ascribe the actual behaviour of an atom to the particular values of its hidden variables, even though it may be impossible to know what these values are, while the latter believes that there is no underlying deterministic reason for the particular behaviour of an individual atom.

It seems at first sight that a hidden variables theory would avoid the problem of non-locality. Consider the correlated pair experiment described in Section 14.3. From the deterministic point of view, a pair is produced with each particle in a unique state, characterized not only by an eigenvalue of some component of spin but also by the values of a set of hidden variables. The values of the hidden variables are correlated, and so are the spin states – in a state of two spin-$\frac{1}{2}$ particles with total spin $S = 0$ the individual spins are always oppositely oriented. In one pair (according to a deterministic theory) they might be oppositely oriented along the z axis, while in the next they could be oppositely oriented along an axis at angle θ to the z axis, and so on. When a particle arrives at filter F_A its spin is oriented in some well-defined direction, but in general this will not be the same as the direction a. So the spin of the particle will flip to either up or down along the a direction, depending on the values of its hidden variables. Since the values of B's hidden variables are correlated with those of A, they will be sufficient to determine whether or not it passes the filter F_B without the necessity of faster-than-light signals from the event at D_A.

The hidden variables picture presented above avoids non-local interactions because all information necessary to determine uniquely the behaviour of each particle in any given situation is carried by the particle itself. The values of its hidden variables and the setting on filter F_B prescribe whether B will score a hit or a miss. This result is correlated with the result at D_A only because B's hidden variables are correlated with A's through the common production process, just as its spin state is, and the result at D_B is not affected by the choice of setting on the distant filter F_A.

But it can be proved that such a local hidden variables theory can never reproduce all the predictions of quantum mechanics. Suppose that, in connection with the two-particle correlation experiment, we make the following locality assumption:

Whether particle B scores a hit or miss at D_B is not affected by the direction chosen for the setting of filter F_A.

This means that the interaction between particle B and the filter F_B is local, and depends only on the state of B (including the values of all hidden variables). This assumption is sufficient to prove the following inequality concerning the correlation functions defined in (14.10):

$$|E(a, b) + E(a, b') + E(a', b) - E(a', b')| \leqslant 2 \qquad (14.18)$$

This is one version of an inequality first derived by John Bell [6].

Although we have presented it in the context of measurements on correlated pairs of spin-$\frac{1}{2}$ particles, the ideas concerning hidden variables, locality and the inequality (14.18) itself are quite general. The problem of non-locality in correlations between the results of measurements on pairs of quantum particles from a common source is the essence of the famous Einstein–Podolsky–Rosen paradox [5]. The locality assumption can be expressed more generally as

The outcome of a measurement on a particle made in one region of space depends only on local conditions and the state of the particle, and is unaffected by the settings of remote instruments.

In the general case of a measurement with two possible outcomes the correlation function is defined by an expression of the form (14.10), provided the results denoted by $+$ and $-$ are given an appropriate interpretation. For example, if the two particles are photons and the filters are polarizers then $+$ and $-$ refer to two states of linear polarization, one parallel to the direction in which the polarizer is set and the other perpendicular to that direction. (The specific form of the quantum mechanical predictions for the four probabilities given in (14.15) only applies to measurements on spin-$\frac{1}{2}$ particles, but corresponding expressions can be derived in other cases.)

The left-hand side of (14.18) contains correlation functions relating to four different sets of measurements, in each of which the settings of the filters F_A and F_B are different. If we substitute the values predicted by

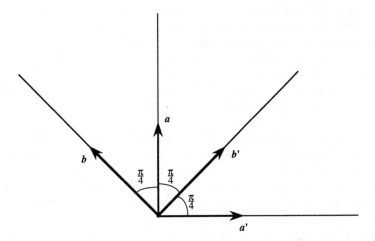

Figure 14.4 Orientations a and a' of F_A, and b and b' of F_B for which the quantum prediction violates Bell's inequality.

quantum mechanics for a pair of spin-$\frac{1}{2}$ particles in a state with total spin 0, (14.16), we obtain

$$|-a \cdot b - a \cdot b' - a' \cdot b + a' \cdot b'| \leqslant 2 \qquad (14.19)$$

But there are some choices of the four directions for which this is not true. For example, if we choose the directions of a, a', b and b' as shown in Figure 14.4 then $a \cdot b = a \cdot b' = a' \cdot b' = \cos\frac{1}{4}\pi = \sqrt{\frac{1}{2}}$ and $a' \cdot b = \cos\frac{3}{4}\pi = -\sqrt{\frac{1}{2}}$. The left-hand side of (14.19) takes the value $2\sqrt{2}$, and the inequality (14.18) is violated. This violation of Bell's inequality by the predictions of quantum mechanics does not occur only for the particular two-particle system we have considered, but is quite general. Since the essential ingredient in the derivation of the inequality is the locality assumption, this means that quantum mechanics cannot be a completely local theory.

The question as to whether inherently non-local correlations between measurements do in fact occur now becomes a question of whether or not quantum mechanics is correct. The expression for the correlation function, (14.16), results directly from very fundamental ideas in quantum theory – the quantization of angular momentum, and the superposition principle with the associated interpretation of expansion coefficients in terms of probabilities. These ideas have given correct predictions in an enormous range of experimental situations, as we have tried to indicate throughout this book, so that it would not be easy to find a new theory that gave the same, correct, results in all other areas, yet predicted a form for the correlation function E that never violated Bell's inequality. But ultimately the answer to the question must be determined by experiment. The correlation function has been measured in two-photon correlation experiments [7], and the left-hand side of the inequality (14.18) determined for polarizer settings a, a', b and b' for which the quantum mechanical prediction violates the inequality. These experiments are very difficult to perform and analyse, but they seem to indicate that the inequality is indeed violated.

If an experiment violates Bell's inequality, its result can never be predicted by a local hidden variables theory. Even though this does not rule out the existence of hidden variables, it means that any satisfactory hidden variables theory would have to include non-local effects. The indeterminacy and non-locality that quantum mechanics introduces reflect real characteristics of the physical universe, and so far no other theory can explain so accurately or elegantly the structure and behaviour of matter in terms of its molecular, atomic and subatomic constituents.

References

[1] Mandl F. (1971). *Statistical Physics*, Section 11.5. London: Wiley.
[2] Omar M.A. (1975). *Elementary Solid State Physics*, Section 10.3. Reading MA: Addison-Wesley.
[3] See Ref. [2], Section 10.7.
[4] Bell J.S. (1987). Bertlmann's Socks and the Nature of Reality. In *Speakable and Unspeakable in Quantum Mechanics*, Chap. 16. Cambridge University Press.
[5] Mermin N.D. (1985). Is the Moon there when nobody looks? Reality and the quantum theory. *Physics Today* **38**, 39.
[6] Bell J.S. (1964). *Physics* **1**, 195 (reprinted as Chap. 2 of the volume cited in [4]).
[7] Aspect A. (1991). *Europhysics News* **22**, 73.

Problems

14.1 For an atom with total angular momentum quantum number $\frac{1}{2}$ find
   (a) the expansion of the state with $j_z = -\frac{1}{2}$ in terms of the eigenstates of (i) $\hat{J}_x$ and (ii) $\hat{J}_y$;
   (b) the expansions of both eigenstates of $\hat{J}_z$ in terms of eigenstates of $\hat{\mathbf{J}} \cdot \mathbf{n}$, where $\mathbf{n}$ is a unit vector at 60° to the z axis in the plane containing the z axis and making an angle of 45° with the x axis.

14.2 Find the relative intensities of the beams emerging from the second SG apparatus in Figure 14.2 if the beam entering the apparatus has $j_z = +\frac{1}{2}$ and the field $\mathcal{B}_2$ is in the direction n specified in Problem 14.1(b). If the field in the third SG apparatus is in the z direction, and the two beams take time $t = \pi/3\omega$ to pass through the second apparatus (where $\pm\hbar\omega$ are their energy eigenvalues), find the probability of finding an atom emerging from the third apparatus in the $j_z = +\frac{1}{2}$ state.

14.3 (a) In an experiment of the type illustrated in Figure 14.3 the following set of 50 results is obtained by an observer monitoring detector D_A, when the setting of filter F_A is a:

A: 0 0 0 1 0 0 0 1 0 0 0 0 0 1 0 1 0 1 1 0 1 0 0 1 1
      1 0 0 0 1 1 1 1 0 1 1 0 1 1 1 0 1 0 1 1 1 1 1 0 0

(0 represents a miss and 1 a hit). Write down the result for the other particle in each pair, obtained by an observer monitoring detector D_B if the setting of filter F_B is a and if it is $-a$.

   (b) Find the value of the correlation function from the following results of observations on 50 pairs of particles:

A: 0 0 1 1 1 0 0 0 0 0 1 1 1 0 0 1 1 0 0 1 1 1 1 0 1
      0 0 1 1 1 0 1 0 0 0 0 0 0 1 1 1 1 1 0 1 1 0 1 0 0

B: 1 1 0 0 0 0 1 1 0 1 0 0 1 1 1 0 0 0 1 0 0 1 1 0 0
      0 0 0 0 1 1 1 0 1 0 1 1 1 1 0 1 0 0 0 1 0 1 1 1 1

Deduce the angle between the settings a and b of the two filters.

14.4 Calculate the value of the left-hand side of (14.19), and determine whether or not Bell's inequality is violated, for the following orientations of the vectors a', b', a and b:

   (a) a' is in the x direction, b and b' are reached from it by anticlockwise rotations in the (x, y) plane through 30° and 120° respectively, and a is in the y direction;

   (b) a' is along the x direction, a and b are reached from it by anticlockwise rotations through 30° and 120° respectively, and b' is in the y direction.

APPENDIX A

The Two-Body Problem in Classical Mechanics

A1 The kinetic energy of a two-particle system

The kinetic energy of a system of particles can always be written as the sum of the kinetic energy associated with the motion of the system as a whole and the kinetic energy associated with the internal motions. We shall consider the particular case of a two-particle system.

Consider two particles, with masses m_1 and m_2 and position vectors r_1 and r_2, as shown in Figure A1. The centre of mass of the system is at

$$R = \frac{m_1 r_1 + m_2 r_2}{m_1 + m_2} \tag{A1.1}$$

and the relative position vector is the vector from the first to the second particle

$$r = r_2 - r_1 \tag{A1.2}$$

The momenta of the particles are

$$p_1 = m_1 \frac{dr_1}{dt}, \qquad p_2 = m_2 \frac{dr_2}{dt} \tag{A1.3}$$

and the momentum of the system as a whole is

$$P = p_1 + p_2 = (m_1 + m_2) \frac{dR}{dt} \tag{A1.4}$$

This is equal to the momentum of a particle of mass $m_1 + m_2$ moving with the centre of mass of the system. Rewriting (A1.3) in terms of the vectors (R, r) instead of (r_1, r_2), we find

$$\left.\begin{aligned}
p_1 &= m_1 \frac{d}{dt}\left(R - \frac{m_2}{m_1 + m_2} r\right) = \frac{m_1}{m_1 + m_2} P - p \\
p_2 &= m_2 \frac{d}{dt}\left(R + \frac{m_1}{m_1 + m_2} r\right) = \frac{m_2}{m_1 + m_2} P + p
\end{aligned}\right\} \tag{A1.5}$$

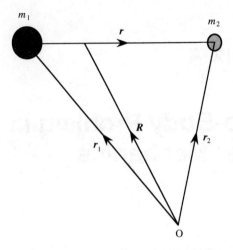

Figure A1 Two-particle system.

where we have defined

$$p = \frac{m_1 m_2}{m_1 + m_2} \frac{dr}{dt} = \frac{m_1}{m_1 + m_2} P - p_1 = -\frac{m_2}{m_1 + m_2} P + p_2 \quad \text{(A1.6)}$$

Note that the positions of the two particles relative to the centre of mass
are

$$r_1 - R = -\frac{m_2}{m_1 + m_2} r, \qquad r_2 - R = \frac{m_1}{m_1 + m_2} r$$

so that $-p$ and p are the momenta of the first and second particles
respectively relative to the centre of mass. (If the centre of mass of the
system is at rest, they are the momenta of the two particles. It is always
possible to transform to a frame of reference moving with the centre of
mass, and this is called the **centre-of-mass frame.**)

If we define

$$\mu = \frac{m_1 m_2}{m_1 + m_2} \quad \text{(A1.7)}$$

(A1.6) shows that the momentum p is the same as that of a particle of
mass μ and velocity equal to the relative velocity dr/dt. The mass μ is
called the **reduced mass** of the two-particle system.

The kinetic energy of the two-particle system can now be expressed in terms of the velocity of the centre of mass and the relative velocity:

$$E_{kin} = \frac{1}{2m_1}(p_1)^2 + \frac{1}{2m_2}(p_2)^2$$

$$= \frac{1}{2m}P^2 + \frac{1}{2\mu}p^2 \qquad \text{(A1.8)}$$

where m is the total mass of the system:

$$m = m_1 + m_2 \qquad \text{(A1.9)}$$

The motion of one particle relative to the other, that is, the **internal motion** of the system, can thus be described in terms of a single particle with mass μ and velocity dr/dt. The total kinetic energy is just the kinetic energy of this (imaginary) particle plus the kinetic energy of the motion of the system as a whole.

A2 Two particles interacting through a central force

Consider a two-particle system with no external forces acting on it, and suppose that the interaction between the two particles can be represented by a potential $V(r)$ that depends only on their separation r. Using (A1.8), the total energy of the system can be written as

$$E = \frac{1}{2m}P^2 + \frac{1}{2\mu}p^2 + V(r) = E_{cm} + E_{rel} \qquad \text{(A2.1)}$$

The total energy associated with the motion of the centre of mass is

$$E_{cm} = \frac{1}{2m}P^2 \qquad \text{(A2.2)}$$

and this is constant if there are no external forces acting on the system, since an external force is defined through

$$F_{ext} = \frac{d(p_1 + p_2)}{dt} = \frac{dP}{dt} \qquad \text{(A2.3)}$$

The centre of mass of the system therefore moves with constant velocity if $F_{ext} = 0$.

From (A2.1), the conservation of total energy and the fact that E_{cm} is constant means that E_{rel} is also constant. The internal motion of the two-particle system can be determined from the energy equation

$$E_{rel} = \frac{1}{2\mu}p^2 + V(r) \qquad \text{(A2.4)}$$

This is the energy equation for a particle of mass μ and position vector r. So the problem of how the two particles move relative to one another is reduced to the problem of how a single particle with the reduced mass moves in the potential $V(r)$.

Since the potential $V(r)$ in (A2.4) is central, the angular momentum about the origin is conserved. As r is the position of particle 2 relative to particle 1, the origin is at the position of particle 1. The angular momentum that is conserved is the angular momentum about particle 1 of a particle with the reduced mass μ situated at the position of particle 2. (We could equally well have defined r as the position vector of particle 1 relative to particle 2, in which case the labels 1 and 2 should be exchanged in the previous sentence.)

The square of the momentum of a particle can always be separated into radial and angular components, since

$$p^2 = \frac{(r \cdot p)^2}{r^2} + \frac{(r \times p)^2}{r^2}$$

By definition, the angular momentum about the origin is

$$M = r \times p$$

and so we can write

$$p^2 = p_r^2 + \frac{M^2}{r^2} \qquad\qquad (A2.5)$$

where $p_r = r \cdot p/r$ is the radial momentum. In terms of the motion of the real particles, M is proportional to the angular momentum of either about the centre of mass. The position of particle 2 relative to the centre of mass is $r_2 - R = (m_1/m)r$ (from (A1.1)), and its momentum relative to the centre of mass is p. So the angular momentum of particle 2 about the centre of mass is

$$(r_2 - R) \times p = \frac{m_1}{m} r \times p = \frac{m_1}{m} M$$

Similarly, the angular momentum of particle 1 about the centre of mass is

$$(r_1 - R) \times (-p) = -\frac{m_2}{m} r \times (-p) = \frac{m_2}{m} M$$

The total angular momentum of the two-particle system about the position of either particle is the sum of these, which is M.

APPENDIX B

Analytical Solutions of Eigenvalue Equations

We present analytical solutions of the eigenvalue equation for orbital angular momentum, (4.41), the energy eigenvalue equation for a one-dimensional simple harmonic oscillator, (5.3), and the radial equation for the bound states of the hydrogen atom, (7.12).

B1 Legendre's equation

The eigenvalue equation for the orbital angular momentum, (4.41), is reduced to Legendre's equation when the eigenvalue of the z component of orbital angular momentum is zero. Legendre's equation can be written as

$$(1 - x^2)\frac{d^2y}{dx^2} - 2x\frac{dy}{dx} + \alpha^2 y = 0 \qquad \textbf{(B1.1)}$$

and $y(x)$ must be finite everywhere in the range $-1 \leqslant x \leqslant 1$, and in particular when $x = \pm 1$. This equation can be solved by the power series expansion method. That is, we assume that the solution can be expanded in the form of an infinite power series in x:

$$y = (a_0 + a_1 x + a_2 x^2 + \ldots) = \sum_{k=0}^{\infty} a_k x^k \qquad \textbf{(B1.2)}$$

This series is differentiated term by term, and the result substituted in the differential equation. The left-hand side of the equation becomes an infinite power series in x, and for this to vanish for all x the coefficients of each power of x must separately be equated to zero.

The first and second derivatives of the expansion (B1.2) are

$$\frac{dy}{dx} = \sum_{k=0}^{\infty} a_k k x^{k-1} \tag{B1.3}$$

$$\frac{d^2 y}{dx^2} = \sum_{k=0}^{\infty} a_k k(k - 1) x^{k-2} \tag{B1.4}$$

If the expansions (B1.2)–(B1.4) are substituted in (B1.1), we obtain

$$\sum_{k=0}^{\infty} a_k[k(k - 1)(x^{k-2} - x^k) - 2kx^k + \alpha^2 x^k] = 0$$

which can be rewritten as

$$\sum_{k=0}^{\infty} a_k\{k(k - 1)x^{k-2} - [k(k + 1) - \alpha^2]x^k\} = 0 \tag{B1.5}$$

If we expand the power series on the left-hand side of (B1.5), we obtain

$$(2a_2 + \alpha^2 a_0)x^0 + [6a_3 - (2 - \alpha^2)a_1]x + [12a_4 - (6 - \alpha^2)a_2]x^2 + \ldots$$
$$+ \{(k + 2)(k + 1)a_{k+2} - [k(k + 1) - \alpha^2]a_k\}x^k + \ldots = 0 \tag{B1.6}$$

In order for (B1.6) to be true for all x, the coefficients of each power of x on the left-hand side must be zero. The coefficient of x^k vanishes for all $k = 0, 1, 2, \ldots$ if

$$a_{k+2} = a_k \frac{k(k + 1) - \alpha^2}{(k + 2)(k + 1)} \tag{B1.7}$$

This is a recurrence relation, from which the values of all the coefficients a_k with $k \geqslant 2$ can be found in terms of a_0 and a_1. Thus we find

$$a_2 = -\tfrac{1}{2}\alpha^2 a_0, \quad a_3 = \tfrac{1}{6}(2 - \alpha^2)a_1, \quad a_4 = -\tfrac{1}{12}(6 - \alpha^2)a_2 = -\tfrac{1}{12}(6 - \alpha^2)\tfrac{1}{2}\alpha^2 a_0$$

and so on. All the coefficients of even powers of x are determined in terms of a_0, and those of odd powers of x in terms of a_1. The coefficients a_0 and a_1 are undetermined, and are the two arbitrary constants that we should expect in the general solution of a second-order differential equation. The general solution is therefore of the form

$$y(x) = a_0 y_0(x) + a_1 y_1(x) \tag{B1.8}$$

where $y_0(x)$ and $y_1(x)$ are independent solutions of the differential equation. $y_0(x)$ is a power series in even powers of x, and $y_1(x)$ is a power series in odd powers of x.

We must now ensure that $y(x)$ satisfies the requirement that it should be finite in the range $-1 \leqslant x \leqslant 1$. The convergence of a power series can be tested by comparing it with the infinite power series expansion of a function whose behaviour is known. If we examine the ratio a_{k+2}/a_k given by (B1.9) in the limit as $k \to \infty$, we find that $a_{k+2}/a_k \to 1$ as $k \to \infty$. This means that if we go to terms with sufficiently high k in the series, we find a group of terms that behaves in the same way as

$$x^k + x^{k+2} + x^{k+4} + x^{k+6} + \ldots = x^k(1 + x^2 + x^4 + x^6 + \ldots)$$

But $1 + x^2 + x^4 + x^6 + \ldots$ is the binomial expansion of $1/(1 - x^2)$ when $|x| < 1$, and this diverges at $x = \pm 1$. So if the power series for $y_0(x)$ or $y_1(x)$ contains an infinite number of terms, it will diverge in the same way. In order to prevent this divergence, the power series must terminate at some finite value of k. In other words, we require the solution to be a polynomial in x rather than an infinite power series. Suppose that the series is terminated at $k = n$. Then we must have a_n non-zero, but the coefficients of all higher powers of x must vanish. This will be true if $a_{n+1} = 0$ and $a_{n+2} = 0$. From (B1.7) we find that $a_{n+2}/a_n = 0$ if

$$\alpha^2 = n(n + 1) \quad (n = 0, 1, 2, 3, \ldots) \tag{B1.9}$$

Unless α^2 takes one of these values, there is no solution to (B1.1) that is finite at $x = \pm 1$. This is the condition that determines the spectrum (4.42).

The requirement (B1.9) will make $y_0(x)$ a polynomial if n is even, and $y_1(x)$ a polynomial if n is odd. To ensure that the general solution (B1.8) is finite at $x = \pm 1$, we must also require

$$a_1 = 0 \quad \text{if } n \text{ is even,} \qquad a_0 = 0 \quad \text{if } n \text{ is odd} \tag{B1.10}$$

That is, the solutions to (B1.1) that are finite in the range $-1 \leqslant x \leqslant 1$ are the finite polynomials in x with coefficients determined by (B1.7), (B1.9) and (B1.10). These are called the **Legendre polynomials** $P_n(x)$, the first five of which are given in Table 4.1.

The eigenvalue equation for orbital angular momentum in the general case, (4.41), is of the form

$$(1 - x^2)\frac{d^2 y}{dx^2} - 2x\frac{dy}{dx} + \left(\alpha^2 - \frac{\beta^2}{1 - x^2}\right)y = 0 \tag{B1.11}$$

The solution of the eigenvalue equation for the z component of angular momentum restricts the values of β to 0, ± 1, ± 2, $\ldots$; (4.37). We shall prove that the solutions to (B1.11) are the **associated Legendre polynomials**

$$P_n^m(x) = (1 - x^2)^{m/2}\frac{d^{m/2}}{dx^{m/2}}P_n(x) \tag{B1.12}$$

where $m = |\beta|$ is an integer, and $P_n(x)$ is the polynomial solution to Legendre's equation (B1.1), of degree n.
$P_n(x)$ satisfies (B1.1) with $\alpha^2 = n(n + 1)$:

$$(1 - x^2)\frac{d^2 P_n}{dx^2} - 2x\frac{dP_n}{dx} + n(n + 1)P_n = 0$$

Differentiating this equation m times (where $m < n$), we obtain

$$(1 - x^2)\frac{\mathrm{d}^{m+2} P_n}{\mathrm{d}x^{m+2}} - 2(m + 1)x\frac{\mathrm{d}^{m+1} P_n}{\mathrm{d}x^{m+1}} + [n(n + 1) - m(m + 1)]\frac{\mathrm{d}^m P_n}{\mathrm{d}x^m}$$
$$= 0 \quad \textbf{(B1.13)}$$

This can be proved by induction: it is easy to prove for $m = 1$, and if it is true for some m then it can be proved for $m' = m + 1$ by differentiating the equation once.

Now we multiply (B1.13) from the left by $(1 - x^2)^{m/2}$ and use

$$(1 - x^2)^{m/2}\frac{\mathrm{d}^{m+2} P_n}{\mathrm{d}x^{m+2}} = (1 - x^2)^{m/2}\frac{\mathrm{d}}{\mathrm{d}x}\left(\frac{\mathrm{d}^{m+1} P_n}{\mathrm{d}x^{m+1}}\right)$$

$$= \frac{\mathrm{d}}{\mathrm{d}x}\left[(1 - x^2)^{m/2}\frac{\mathrm{d}^{m+1} P_n}{\mathrm{d}x^{m+1}}\right]$$

$$+ \frac{mx}{1 - x^2}(1 - x^2)^{m/2}\frac{\mathrm{d}^{m+1} P_n}{\mathrm{d}x^{m+1}}$$

$$(1 - x^2)^{m/2}\frac{\mathrm{d}^{m+1} P_n}{\mathrm{d}x^{m+1}} = (1 - x^2)^{m/2}\frac{\mathrm{d}}{\mathrm{d}x}\left(\frac{\mathrm{d}^m P_n}{\mathrm{d}x^m}\right)$$

$$= \frac{\mathrm{d}}{\mathrm{d}x}\left[(1 - x^2)^{m/2}\frac{\mathrm{d}^m P_n}{\mathrm{d}x^m}\right]$$

$$+ \frac{mx}{1 - x^2}(1 - x^2)^{m/2}\frac{\mathrm{d}^m P_n}{\mathrm{d}x^m}$$

The result is

$$\left\{(1 - x^2)\frac{\mathrm{d}^2}{\mathrm{d}x^2} - 2x\frac{\mathrm{d}}{\mathrm{d}x} + \left[n(n + 1) - \frac{m^2}{1 - x^2}\right]\right\}(1 - x^2)^{m/2}\frac{\mathrm{d}^m P_n}{\mathrm{d}x^m} = 0$$

This shows that $P_n^m(x) = (1 - x^2)^{m/2}\, \mathrm{d}^m P_n/\mathrm{d}x^m$ satisfies (B1.11) with $\alpha^2 = n(n + 1)$ and $\beta^2 = m^2$. Since $P_n(x)$ is a polynomial of degree n, all its derivatives vanish for $m > n$, and the associated Legendre polynomial $P_n^m(x)$ is defined only for $m \leqslant n$.

B2 The energy eigenvalue equation for the simple harmonic oscillator

The differential equation (5.3) is of the form

$$\frac{\mathrm{d}^2 \psi}{\mathrm{d}x^2} + (2\epsilon - x^2)\psi = 0 \quad \textbf{(B2.1)}$$

As it stands, this equation cannot be solved by the power series expansion method, but it can be transformed to a suitable form by the substitution

$$\psi(x) = y(x)e^{-x^2/2} \tag{B2.2}$$

Substituting (B2.2) in (B2.1), we obtain **Hermite's equation**

$$\frac{d^2y}{dx^2} - 2x\frac{dy}{dx} - (1 - 2\epsilon)y = 0 \tag{B2.3}$$

If we assume a power series solution of the form (B1.2), and substitute (B1.2)–(B1.4) in (B2.3), it becomes

$$\sum_{k=0}^{\infty} a_k[k(k-1)x^{k-2} - (2k+1-2\epsilon)x^k] = 0 \tag{B2.4}$$

The recurrence relation ensuring that the coefficient of x^k vanishes for any $k \geqslant 0$ (corresponding to (B1.7)) is

$$a_{k+2} = a_k\frac{2k+1-2\epsilon}{(k+2)(k+1)} \tag{B2.5}$$

Once again the general solution is of the form (B1.8), with only even powers of x appearing in $y_0(x)$, and only odd powers in $y_1(x)$.

In the present case the boundary condition is that $\psi(x)$, given by (B2.2), should be finite as $x \to \pm\infty$. The behaviour of $y(x)$ as $x \to \pm\infty$ is determined by the terms involving high powers of x in the expansion of the form (B1.2), and we can investigate this behaviour by examining the expansion coefficients a_k in the limit as $k \to \infty$. From (B2.5) we find that $a_{k+2}/a_k \to 2/k$ as $k \to \infty$. This is the same behaviour as that of the power series for e^{x^2}, since

$$e^{x^2} = 1 + x^2 + \frac{1}{2!}x^4 + \frac{1}{3!}x^6 + \ldots = \sum_{k=0}^{\infty}\frac{1}{k!}x^{2k}$$

The ratio of the coefficients of x^{k+2} and x^k in this series is

$$\frac{(\frac{1}{2}k)!}{(\frac{1}{2}k+1)!} = \frac{2}{k+2} \to \frac{2}{k} \quad \text{as } k \to \infty$$

Therefore both $y_0(x)$ and $y_1(x)$ behave in a similar way to e^{x^2} at large $|x|$, and so, from (B2.2), $\psi(x)$ diverges like $e^{x^2/2}$ as $x \to \pm\infty$.

The divergent behaviour of $\psi(x)$ can be avoided only if $y(x)$ is a finite polynomial rather than an infinite power series. From (B2.5), the condition for the series to terminate at $k = n$ is

$$2n + 1 - 2\epsilon = 0$$

or

$$\epsilon = n + \tfrac{1}{2} \quad (n = 0, 1, 2, 3, \ldots) \tag{B2.6}$$

This gives the energy eigenvalue spectrum (5.15).

As in the case of the solution to the Legendre equation, the condition (B2.6) terminates $y_0(x)$ if n is even, or $y_1(x)$ if n is odd. So again we must choose $a_1 = 0$ (so that $y_1(x) = 0$) if n is even, or $a_0 = 0$ (so that $y_0(x) = 0$) if n is odd. The polynomial solutions to (B2.3) are the **Hermite polynomials** $H_n(x)$, the first five of which are given in Table 5.1.

B3 The radial equation for the hydrogen atom

The radial equation for hydrogen is the differential equation (7.12), which can be written as

$$\frac{d^2 R}{dx^2} + \frac{2}{x}\frac{dR}{dx} + \left[\epsilon + \frac{2}{x} - \frac{l(l+1)}{x^2}\right]R = 0 \qquad \textbf{(B3.1)}$$

where for bound states ϵ must be negative: $\epsilon = -|\epsilon|$. Again this equation cannot be solved by the power series method until an appropriate substitution has been made. In this case we write

$$R(x) = e^{-\alpha x}x^l y(x) \qquad \textbf{(B3.2)}$$

Substituting this in (B3.1), we obtain

$$\frac{d^2 y}{dx^2} + 2\left(\frac{l+1}{x} - \alpha\right)\frac{dy}{dx} + \left\{\alpha^2 - |\epsilon| + \frac{2}{x}[1 - \alpha(l+1)]\right\}y = 0$$

$$\textbf{(B3.3)}$$

This can be simplified by the choice

$$\alpha^2 = |\epsilon| \qquad \textbf{(B3.4)}$$

and a power series solution of the form (B.1.2) can then be assumed. Substituting (B1.2), (B1.3) and (B1.4) in (B3.3), we obtain

$$\sum_{k=0}^{\infty} a_k\{k(k + 2l + 1)x^{k-2} - [2\alpha(k + l + 1) - 2]x^{k-1}\} = 0$$

The recurrence relation guaranteeing that this equation is satisfied for all x is

$$a_{k+1} = a_k \frac{2\alpha(k + l + 1) - 2}{(k + 1)(k + 2l + 2)} \qquad \textbf{(B3.5)}$$

In this case all coefficients can be expressed in terms of a_0, so we have only found one of the two independent solutions to the differential equation. This is because we assumed that the power series started with the constant term. In fact, there is another power series solution, which starts with a term proportional to $x^{-(2l+1)}$. However, this would give a term in $R(x)$ proportional to $e^{-\alpha x}x^{-(l+1)}$, which diverges at $x = 0$ for all allowed values of l. $R(x)$ is required to be finite for all x in the range 0 to ∞, so only the power series starting with the constant term can satisfy the boundary conditions at $x = 0$.

The behaviour of $y(x)$ as $x \to \infty$ is determined, as in the previous case, by the behaviour of the terms in the power series for which $k \to \infty$. From the recurrence relation (B3.5) we find that $a_{k+1}/a_k \to 2\alpha/k$ as $k \to \infty$. This should be compared with the corresponding ratio for the series

$$e^{2\alpha x} = \sum_{k=0}^{\infty} \frac{1}{k!}(2\alpha)^k x^k$$

Here the ratio of the coefficients of x^{k+1} and x^k is

$$\frac{(2\alpha)^{k+1}}{(k+1)!}\frac{k!}{(2\alpha)^k} = \frac{2\alpha}{k+1} \to \frac{2\alpha}{k} \quad \text{as } k \to \infty$$

So $y(x)$ diverges as $e^{2\alpha x}$ as $x \to \infty$, and consequently $R(x)$ diverges as $e^{\alpha x}x^l$. Once again, the divergence is avoided by terminating the power series, so that $y(x)$ is a polynomial. From (B3.5), the condition for the series to terminate at the term in x^p is

$$2\alpha(p + l + 1) - 2 = 0 \quad (p = 0, 1, 2, \ldots)$$

or

$$\alpha = \frac{1}{p + l + 1} \qquad\qquad \textbf{(B3.6)}$$

Because of the identification of α^2 with $|\epsilon|$, (B3.4), the eigenvalue spectrum determined by (B3.6) is

$$\epsilon = -\frac{1}{n^2} \quad (n = p + l + 1) \qquad\qquad \textbf{(B3.7)}$$

If l is identified as the orbital angular momentum quantum number, with eigenvalues 0, 1, 2, ..., the allowed values for n are 1, 2, 3, ... Equation (B3.7) leads to the energy eigenvalue spectrum for hydrogen given in (7.13).

The polynomial determined by the condition (B3.6) is of degree p, but, according to the recurrence relation (B3.5), its coefficients depend on l as well as on p. These are the **Laguerre polynomials** $L_p^q(x)$, where we have defined $q = 2l + 1$. There is thus a set of different polynomials of the same degree p that satisfy (B3.3), each associated with a different value of l. Equation (B3.7) indicates that the eigenvalue of ϵ depends only on n. So polynomials corresponding to the same eigenvalue $-1/n^2$ are of different degrees, ranging from $p = 0$ with $l = n - 1$ to $p = n - 1$ with $l = 0$.

APPENDIX C

The Computer Demonstrations

The disk contains programs to demonstrate, interactively, some of the fundamental ideas of quantum mechanics. It will run on any IBM-compatible pc, with at least a VGA monitor. Some of the programs will be rather slow on machines without a mathematical co-processor.

The full menu of programs, each of which is described below, is as follows:

(1) The one-dimensional Schrödinger equation
  (a) Infinite square well
  (b) Finite square well
  (c) Harmonic oscillator
  (d) Anharmonic oscillator
  (e) Triangular well

(2) The Kronig–Penney model

(3) The Schrödinger equation: central potentials
  (a) Coulomb potential
  (b) Harmonic oscillator (three-dimensional)

(4) Orbital angular momentum

(5) Transmission
  (a) Plane wave, one-dimensional square well
  (b) Plane wave, one-dimensional square barrier
  (c) Gaussian wave packet, one-dimensional square barrier

(6) Wave packets (one-dimensional)
  (a) Infinite square well
  (b) Harmonic oscillator

C1   The Schrödinger equation in one dimension

This program solves the Schrödinger equation

$$\left[-\frac{\hbar^2}{2m}\frac{d^2}{dx^2} + V(x)\right]\psi(x) = E\psi(x) \qquad \text{(C1.1)}$$

for each of five different potentials $V(x)$. The user may choose the mass m (in units of the free electron mass m_e) and parameters that define the potential. (The programs can accept a wide range of masses: of particular interest, apart from $m = m_e$, are the proton mass $m = 1836m_e$ and the effective mass of an electron in a semiconductor – in GaAs $m = 0.067m_e$.) For any chosen value of the energy E (in eV) the options *Draw psi* and *Draw psi²* plot the corresponding solution to (C1.1), $\psi(x)$, and the probability distribution $|\psi(x)|^2$. Except in the case of the triangular well, the solutions plotted are either symmetric or antisymmetric about the centre of the well, depending on whether the user has selected *Even* or *Odd*. (Default values for all variables are incorporated in the programs.) In general, as will be evident from the displayed function, the solution $\psi(x)$ does not obey the physical boundary conditions. The eigenvalues (those values of E for which $\psi(x)$ satisfies the boundary conditions) can be found by a process of trial and error, by observing the behaviour of the wave function outside the (marked) classically allowed region for different values of E. Alternatively, the option *Find E* or *Shoot for E* will calculate the energy eigenvalues, display them correct to five significant figures, and plot the eigenfunction.

(a) Infinite square well

Potential:

$$V(x) = \begin{cases} 0 & (0 \leqslant x \leqslant L) \\ \infty & (x < 0, x > L) \end{cases} \qquad \text{(C1.2)}$$

Boundary conditions:

$$\psi(x) = 0 \quad (x = 0, L) \qquad \text{(C1.3)}$$

Choice of parameters: m/m_e, L in nm.

Select *Parity* (even or odd) and E in eV.

If the chosen value of E is close enough to an eigenvalue E_n (and the selected parity is correct), *Draw psi* shows the corresponding energy eigenfunction, which satisfies the boundary conditions (C1.3) exactly, and indicates the level n concerned.

The option *Find E* calculates the eigenvalue E_n for a selected level n using (3.39), and displays the corresponding eigenfunction.

(b) Finite square well

Potential:

$$V(x) = \begin{cases} 0 & (0 \leqslant x \leqslant L) \\ V_0 & (x < 0, x > L) \end{cases} \tag{C1.4}$$

Boundary conditions:

$$\psi(x) \to 0 \quad (x \to \pm\infty) \tag{C1.5}$$

Choice of parameters: m/m_e, L in nm, V_0 in eV. The value of the well parameter $w = \sqrt{(8mV_0L^2/h^2)}$ is displayed.

Select *Parity* (even or odd) and E (less than V_0) in eV.

In general it is impossible to select a value of E for which *Draw psi* shows a wave function that satisfies the boundary conditions (C1.5) exactly, because of the finite accuracy of the computer. The boundary conditions are not incorporated in the drawing program, except when the option *Find E* is used (or when E is very close to an eigenvalue).

The option *Find E* calculates the eigenvalue E_n for a selected level n using the equations given in Problem 3.7, and displays the corresponding eigenfunction.

(c) Harmonic oscillator

Potential:

$$V(x) = \tfrac{1}{2}m\omega^2 x^2 = \tfrac{1}{2}\hbar\omega\xi^2 \tag{C1.6}$$

where ξ is given in (5.2).

Boundary conditions:

$$\psi(\xi) \to 0 \quad (\xi \to \pm\infty) \tag{C1.7}$$

There is no choice of parameters in this case.

Select *Parity* (even or odd) and E in units of $\hbar\omega$.

Draw psi and *Draw psi²* plot $\psi(\xi)$ and $|\psi(\xi)|^2$ as functions of the dimensionless variable ξ, and the limits of the classically allowed region, (5.32), are indicated. If the chosen value of E is sufficiently close to an eigenvalue, the corresponding value of n will be displayed (provided the correct parity has been selected).

The option *Find E* calculates an eigenvalue E_n in units of $\hbar\omega$, using (5.15). The value returned is the eigenvalue, for a level with the chosen parity, that is closest to the value of E currently displayed.

(d) Anharmonic oscillator

Potential:

$$V(x) = \tfrac{1}{2}\hbar\omega(\xi^2 + c\xi^4) \qquad\qquad (\textbf{C1.8})$$

where ξ is given in (5.2).

Boundary conditions:

$$\psi(\xi) \to 0 \quad (\xi \to \pm\infty) \qquad\qquad (\textbf{C1.9})$$

Choice of parameters: c, the dimensionless constant appearing in (C1.8). The program will not accept values of c greater than 1, since it becomes too slow, even on a machine with a maths co-processor.

Select *Parity* (even or odd) and E in units of $\hbar\omega$. If the chosen value of E is sufficiently close to an eigenvalue, the corresponding value of n will be displayed (provided the correct parity has been selected).

The option *Shoot for E* uses a simple successive approximation method to calculate the eigenvalue E_n, for a level with the selected parity, closest to the value of E currently displayed. Five successive estimates are displayed on the screen, and if this set of values is not sufficiently convergent then the process may be repeated. (Since the computer works to a larger number of significant figures than it displays on the screen, a wave function with improved behaviour at large ξ may sometimes be obtained by repeating *Shoot for E*, even though the calculated value for the energy does not appear to change.)

(e) Triangular well

Potential:

$$V(x) = \begin{cases} -cx & (x \leqslant 0) \\ \infty & (x > 0) \end{cases} \qquad\qquad (\textbf{C1.10})$$

Boundary conditions:

$$\psi(x) \to 0 \quad (x \to -\infty), \qquad \psi(x) = 0 \quad (x = 0) \qquad (\textbf{C1.11})$$

Choice of parameters: m/m_e, c in $\mathrm{eV\,nm}^{-1}$ (the negative slope of the potential in the region $x \leqslant 0$).

Select E in eV.

Draw psi and *Draw psi²* plot $\psi(x)$ and $|\psi(x)|^2$ as functions of the dimensionless variable $\xi = (2mc/\hbar^2)^{1/3}$. The width of the classically allowed region, from the point where $E = V(x)$ to $x = 0$, is indicated.

The option *Shoot* for E uses the same approximation method as in the case of the anharmonic oscillator to calculate the eigenvalue E_n closest to the value of E currently displayed.

C2 The Kronig–Penney model

This program solves the equation (derived in Problem 7.13) for the Kronig–Penney model of an electron in a one-dimensional periodic lattice:

$$\cos k\mathbb{L} = \cos \alpha\mathbb{L} + \frac{C}{2\alpha\mathbb{L}} \sin \alpha\mathbb{L} \qquad \text{(C2.1)}$$

where $\alpha^2 = 2mE/\hbar^2$ and $\mathbb{L}$ is the well width. C (dimensionless) is a measure of the strength of the interaction.

Choice of parameters: m/m_e, $\mathbb{L}$ in nm, C.

Select N, the number of unit cells in the lattice.

The option *Plot* shows the right-hand side of (C2.1) as a function of α, intersected by horizontal lines indicating values of $\cos k\mathbb{L}$. If the selected $N \leqslant 10$, lines corresponding to all the allowed values of $\cos k\mathbb{L}$ are shown. (These values are calculated in Problem 7.14.) The discrete intersection points are marked, and the corresponding values for E (in eV) are displayed, for the first five allowed bands of energies. If the selected $N > 10$, only the lines $\cos k\mathbb{L} = \pm1$ are shown, and the ranges on the horizontal axis over which the right-hand side of the equation lies between these two values is indicated by shading. The energy ranges in the first five bands (in eV) are displayed.

C3 The Schrödinger equation: central potentials

This program integrates a radial equation of the form

$$\left\{ -\frac{\hbar^2}{2m} \left[\frac{d^2}{dr^2} + \frac{2}{r}\frac{d}{dr} - \frac{l(l+1)}{r^2} \right] + V(r) \right\} R(r) = ER(r) \quad \text{(C3.1)}$$

for two central potentials. In each case the user may select a value for l and E, and the options *Draw psi* and *Draw psi²* plot the corresponding solution $R(r)$ and the probability distribution $r^2|R(r)|^2$ (see Section 7.2). Unless E is an eigenvalue, $R(r)$ does not satisfy the boundary conditions

$$R(r) \to 0 \quad (r \to \infty), \qquad R(r) \quad \text{finite at} \quad r = 0 \qquad \text{(C3.2)}$$

(a) The Coulomb potential

$$V(r) = -\frac{e^2}{4\pi\varepsilon_0 r} \qquad \text{(C3.3)}$$

The option *Find E* uses (7.13) to calculate the energy, in units of the Rydberg energy, for any chosen value of n.

(b) The three-dimensional harmonic oscillator

$$V(r) = \tfrac{1}{2}m\omega^2 r^2 \qquad \text{(C3.4)}$$

The option *Find E* uses the equation $E_n = (n + \tfrac{3}{2})\hbar\omega$ to calculate the energy, in units of $\hbar\omega$. The calculated eigenvalue is that for the level (with n not less than the selected value of l) closest to the value of E currently displayed.

C4 Orbital angular momentum

This program will plot the square root of the probability distribution, $|Y_{lm_l}(\theta, \phi)|$, for a particle in an orbital angular momentum eigenstate. The distribution is symmetric about the z axis (that is, independent of ϕ).

Select values for l (zero or integer) and m_l (in the range $-l \leq m_l \leq l$).

The option *Outline* draws the plane curve $r = |Y_{lm_l}(\theta, \phi)|$ (independent of ϕ), where r is the distance of the curve from the origin at the angle θ. The options *Wireframe* and *Solid* redraw the curve, and rotate it about the z axis to produce a three-dimensional representation.

C5 Transmission

(a) Plane wave, well

This program plots the transmission coefficient for a particle in the one-dimensional square well (C1.4), with $E \geq V_0$.

Choice of parameters: m/m_e, L in nm, V_0 in eV. The value of the well parameter $w = \sqrt{(8mV_0L^2/h^2)}$ is displayed.

The option *Plot T* draws the transmission coefficient (3.57) as a function of $E/V_0 \geqslant 1$.

(b) Plane wave, barrier

This program plots the transmission coefficient for a particle incident on the one-dimensional square barrier

$$V(x) = \begin{cases} V_0 & (0 \leqslant x \leqslant L) \\ 0 & (x < 0, x > L) \end{cases} \tag{C5.1}$$

Choice of parameters: m/m_e, L in nm, V_0 in eV. The value of the barrier parameter $w = \sqrt{(8mV_0L^2/h^2)}$ is displayed.

The option *Plot T* draws the transmission coefficient T as a function of E/V_0. The expressions used for T in the two ranges $0 \leqslant E/V_0 \leqslant 1$ and $E/V_0 \geqslant 1$ are derived in Problem 3.13.

(c) Gaussian wave packet

This program shows the behaviour of a Gaussian wave packet incident on the one-dimensional square barrier (C5.1).

Select the ratio V_0/E (which may be less than or greater than 1) and the width L in arbitrary units.

The option *Display* shows the wave packet propagating towards the barrier, and being partially reflected and partially transmitted.

C6 Wave packets

This program allows construction of wave packets, and investigation of their behaviour as a function of time.

The wave packet, of the form

$$\Psi(x, t) = \sum_n c_n \psi_n(x) e^{-iE_n t/h} \tag{C6.1}$$

is a superposition of energy eigenfunctions $\{\psi_n(x)\}$ for either

(a) a particle in an infinitely deep one-dimensional square well, or

(b) a one-dimensional simple harmonic oscillator.

The user may choose values for the first 13 coefficients in the expansion
(The wave packet need not be normalized.)

The options *Plot psi(0)* and *Plot psi²(0)* show $\Psi(x,0)$ and $|\Psi(x,0)|^2$ as
functions of x at time $t = 0$, while the option *Plot psi²(t)* produces a
dynamic display showing the development of the probability distribution
with time.

Index

References to sections are in bold, and references to problems are indicated by 'p' after the page number.